Lecture Notes in Earth Sciences

Vol. 1: Sedimentary and Evolutionary Cycles. Edited by U. Bayer and A. Seilacher. VI, 465 pages. 1985.

Vol. 2: U. Bayer, Pattern Recognition Problems in Geology and Paleontology. VII, 229 pages. 1985.

Vol. 3: Th. Aigner, Storm Depositional Systems. VIII, 174 pages. 1985.

Vol. 4: Aspects of Fluvial Sedimentation in the Lower Triassic Buntsandstein of Europe. Edited by D. Mader. VIII, 626 pages. 1985.

Vol. 5: Paleogeothermics. Edited by G. Buntebarth and L. Stegena. II, 234 pages. 1986.

Vol. 6: W. Ricken, Diagenetic Bedding. X, 210 pages. 1986.

Vol. 7: Mathematical and Numerical Techniques in Physical Geodesy, Edited by H. Sünkel. IX, 548 pages. 1986.

Vol. 8: Global Bio-Events. Edited by O.H. Walliser. IX, 442 pages. 1986.

Vol. 9: G. Gerdes, W.E. Krumbein, Biolaminated Deposits. IX, 183 pages. 1987.

Vol. 10: T.M. Peryt (Ed.), The Zechstein Facies in Europe. V, 272 pages. 1987.

Vol. 11: L. Landner (Ed.), Contamination of the Environment. Proceedings, 1986. VII, 190 pages. 1987.

Vol. 12: S. Turner (Ed.), Applied Geodesy. VIII, 393 pages. 1987.

Vol. 13: T.M. Peryt (Ed.), Evaporite Basins. V, 188 pages. 1987.

Vol. 14: N. Cristescu, H.I. Ene (Eds.), Rock and Soil Rheology. VIII, 289 pages. 1988.

Vol. 15: V.H. Jacobshagen (Ed.), The Atlas System of Morocco. VI, 499 pages. 1988.

Vol. 16: H. Wanner, U. Siegenthaler (Eds.), Long and Short Term Variability of Climate. VII, 175 pages. 1988.

Vol. 17: H. Bahlburg, Ch. Breitkreuz, P. Giese (Eds.), The Southern Central Andes. VIII, 261 pages. 1988.

Lecture Notes in
Earth Sciences

Edited by Somdev Bhattacharji, Gerald M. Friedman,
Horst J. Neugebauer and Adolf Seilacher

17

H. Bahlburg Ch. Breitkreuz
P. Giese (Eds.)

The Southern Central Andes

Contributions to Structure and Evolution
of an Active Continental Margin

Springer-Verlag
Berlin Heidelberg GmbH

Editors

Dr. Heinrich Bahlburg
Priv. Doz. Dr. Christoph Breitkreuz
Institut für Geologie und Paläontologie
Technische Universität Berlin
Ernst-Reuther-Platz 1, D-1000 Berlin 10, FRG

Prof. Dr. Peter Giese
Institut für Geophysikalische Wissenschaften
Freie Universität Berlin
Rheinbabenallee 49, D-1000 Berlin 33, FRG

ISBN 978-3-540-50032-2   ISBN 978-3-540-45904-0 (eBook)
DOI 10.1007/978-3-540-45904-0

© Springer-Verlag Berlin Heidelberg 1988
Originally published by Springer-Verlag Berlin Heidelberg New York in 1988

2132/3140-543210

Escarpment of the Atacama Fault Zone near Antofagasta (Northern Chile):
Recent uplift of the Coastal Cordillera (view towards the north).

PREFACE

The suggestion to compile and publish this volume dealing with some geoscientific problems of the Central Andes came up during a conference on "Mobility of Active Continental Margins" held in Berlin, February 1986. At this international conference, organized by the Berlin Research Group "Mobility of Active Continental Margins", colleagues from Europe, Southern and Northern America reported on their current investigations in the Central Andes.

The Central Andes claim a special position in the 7000 km long Andean mountain range. In Northern Chile, Southern Bolivia and Northwest Argentina the Central Andes show their largest width with more than 650 km and along a Geotraverse between the Pacific coast and the Chaco all typical Andean morphotectonic units are well developed. Here, the pre-Andean evolution is documented by outcropping of Paleozoic and pre-Cambrian rocks. The characteristic phenomena of the Andean cycle can be studied along the entire geotraverse. The migration of the tectonic and magmatic activity starting in Jurassic and being active till Quaternary is clearly evidenced. Besides the Himalaya, the Central Andes show with 70-80 km and -400 mgal the largest crustal thickness known in mountain ranges. These and many other interesting and exciting geoscientific features encouraged a group of geoscientists from both West-Berlin universities (Freie Universität and Technische Universität) to focus their studies along a geotraverse through the Central Andes. The realization of these studies would not have been possible without the active assistance and close cooperation of our colleagues from the geoscientific institutions in Salta (Argentina), La Paz and Santa Cruz (Bolivia) and Antofagasta and Santiago (Chile). Concerning the German participation, this joint and interdisciplinary project is financially supported since 1982 as Reserach Group" Mobility of Active Continental Margins" by the German Research Society and by the West-Berlin universities as well. A number of colleagues from universities in West Germany take part in this project, too.

The papers presented here deal with the period from Late Precambrian up to the youngest phenomena in Quaternary. The contributions cover the whole spectrum of geoscientific research, geology, paleontology, petrology, geochemistry, geophysics and geomorphology.
In conclusion, the data published here may help to improve the picture of Andean structure and evolution. The detailed investigations carried out in the past years show, that the first simple plate tectonic models proposed in the beginning of the seventies have to improved and modified. Furthermore, the results can be seen as contribution to the international Lithospheric Project and as a useful data base for the construction of a Central Andean Transect.

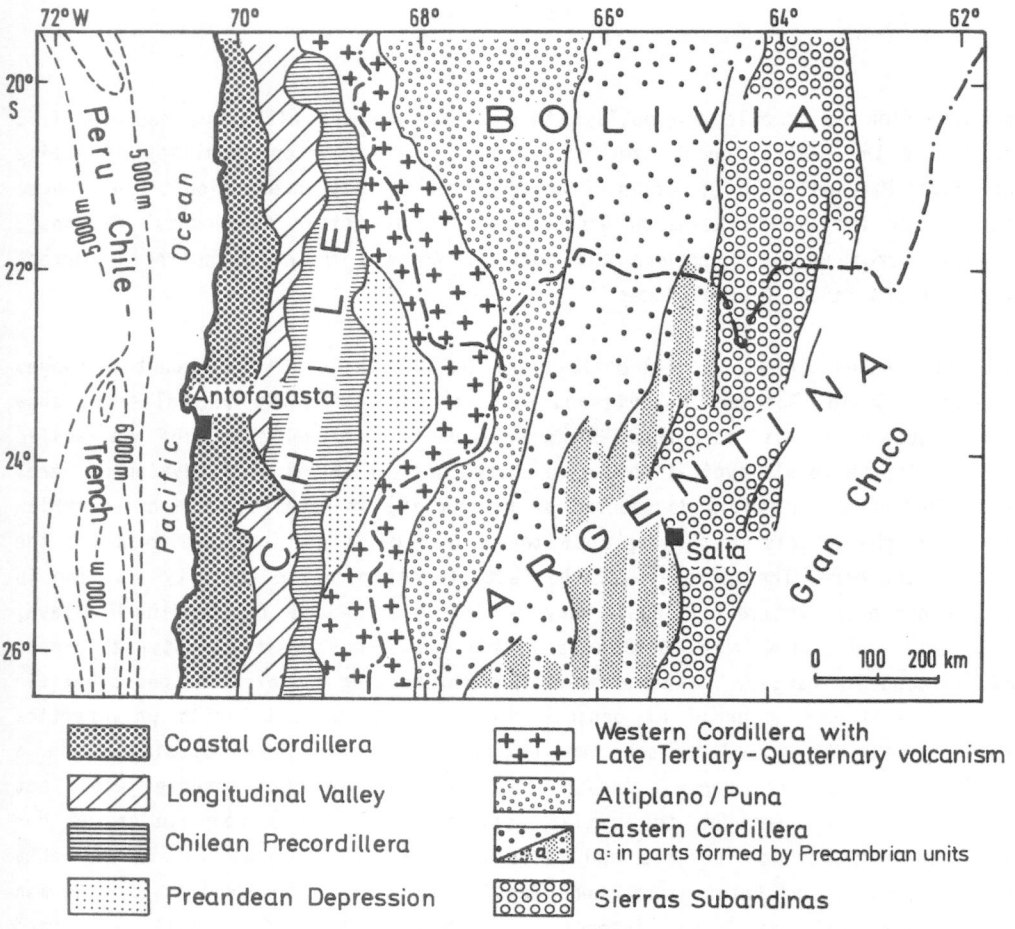

Simplified map of the main morphostructural units of the southern Central Andes
between 20° and 26° S

We are grateful to B. Dunker, H. Glowa, F. Jurtan, B. Kleeberg, and K. Zeschke,
Berlin, for their technical assistance in the preparation of this volume.

The editors

CONTENTS

A: PRE-ANDEAN EVOLUTION..1

I. Basement and Paleozoic Basins

BAEZA, L. & PICHOWIAK, S.:
Ancient crystalline basement provinces in the North Chilean Central Andes -
relics of continental crust development since the Mid-Proterozoic..................3

ACENOLAZA, F.G., MILLER, H. & TOSELLI, A.J.:
The Puncoviscana Formation (Late Precambrian - Early Cambrian).- Sedimentology, tectonometamorphic history and age of the oldest rocks of NW Argentina......25

KUMPA, M. & SANCHEZ, M.C.:
Geology and sedimentology of the Cambrian Grupo Mesón (NW Argentina)..............39

MOYA, C.:
Lower Ordovician in the southern part cf the Argentine Eastern Cordillera........55

BAHLBURG, H., BREITKREUZ, C. & ZEIL, W.:
Geology of the Coquena Formation (Arenigian-Llanvirnian) in the NW Argentine
Puna: Constraints on geodynamic interpretation....................................71

BREITKREUZ, C., BAHLBURG, H. & ZEIL, W.:
The Paleozoic evolution of Northern Chile: Geotectonic implications..............87

B: ANDEAN EVOLUTION..103

I. Mesozoic-Cenozoic Basins

GRÖSCHKE, M., HILLEBRANDT, A.v., PRINZ, P., QUINZIO, L.A. & WILKE, H.-G.:
Marine Mesozoic Paleogeography in Northern Chile between 21° - 26°S..............105

MARQUILLAS, R. & SALFITY, J.A.:
Tectonic framework and correlations of the Cretaceous-Eocene Salta Group;
Argentina...119

CHONG D., G.:
The Cenozoic saline deposits of the Chilean Andes between 18°00' and 27°00'
South Latitude..137

ABELE, G.:
Geomorphical west-east section through the North Chilean Andes near Antofagasta...153

II. Mesozoic-Cenozoic magmatism and tectonics...................................169

BUCHELT, M. & TELLEZ C., C.:
The Jurassic La Negra Formation in the area of Antofagasta, Northern Chile
(lithology, petrography, geochemistry)...171

SCHMITT-RIEGRAF, C. & PICHLER, H.:
Cenozoic ignimbrites of the Central Andes: A new genetic model...................183

GÖTZE, H.-J., SCHMIDT, S. & STRUNK, S.:
Central Andean gravity field and its relation to crustal structures..............199

WIGGER, P.:
Seismicity and crustal structure of the Central Andes...........................209

REUTTER, K.-J., GIESE, P., GÖTZE, H.-J., SCHEUBER, E., SCHWAB, K., SCHWARZ,
G. & WIGGER, P.:
Structures and crustal development of the Central Andes between 21° and 25°S.....231

A: PRE-ANDEAN EVOLUTION

I: Basement and Paleozoic Basins

ANCIENT CRYSTALLINE BASEMENT PROVINCES IN THE NORTH CHILEAN CENTRAL ANDES - RELICS OF CONTINENTAL CRUST DEVELOPMENT SINCE THE MID PROTEROZOIC

L. Baeza* & S. Pichowiak**
* Departamento de Geociencias, Universidad del Norte,
Casilla 1280. Antofagasta, Chile
** Institut für Geologie, Freie Universität, Altensteinstr. 34a,
1000 Berlin 33, Federal Republic Germany.

ABSTRACT

Ancient crystalline basement provinces are found only rarely in the North Chilean parts of the Central Andes and they show only diffuse structural similarities due to multiphase tectonic and metamorphic reworking. Radiometric dating of supracrustal series and intrusive cycles point to a crustal creation history reaching back as far as the Mid-Proterozoic (1460 ± 448 Ma: BELÉN metabasalts; $1254^{+97}/_{-94}$ Ma: CHOJA migmatite; $1213^{+28}/_{-25}$; CHOJA orthogneiss). The oldest rock records are of mostly low- to medium-grade metamorphic volcanic-clastic sequences with intrusive granitic pulses. High-grade metamorphic rocks occur only in few sites as granulitic gneisses and migmatites. Most of the radiometric data suggest various phases of metamorphic formation (Brazilian, Caledonian and Variscan events). Primary ages shown to be Mid-Proterozoic by the depositional events of basic to intermediate volcanic and volcanic-clastic rocks. Amphibolitic metabasalts are of primitive type, comparable to modern tholeiitic rocks of CFB (continental flood basalts) affinities. Shallow seated granitic bodies intruded the supracrustals during syn- to postkinematic stages of various orogenic cycles. The oldest intrusive cycles can be identified as "Brazilian", "Caledonian" and "Variscan" and vary in different provinces. The "Andean cycle" is responsible for the main tectonic kinematics and continous magmatic pulses. The characteristics of granitic rocks vary only slightly. Geochemical data indicate complex magmatic systems with possible crustal contaminants (transition from WPG to VAG trace element affinities; Sr_i values from 0.7051 to 0.7062).

INTRODUCTION

Andean mountain building processes have been well investigated and interpreted - as a classical example of a young orogen in the Circumpacific region. Kinematics are understood as triggered by the subduction of oceanic lithosphere at an active continental margin at least since the Triassic-Jurassic transition (e.g. FARRAR et al., 1970; McNUTT et al., 1975; PICHOWIAK et al., 1988). Most of the research published on the Central Andes region has focused on Mesozoic and Cenozoic processes

Lecture Notes in Earth Sciences, Vol. 17
H. Bahlburg, Ch. Breitkreuz, P. Giese (Eds.),
The Southern Central Andes
© Springer-Verlag Berlin Heidelberg 1988

Fig. 1: Sketch-map of the South American basement units, referring to the North Chilean Central Andes.
Modified, according to COBBING et al. (1977) and LITHERLAND et al. (1985).

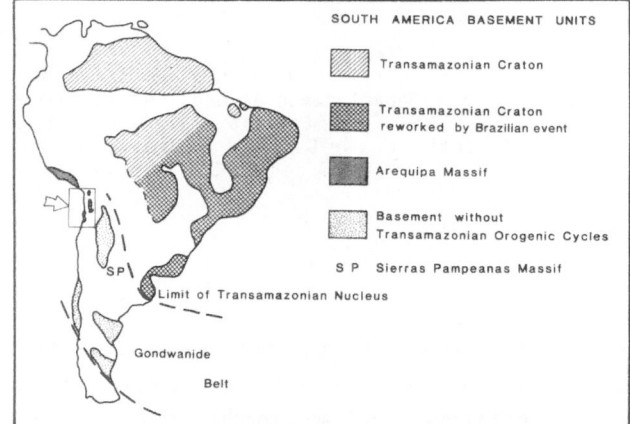

Fig. 2: Sketch-map of the Central Andes crystalline basement provinces. Cross-hatched areas are complex amphibolite-micaschist-gneiss terraines with medium- to high-grade regional meta-morphism of Proterozoic to Early Paleozoic formational events (see text). Areas with crosses are Late Proterozoic to Late Paleozoic granitic domaines.

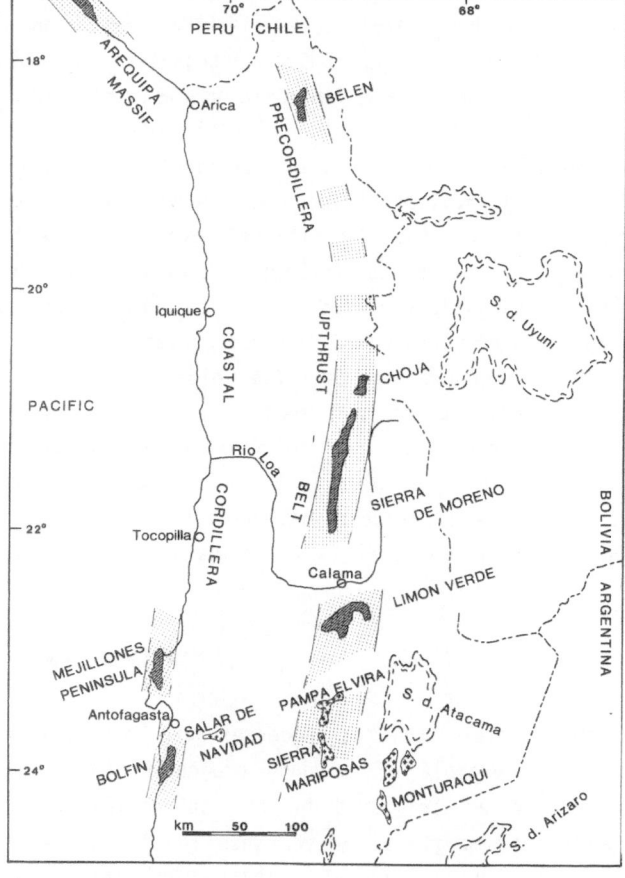

of subduction related tectonism and magmatism - a consequence of the seismic data gained by the geophysicists in the sixties and seventies which suggested crustal thicknesses in the Central Andes of up to 70 kms (ISACKS et al., 1968), quite a contrast to the situation in the Western Pacific where island arc regimes dominate modern geodynamic history. Another contrast was obvious when geochemical data of young Andean volcanic rocks were published which could be interpreted satisfactorily only with the help of models that included crustal influxes on the magmatic systems (e.g. KLERKX et al., 1977). This was quite understandable as the magmas had to pass an extremely long way through the continental crust and so the next step was to deal with isotopic modelling on crustal contamination and magma mixing (e.g. JAMES, 1981). After that research on Pre-Andean geological records as the source of crustal contamination in magmatic systems became popular. In fact, there were not many existing investigations on crystalline basement rocks outcropping in the Central Andean ranges which dealt with the lithological record on metamorphic conditions and geochemical characteristics of rocks. Most of the information for the Central Andes region of North Chile - the area focussed on in this review and an area with only few basement provinces - was compiled during mapping expeditions e.g. of the IIGC (Instituto de Investigaciones Geológicas, Chile) and some more detailed regional investigations were based on them. It was not merely the possibility of these crystalline basement rocks being a source for crustal contamination that made them interesting objects of study. These rocks alone were the key to more information on the "Andean Foundation", as it is obvious that Paleozoic and probably older supracrustal rocks form the visible basement of Andean developments in various regions of the Coastal Cordillera in Chile and Peru, also the Pre-Cordillera, then the Puna and the Pampean Ranges of Argentina, and the East-Cordillera of Bolivia and Argentina - to mention only the Central Andes segment. Another curious aspect is that the importance of Pre-Andean intrusive events grew with the rise of results from radiometric dating in the last decade. It thus became quite clear that several core structures of the Coastal Cordillera and the Pre-Cordillera of Northern Chile are the sites of large plutonic bodies of Late- to Early Paleozoic pulses (see e.g. BAEZA & PICHOWIAK 1988), which were earlier suspected to be part of the "Andean Batholith". It was in fact the detailed and refined time-scale of magmatic and metamorphic dating which helped to devide the "Pre-Andean" into distinct orogenic and anorogenic cycles.

At present the known timespan of intrusive events in the Central Andes region of North Chile ranges from 583 Ma to 43 Ma. Metamorphic ages cover a smaller span of time and range from 466 Ma to 175 Ma with culminations of Ordovician ($\sim$ 430 Ma) and Variscan pulses ($\sim$ 280 Ma) (COIRA et al., 1982; HERVÉ et al., 1981; DAMM et al., 1988). Indeed it was usually difficult to determine the metamorphic events with radiometric dating. Most of the "metamorphic" data therefore were gained from intrusive ages of synorogenic bodies. The current existing amount of very systematic data for the Central Andes region certainly leaves no room for serious doubt of the existence of an important Variscan orogenic cycle, a likewise important Caledonian event, and last not least there is also evidence of Mid- to Late Proterozoic kinematics with some yet unsolved questions:

- Did the Andean development of the Central Andes region take place completely on older sialic crust connected to the Transamazonian cratonic units or to provinces of a "Gondwanide Belt" respectively (see Fig. 1) ?

- How far back into the Precambrian do the roots of continental crust-creation cycles in the Andean basement reach and do we know all the successive cycles of ensialic processes prior to the development of the Andean ?

The aim of this paper is not to seak final solutions to these questions as the state of data today is such that the problems have just begun to emerge. Therefore we will here content ourselves with trying to give a representative synthesis of the data record gained in the North Chilean basement provinces.

GEOLOGICAL OUTLINE

The more extended and complete North Chilean crystalline basement areas dealt with here (Fig. 2) are the BELÉN region ($18^0 30'$ lat.S., $71^0 30'$ long.W), the QUEBRADA CHOJA ($21^0 05'$lat.S., $68^0 55'$ long.W), the CERROS LIMON VERDE ($22^0 40$ lat.S, $68^0 50'$ long.W), and the MEJILLONES PENINSULA ($23^0 00'$-$23^0 50'$lat.S, $70^0 30'$-$70^0 35'$long.W). Older basement areas mainly consisting of granitic complexes are the MONTURAQUI- the PAMPA ELVIRA- and the SIERRA MARIPOSAS region of the Pre-Cordillera and the SALAR DE NAVIDAD complex sited to the east of the Coastal Cordillera near the town of Antofagasta. The BOLFIN granulitic terrain in the Coastal Cordillera south of Antofagasta has recently been investigated by RÖSSLING (1988) (see also REUTTER et al., this volume).

In general it is possible to compare the more complex regions as they are similar in phenomenology. Low- to medium-grade metamorphic micaschists, micagneisses and amphibolites are the products of greenschist- to amphibolite-facies metamorphism of volcanic and clastic deposits. More or less intensive folding is obvious in all of the regions. Stress and strain relations are highly variable, but show some systematic aspects that can be correlated from one region to the other. Rocks indicative for granulite-facies conditions are only found in the CHOJA and BOLFIN areas. It is not solved yet if those of the BOLFIN area refer to an autochtonous or allochtonous area of regional hyperstene facies metamorphism.

Synintrusives are commonly found as apophyses and aplitic veins but also as foliated bodies of various dimensions. Postkinematic intrusives sometimes cut through all other structures. Brittle deformation is very common and surely the most longstanding regime. Most of this deformation type may be linked to Mesozoic shear systems - the overall importance of these shear regimes in the Coastal Cordillera connected with subduction processes has been pointed out by some authors (e.g. SCHEUBER, 1987; ARABASZ, 1971). N-S striking sinistral and dextral movements are responsible for "stockwork-tectonics" with intracrustal uprises of tens of kilometers and the patchy outcropping of the crystalline basement. The basement areas are generally flanked or separated from the surroundings by steeply dipping faults, there are no obvious hints as to nappe-tectonics.

BELEN, CHOJA and LIMON VERDE are sited in the N-S strike of the Pre-Cordillera - distributed at least over an N-S extension of nearly 1000 kms . These are not the only basement outcrops known from the Pre-Cordillera - the CHISMISA area between BELEN and CHOJA (19^040' lat.S, 69^015' long.W) and the SIERRA de MORENO between CHOJA and LIMON VERDE which are also parts of the Pre-Cordillera, were first mentioned by GONZALEZ-BONORINO & AGUIRRE (1970). The SIERRA De MORENO basement, east of the village of Quillagua which later was briefly described by SKARMETA & MARINOVIC (1981), was even the region in which possible Precambrian carbonaceous metasediments were found, reported by VENEGAS & NIEMEYER (1982). South of the LIMON VERDE mountains - as a structural elongation - several plutonic bodies of Paleozoic (Carboniferous-Permian) setting occur, such as the CERRO CATORCE De FEBRERO complex in the PAMPA ELVIRA (BAEZA & PICHOWIAK, 1988) and the SIERRA De MARIPOSAS intrusives (RAMIREZ & GARDEWEG, 1982). These rocks are not found in contact with any metasedimentary country rocks but show quite remarkable similarities in their compositional range, structural style of emplacement and age relations to those intrusives found in the complex regions.

Even if there is no direct connection between the individual outcropping areas there can be no doubt that they are part of the same structural unit - the Pre-Cordilleran anticlinorium or the PRE-CORDILLERAN UPTHRUST BELT (see Fig. 2), which was formed during the Late Mesozoic. The metamorphic provinces represent the most deepseated core complexes of this structure. Their formation, deformation and metamorphism was polyphase - with Pre-Andean activity and even some roots in the Mid-Proterozoic. Deformational-, shear- and thrust kinematics were active at least until the Eocene, but the Andean cycle may not have attributed more than a final very-low-grade retrograde overprint to the complex metamorphism of the outcropping basement series.

The MEJILLONES PENINSULA area is surely the most exotic of the crystalline basement provinces. It occurs quite isolated in the Coastal Cordillera, resembling rocks do not crop out further to the south of the Coastal Cordillera before the Central Chile region (the first basement-patches occur in the Los Chorros area: 29^010' lat. S,

71°30' long. W, south of La Serena the basement starts to crop out more frequently, then the Coastal Cordillera is finally dominated by crystalline basement rocks streching from the city of Valparaiso to the island of Chiloe. However, even if these rocks are similar in their phenomenology to the MEJILLONES basement, they suffered exclusively Mid- to Late Paleozoic metamorphism - see e.g. HERVÉ et al., 1981). To the north of the Mejillones Peninsula no more metamorphic rocks are seen, prior to reaching the AREQUIPA MASSIF in the Coastal Cordillera of southern Peru (Fig. 2). Fundamental works on the AREQUIPA MASSIF have confirmed the Precambrian setting of the metamorphic series (e.g. JENKS, 1948, COBBING & PITCHER, 1972, STEWART et al., 1974, COBBING et al., 1977, SHACKLETON et al., 1979). Transamazonian and Brazilian events (2000 - 1800 Ma / 650 - 500 Ma) were recognized as well as also Paleozoic thermal regimes and low- to medium-grade metamorphism which affected these rocks during Devonian to Mid-Permian times (MÉGARD et al., 1971). Mid-Devonian to Mid-Permian ages are reported likewise from the Coastal Cordillera in Central Chile (GONZALEZ-BONORINO & AGUIRRE, 1970, HERVÉ et al., 1981), but no hints as to older formation ages and metamorphic events have been discovered so far.

The BELÉN region

The "Formación Esquistos de Belén" was first mentioned by MONTECINOS (1963) as a type of basement province with some affinities to other South American regions of Pre-Andean settings. It is the northernmost outcropping basement province of the Pre-Cordilleran Upthrust Belt (Fig. 2). PACCI et al. (1980) tried to date these series by using Rb/Sr-WR and K/Ar methods. These authors were the first to report possible Precambrian formation events in Northern Chile. Finally DAMM et al. (1986, 1988) presented more detailed geochemical and geochronological data.

The somewhat patchy outcropping basement occurs along a NNW-SSE trending core structure, bordered by steeply eastward dipping faults. The lithology of this basement complex can be described appropriately as an amphibolite series with intercalated micaschists of highly variable thicknesses. Micaschists are sometimes transitional to micagneisses. The whole sequence is frequently penetrated by serpentinite stocks or lenticular to dykelike bodies. Final granodioritic intrusions are of postkinematic style.

Foliation and lepidoblastic textures and a transition to nematoblastic /lepidoblastic arrangements characterize the medium- to coarse-grained micaschists and micagneisses. The amphibolites that occur are medium- to fine-grained, nematoblastic and partly fibroblastic. Two types of hornblende (tschermakite, actinolite, after DAMM et al., 1988), almandine, plagioclase ± biotite, ± ilmenite and ± quartz are the main assemblages of the amphibolites, whereas the schists and gneisses contain quartz, plagioclase, biotite, muscovite, cordierite, garnet and $\pm$ sillimanite.

The regional metamorphism of this series was defined by DAMM et al.(1988) from the mineral assemblage of the amphibolitic rocks, reaching the medium-grade with 600^0 to 750^0C at 3 to 5 kb. A retrogressive low-grade overprint was also determined, caused by a strong shear-deformation regime.

Some preliminary age determinations of PACCI et al. (1980) left some doubts as to whether the given "reference isochrone age" of 1000 Ma was an accurate protolith age or an apparent age, caused by metamorphic disturbances in the Rb/Sr system. Nd/Sm-WR dating attempts on amphibolite samples resulted in an "errorchrone age" of 1460 ± 448 Ma by DAMM et al. (1986) and could at least establish the assumption of Precambrian formation. The same authors also estimated the timing of metamorphic events by a recalculation of PACCI's data, to a culminating phase from 500 to 440 Ma, similar results were also summarized by MPODOZIS et al. (1983).

The QUEBRADA CHOJA region

Following the same Precordillean upthrust stucture nearly 200 kms to the South, the QUEBRADA CHOJA basement province is exposed in a quite similar tectonic situation as the BELÉN region (Fig. 2). VERGARA & SATO (1978) gave first descriptions and geological maps. Detailed petrographical and geochemical work were published by DAMM et al.(1986, 1988), who were once more able to prove traces of a Precambrian cycle by radiometric dating on migmatitic and gneissic rocks.

Micaschists-micagneisses and amphibolites occur similarly to the BELÉN region with comparable textural patterns and mineral assemblages of the main phases. There are some differences in the thicknesses and the arrangements of the strata. Some of the gneissic parts especially form thicker "layers" and exhibit a more coarse, plutonic type of fabric. Those can be considered to be orthogneisses, whereas the banded gneissic parts and the thinner gneissic strata here and in the BELÉN region were interpreted as para-rocks. Controlled by complex tectonics, in some places the same sequence crops out transformed to an amphibolite-migmatite series. This includes some smaller intrusives of an in-situ anatectic granite. Again the whole assemblage is intruded by postkinematic granodiorites of a Pre-Andean event, finally magmatic pulses of the Andean cycle crosscut the whole structure.

A two stage model of the deformational and metamorphic history of this region was drawn up by DAMM et al.(1988). Barrowian-type metamorphism with a high-grade (amphibolite facies) culmination is followed by a retrogression, probably related to similar shear deformational pulses as in the BELÉN region. The high-grade critical parageneses are: calcic plagioclase - Mg-hornblende - diopsidic clinopyroxene - ilmenite $\pm$ quartz $\pm$ biotite, found in the amphibolitic parts of the amphibolite-migmatite suite. At this point the maximum degree of metamorphism reached 2.5 to 4.5 kb and up to 850^0C.

Age determinations mentioned above (DAMM et al. 1986, 1988) on magmatic zircon populations with the U/Pb method yielded ages of $1254^{+97}/_{-94}$ Ma and $1213^{+28}/_{-25}$ Ma for a migmatitic rock and an orthogneiss, respectively. Both data were obtained as discordia upper-intercepts and interpreted as being due to primary intrusive event, while a lower-intercept at $415^{+36}/_{-38}$ Ma was sufficiently synchronous with the U/Pb dating of a second zircon population from the same migmatite sample with an upper intercept at $466^{+8}/_{-7}$ Ma. That at least gave the time limit for the high-temperature metamorphic phase, including the migmatisation processes.

The LIMON VERDE region

The metamorphic series of LIMON VERDE located south of the town of Calama - a direct continuation of the line of the SIERRA De MORENO basement (Fig. 2) - consist of micaschists, amphibolites and subordinate strata of quartzites, outcropping in the western part of the Sierra Limón Verde area. These series were first mentioned briefly by HARRINGTON (1961); detailed mapping, petrographic and geochemical descriptions were given by BAEZA (1984). Some investigations on isotope geochemistry are to be found in HERVÉ et al. (1985), ROGERS (1985) and DAMM et al. (1986).

The micaschists exhibit strongly foliated lepidoblastic textural features with alternating parts of mica enrichments and xenoblastic quartz bands. The most common rock type is a garnet-mica-schist assemblage of: quartz - white mica - biotite - garnet $\pm$ plagioclase $\pm$ sphene. Amphibolites exhibit medium to coarse nematoblastic textures with interweaved amphibole prisms. Garnet-amphibolites and common amphibolites occur most frequently. The main assemblage consists of: green hornblende - oligoclase - zoisite - sphene $\pm$ garnet $\pm$ rutile, biotite is sometimes present. Some rarely intercalated quartzites with quartz - white mica - biotite - garnet - assemblages are of minor importance.

The conditions of metamorphism are characterized by critical mineral assemblages for micaschists and amphibolites respectively, with: quartz - white mica - red biotite - almandine-rich garnet $\pm$ oligoclase $\pm$ sphene; green hornblende - oligoclase - sphene - $\pm$ clinozoisite - quartz $\pm$ almandine-rich garnet $\pm$ biotite. The presence of: hornblende - garnet - oligoclase is sufficient reason for assuming metamorphism to be limited to the almandine - amphibolite facies (TURNER, 1981: transitional almandine zone between greenschist- and amphibolite facies). This also corresponds to the medium-grade metamorphism of WINKLER (1979). BAEZA (1984) estimates the p/T metamorphic conditions to be intermediate with approximately 4 - 5 kb and a maximum of 550 ^{O}C.

The age relations of the metamorphic series are once more of special interest - the timing of metamorphic events as well as the protolith ages. Various attempts undertaken by different investigators succeeded only in dating intrusive events that superimposed the metaseries during a Variscan period. HERVÉ et al.(1985) interpreted Rb/Sr-WR and K/Ar dates in minerals of the metaseries, as well as K/Ar WR dates in granitic rocks, as pointing to an orogenic cycle with its metamorphic culmination at around 300 Ma and terminating in postkinematic granitic intrusions lasting until 220 Ma. Granitic intrusion pulses were also determined by ROGERS (1985) with Rb/Sr-WR dating at 266±42 Ma, and by DAMM et al.(1986, 1988) with U/Pb zircon dating at 289±1.5 Ma. None of the age determination attempts resulted in accurate protolith ages, and aspects of calculations vary greatly according to the authors - HERVÉ et al. postulated 405 Ma, ROGERS calculated Precambrian formation in the range from 913 to 1308 Ma by Sm/Nd isotopic studies and DAMM et al. discovered inherited components with U/Pb dating on zircons of gneissic rocks which gave some hints as to a formation age for protoliths around 777±36 Ma.

MEJILLONES PENINSULA

Preliminary descriptions of the metamorphic series north and south of Antofagasta were given in some detail by FERRARIS & DI BIASE (1978) with first maps of the area. Re-examinations and intensified work on mapping and lithological records were published by VENEGAS (1979) and BAEZA & VENEGAS (1985). Geochemistry mainly of the amphibolitic rocks of the metamorphic rhythmites were the subject of investigations by BAEZA (1984) and DAMM et al. (1986). First radiometric ages were reported by FERRARIS & DI BIASE (1978) from Pb-alpha determinations, however, some problems as to interpretation remained. At the same time HALPERN (1978) submitted some Rb/Sr-WR data pointing to Early Jurassic magmatic activity. More systematic approaches towards determining radiometric ages were presented by DIAZ et al. (1985) and DAMM et al. (1986).

Two areas with somewhat different metamorphic characteristics can be distinguished on MEJILLONES PENINSULA. They were recently defined by BAEZA & VENEGAS (1988) as "Formación Punta Angamos", outcropping exclusively in the north of the peninsula and the "Formación Jorgino" (redefined Formación Jorgino by FERRARIS & DI BIASE, 1978) which is restricted to the central region.

Formación Punta Angamos: The series consists of micaschists, quartzschists, migmatitic gneisses, quartzites and subordinate amphibolites. Micaschists and quartzschists are fine-grained, foliated and typically banded; a lepidoblastic arrangement of white mica and chlorite is dominant. Porphyroblasts of garnet are common. In the vicinity of contacts to tonalitic-granodioritic intrusives of the Variscan cycle andalusite and sillimanite are evidence of contact metamorphism. Quartzites are fine-grained and weakly foliated. They are irregularly intercalated in the micaschists. Amphibolitic rocks occur sporadically in distinct zones as lenses or blocklike relics. Migmatitic gneisses of coarse- to medium grain sizes with various migmatitic textures are to be found only in the innermost core of the outcropping area. These rocks are derived from the micaschists by higher degrees of metamorphism that have even reached minimum-melt conditions.

The lithology demonstrates a type of regional metamorphism of low-grade greenschist facies in the chlorite zone. A superimposed contact metamorphism by the above mentioned intrusion caused a typically zoned contact aureole (biotite/andalusite-sillimanite/ orthoclase- sillimanite-zones, after BAEZA, 1984 and BAEZA & VENEGAS 1985).

Formación Jorgino: This series which forms the central parts of the MEJILLONES PENINSULA consists of gneisses, micagneisses and amphibolites. Gneisses and micagneisses are coarse-grained granoblastic. They exhibit a typical banding with lenticularly arranged cumulates of leucocratic (quartz-feldspars aggregates) and melanocratic (biotite - garnet $\pm$ hornblende) minerals and sometimes a marked schistosity. The amphibolites show fine- to medium-grained nematoblastic textures, they are generally foliated. Green hornblende and plagioclase (andesine - labradorite) in varying amounts are the major assemblages. Sphene $\pm$ garnet $\pm$ biotite also occur.

Regional metamorphism of intermediate Barrowian-type is indicated by metamorphic zonation. Three zones were distinguished by BAEZA (1984): biotite-, garnet- and kyanite-zone, representating a prograde metamorphism that reached estimated conditions of 4 - 6 kb and 400^0 - 600^0C.

At the same time, however, there are a number of radiometric age-determinations for Early Mesozoic intrusive pulses of the MEJILLONES PENINSULA and the adjacent regions. Those results give a consistent view in real terms of the magmatic development (see e.g. DIAZ et al., 1985; PICHOWIAK et al., 1988) of Early Andean stage culminating in the activity in the present Coastal Cordillera from 200 to 150 Ma. However, final clues as to Pre-Andean developments are still lacking. Up to now the oldest determinable events are an assumed protolith age for orthoamphibolitic rocks of the central MEJILLONES PENINSULA: 521$\pm$55 Ma (Nd/Sm-WR dating by DAMM et al., 1986). A reasonably comparable age calculation of 530 Ma for micaschists from the northern MEJILLONES PENINSULA was supplied by DIAZ et al. by Rb/Sr-WR dating, and an U/Pb zircon dating for the granodioritic-tonalitic intrusive body of the northern MEJILLONES PENINSULA of 561+12/-14 Ma, cefined by an upper intercept in the concordia diagram, was also given by DAMM et al. On the other hand, the same intrusive yielded an apparent Rb/Sr-WR age of approximately 200 Ma (DIAZ et al.). Geological record verifies that this plutonic phase is obviously linked to the orogenic cycle of deformation and metamorphism of the micaschist-amphibolite sequence - as a late kinematic intrusion. The maximum age therefore probably results from inherited components, whereas the minimum age represents a thermal aspect of the very important initial magmatic cycle of the Early Andean activity. The final statement that can be made regarding the culmination of metamorphic conditions in the micaschist-amphibolite sequence is that it must be hidden somewhere between 530 and 200 Ma.

Geochemistry

The orthoamphibolitic characteristics of nearly all of the samples dealt with from the amphibolite - schist - gneiss sequences has been mentioned by various researchers. BAEZA (1984) and DAMM et al. (1986) used some discrimination methods as proposed by LEAKE (1964) or MISRA (1971) (Fig. 3). Even the geological record favorized an interpretation of these series as sequences of metabasalts with variable intercalations of clastic rocks. Investigations on geochemistry focussed especially

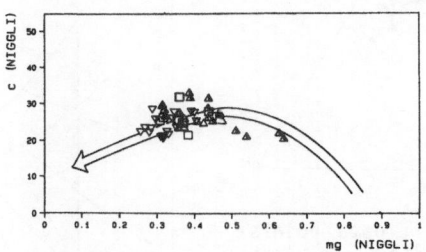

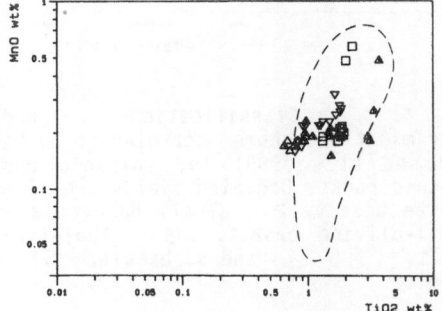

Fig. 3: Para- and orthoamphibolite discrimination diagrams. Upper: *c* versus *mg* (NIGGLI values) diagram according to LEAKE (1964).The arrow indicates a differentiation path of basaltic rocks. Lower: MnO versus TiO_2 diagram according to MISRA (1971). Common orthoamphibolitic rocks plot into the field as indicated by the broken line (for symbols c.f. Fig. 4).

on the characteristics of metabasaltic rocks - to gain some knowledge on the comparability of one region to the other, and on the type of tectonic settings related to the crustal formation of Pre-Andean developments. A brief description and re-interpretation of the results from the mentioned authors follows:

To start with the general features of rock compositions and classification it is remarkable that the compositional range do not show pronounced differences in the four regions. The spectrum of compositions is quite broad and reaches from andesitic

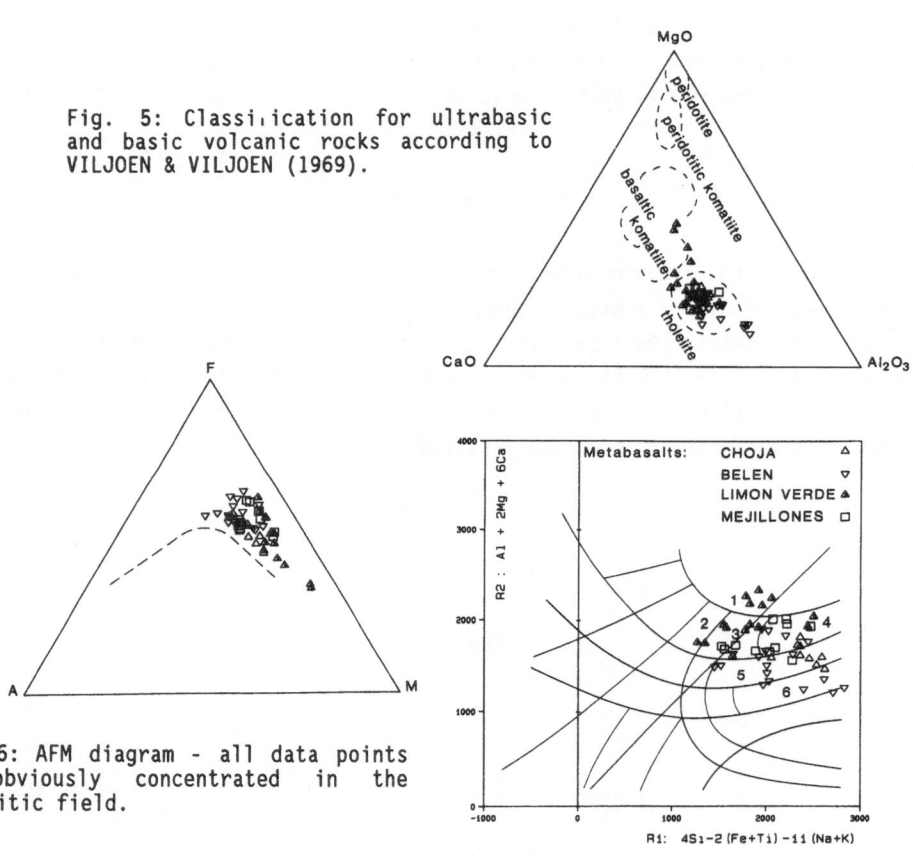

Fig. 5: Classification for ultrabasic and basic volcanic rocks according to VILJOEN & VILJOEN (1969).

Fig. 6: AFM diagram - all data points are obviously concentrated in the tholeiitic field.

Fig. 4: Classification and discrimination scheme according to De La ROCHE et al. (1980) for volcanic and plutonic rocks. Occupied fields are: 1 - picrite basalt, 2 - alkali basalt, 3 - alkali-olivine basalt, 4 - tholeiite basalt, 5 - andesi-basalt, 6 - andesite.

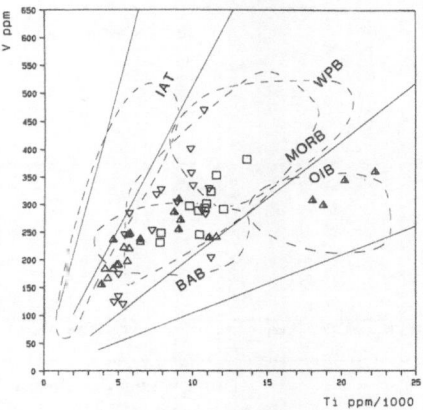

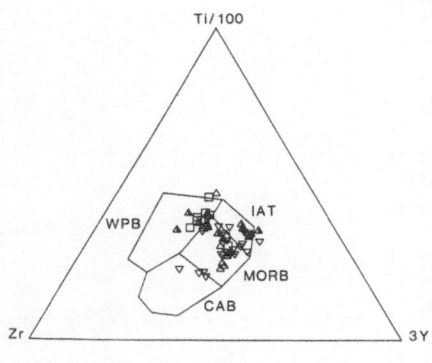

Fig. 7: Setting-type diagram for basaltic rocks according to SHERVAIS (1982). Data points are scattered between various discrimination fields. Abbreviations: IAT - island arc tholeiite, BAB - back arc basalt, MORB - mid ocean ridge basalt, WPB - within plate basalt, OIB - oceanic island basalt (for symbols see Fig. 4).

Fig. 8: Setting-type diagram for basaltic rocks according to PEARCE & CANN (1973). Data points do not show special concentration towards one of the discrimination fields.Same abbreviations as in Fig. 7 (CAB - calc alkali basalt).

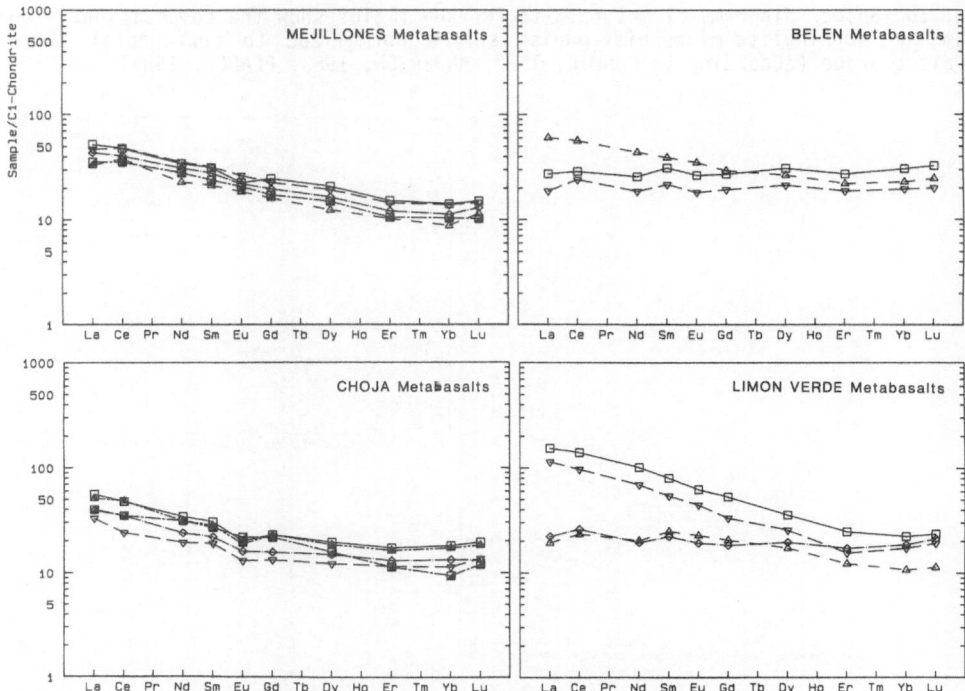

Fig. 9: Chondrite normalized REE spectra of metabasalts from the amphibolite-micaschist-gneiss series. For explanation see text.

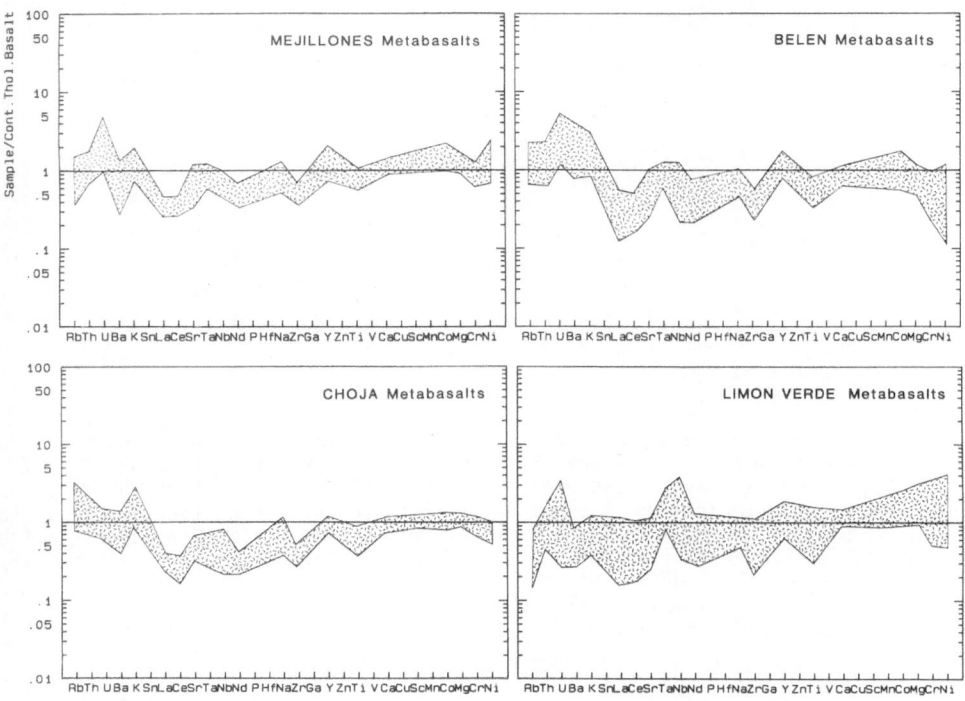

Fig. 10: Spider diagrams of metabasalts (marked fields show the compositional ranges) from the amphibolite-micaschist-gneiss series normalized to continental tholeiite basalt average (according to CONDIE, 1982, ANDERSON, 1983, PEARCE, 1983).

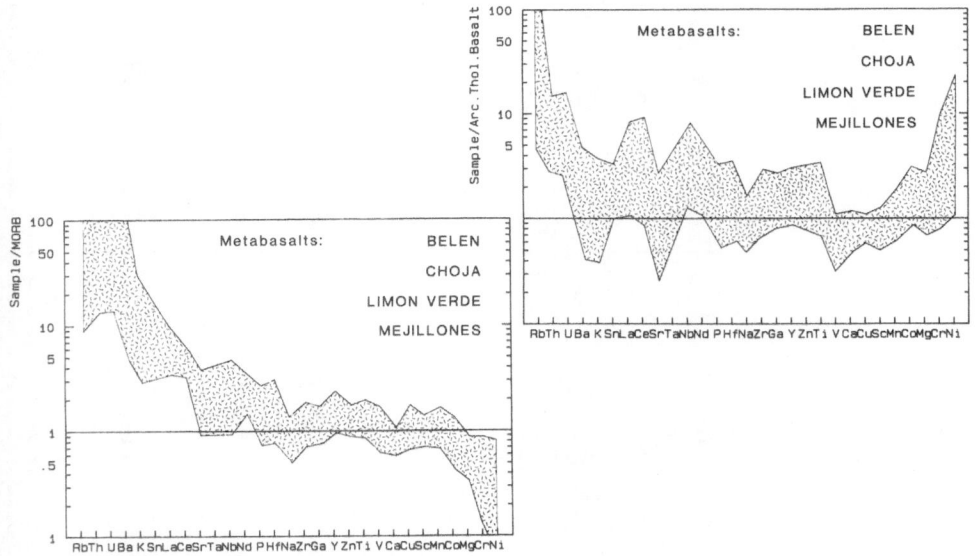

Fig. 11: Spider diagrams of metabasalts from the amphibolite-micaschist-gneiss series normalized to island arc tholeiite basalt- and mid ocean ridge tholeiite basalt averages (according to CONDIE, 1982, ANDERSON, 1983, PEARCE, 1983).

to picritic extremes. Most of the rock compositions investigated show silica
saturation with the exception of the rocks from the LIMON VERDE region which are
concentrated to the more primitive range. More developed rock compositions occur in
the BELÉN region (See Fig. 4, diagram after De La ROCHE et al., 1980, and Fig. 5,
triangle diagram after VILJOEN & VILJOEN, 1969). As a whole all the rocks follow a
tholeiitic differentiation trend in the AFM triangle (Fig. 6). However, one should
keep in mind that these are metamorphic rocks and the degree of metamorphism in
several cases reached conditions which could have caused changes in the primary
compositional balance. Mobilization and migration of the highly incompatible LIL
elements and to some extend also the less mobile HFS elements may have affected the
major compositions and really limit the use of binary and ternary minor- and trace
element discrimination methods. It is therefore obvious from the diagrams of SHERVAIS
(1982) and PEARCE & CANN (1973) that no distinct discrimination of the tectonic
setting is to be found - the data points constitute characteristic features of
tholeiitic rocks from within-plate- via island-arc- to mid-oceanic-ridge type (Figs.
7 & 8). The use of REE spectra (Fig. 9) and "spiderdiagrams" (Figs. 10 & 11) is
somewhat more satisfactory in the discrimination procedures. Again a characteristic
aspect of all the analyzed samples is the similarity seen here in the REE spectra.
The overall primitive characteristics of the rocks are documented by generally flat
spectra with only a low total concentration of REE - a type of REE pattern which is
generally found in primitive tholeiitic basalts of MORB settings, as well as in WPB
environments. Weak enrichment of the LREE against the HREE occurs in all samples
analyzed of the MEJILLONES PENINSULA rocks, all CHOJA rocks, and one of the BELÉN
rocks also shows that trend. Taking into account the assumption that the REE
characteristics are not essentially affected by secondary processes, this behaviour
can result from low degrees of olivine and pyroxene dominated fractionation or low to
medium degrees of partial upper mantle melting of a garnet-free source. Fractionation
(plagioclase dominated) is also indicated by some tendencies of a negative Eu-anomaly
in the CHOJA rocks and one of the MEJILLONES PENINSULA samples. Two of the LIMON
VERDE spectra show a more pronounced LREE enrichment. There have been some
assumptions of BAEZA (1984) about a bimodal character for the LIMON VERDE metabasalts
with one tholeiitic unit and another unit with characteristics transitional to
alkalic rocks. This was demonstrated by discrimination methods, but it is not
definitely clear whether these rock types represent some cumulate phases of the
tholeiitic series (Ti- and P fractionation due to cumulate enrichments). At least the
behaviour of the HREE would correspond with the last assumption, as one could expect
a more pronounced depletion in HREE with a real alkaline trend, related to a
different source type or different degrees of mantle partial melting processes. The
set of spiderdiagrams (Figs. 10 & 11) should document the comparability or diversity
between the rocks dealt with here and to some referring rock types. **Island arc low-K
tholeiite, continental tholeiite basalt and tholeiitic mid ocean ridge basalt** average
values were chosen as reference point and compared to the range of basement

metabasalts (normalization values according to CONDIE, 1982, ANDERSON, 1983, and PEARCE, 1983). On all the cases examined the continental tholeiite basalt example best corresponds with reference line.

Only few investigations on isotope systems of the metabasalts dealt with here exist. DAMM et al. (1988) recently reported on the systematics of epsilon$_{Nd}^T$ values. Samples from all four areas revealed positive values ranging from +2 to +8 and thus, on the

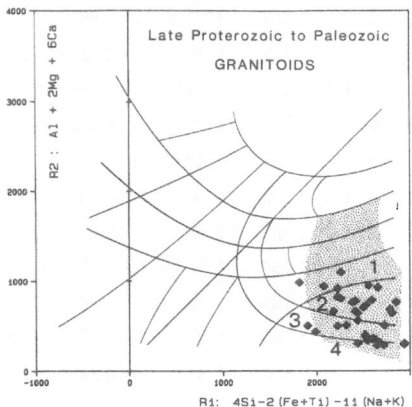
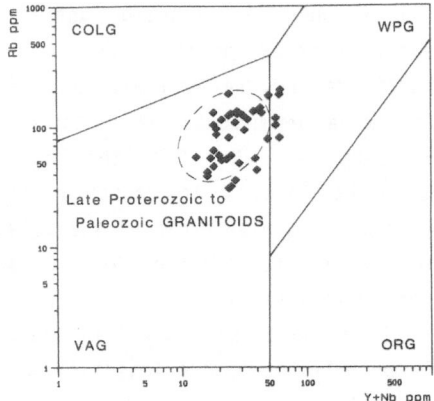

Fig. 12: De La ROCHE et al. (1980) diagram for classification of volcanic and plutonic rocks. Occupied fields are: 1 - tonalite, 2 - granodiorite, 3 - granite, 4 - alkali granite. The shaded space indicates calc alkaline differentiation trends.

Fig. 13: Setting type diagram for granitic rocks according to PEARCE et al. (1984). Most of the data points plot into the space (broken line) of continental/calc alkaline granites according to BALDWIN & PEARCE (1982). (Abbreviations: ORG - oceanic ridge granite, VAG - volcanic arc granite, COLG - collision type granite, WPG - within plate granite.)

Fig. 14: Spider diagram of Late Proterozoic to Paleozoic granites (marked field) - normalized to oceanic ridge granite (ORG) after PEARCE et al. (1984).
Two reference lines are given: Dark squares - Andean type granitoid, light squares - attenuated crust Skaergaard type granitoid.

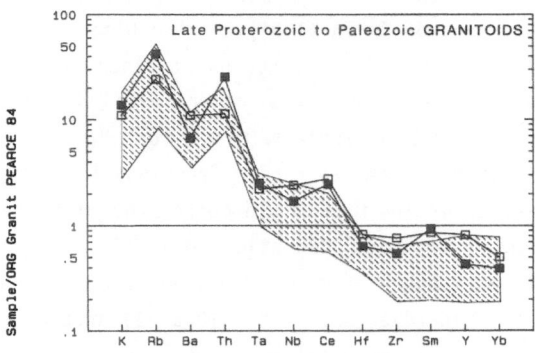

one hand closely follow the depleted mantle evolution, on the other hand also bear possibilities of mixing of inhomogeneously enriched mantle- and crustal-contamination sources with those considerable variations of $epsilon_{Nd}^T$ values.

A striking uniformity in Late Proterozoic to Late Paleozoic granitic rocks of the Central Andes makes it relatively easy to discuss them together as regards their geochemical aspects. Nevertheless a quite significant spread in modal compositions and textural features is to be mentioned in these granites, which, up to now have been dealt with by many authors (e.g. DAMM & PICHOWIAK, 1981, BERG & BAUMANN, 1985, MPODOZIS et al., 1983, DAVIDSON et al., 1981). The De La ROCHE diagram might give an appropriate idea of the compositional range (Fig. 12). Although most of the rocks have a granodioritic to granitic composition, transitions to tonalites and alkali-granites are also common. No true alkalic rocks occur but sometimes monzonitic affinities can result from metasomatic alteration which is a widespread phenomenon in the region, linked to various succeeding pulses of magmatic and shear-zone activities (e.g. ROGERS, 1985, SCHEUBER, 1987). The trend towards calc alkaline differentiation is also a general feature - there is quite a good correlation of that trend in the De La ROCHE diagram and the setting-type diagram by PEARCE et al.(1984), which discriminates most of the granites to the transition of VAG (volcanic arc granites) to WPG (within plate granites) (Fig. 13) - most of the data points even plot into the the field of "continental calc alkaline rocks" according to BALDWIN & PEARCE (1982). Finally it is of some importance that the element spectrum range of these granitic rocks is comparable to reference compositions of continental settings as given by PEARCE et al. (1984) (Fig. 14) - the Andean type granitoid and the attenuated crust Skaergaard type granitoid. Even if these two types of settings are quite different - a convergent active continental margin and a tensional pre-rift stage during intracrustal thinning and stretching - the element patterns reflect the same influx of continental crust components involvement. Although there have been only very few hopeful attempts of isotopic studies on metabasalts, the subject of Pre-Andean granitic rocks was dealt with the topics of magmagenesis more successfully by investigators such as HALPERN (1978), DAMM & PICHOWIAK (1981), SHIBATA et al. (1984), DIAZ et al. (1985), BERG & BAUMANN (1985), HERVÉ et al. (1985), ROGERS (1985), DAMM et al. (1986), BAEZA & PICHOWIAK (1988), DAMM et al. (1988).

Mostly Sr-isotopic data were published by these authors and the thereby gained Sr_i values are quite revealing. Granitic rocks with rather high initial values referring to "S-type" derivation are of an important occurrence only in the Chañaral-Taltal area in the Coastal Cordillera which is directly adjacent to the south of the segment mentioned here shown in Fig. 2 (DAMM & PICHOWIAK reported that initial Sr-values were generally higher than 0.7100 in Carboniferous to Triassic rocks. Similar results were gained by BERG & BAUMANN, but a few Sr_i values ranging from 0.7064 to 0.7046 for Early Permian to Late Triassic granites were also presented by SHIBATA et al.). In general the Sr_i characteristics remain somewhat transitional and "hybrid" between "S-" and "I-type" - at 0.7062 for the LIMON VERDE pluton with 266±42 Ma (ROGERS, 1985), 0.7051 for the CATORCE De FEBRERO pluton in the PAMPA ELVIRA with 285±32 Ma (BAEZA & PICHOWIAK) and approximately 0.7060 for the MEJILLONES PENINSULA northern plutonic complex with an apparent age of 530 Ma, presented by DIAZ et al.(1985) (this age was verified by Nd/Sm WR dating leading to a value of 521±55 Ma by DAMM et al., 1986 - see above).

CONCLUSIONS

The exceptional role of Precambrian and Paleozoic rocks as parts of the basement and precursors of Andean development has been mentioned by various authors (e.g. COBBING et al., 1977, COIRA et al., 1982, ZEIL, 1983, COBBING, 1985, PALMA et al., 1986, BREITKREUZ, 1986, DAMM et al., 1988). The synthesis presented in this paper for a region of the Central Andes in which ancient basement rocks occur quite rarely shows that various cycles of orogenic activity can be defined with the results of radiometric dating. The history of the North Chilean Central Andes of "crustal creation" can be traced back to the Mid-Proterozoic (1250 Ma). These ages are restricted to the northern parts of the Pre-Cordilleran Upthrust Belt (BELEN and CHOJA areas) and can be compared with similar results in the AREQUIPA MASSIF of southern Peru. Evidence of a Brazilian regeneration (650-500 Ma) which is pronounced in the AREQUIPA region occurs only in the MEJILLONES PENINSULA basement and in the adjacent granitic complex of the SALAR De NAVIDAD. On the other hand, a Caledonian regeneration (450-400 Ma) is to be seen exclusively in the Pre-Cordilleran basement series and not in the LIMON VERDE area. Evidence of Paleozoic events that took place 350 to 250 Ma ago, including a Variscan culmination, are most widespread. Medium- to high-grade conditions (mostly amphibolite facies) characterise a large part of regional metamorphism of the amphibolite-micaschist-gneiss domains. Barrowian-type metamorphic zonation only occurs in the MEJILLONES area and to some extent in the CHOJA region. No hints as to paired metamorphic belts as recognized in the Paleozoic of the Coastal Cordillera of Central Chile occur (GONZALEZ-BONORINO & AGUIRRE, 1970, HERVÉ et al., 1981).

While searching for appropriate interpretations of the tectonic setting of the crystalline basement areas it has been quite tempting to start with the use of terms from well known Archean and Proterozoic provinces. However, one could neither talk about related Granit-Greenstone Provinces here, nor do these basement outcrops support any interpretations of them as relics of a Mobile-Belt. Although the similarities in compositional- and age relations of at least the Pre-Cordilleran basement provinces seem to suggest a similar setting during formation and a more or less autochtonous position during all following orogenic cycles, it is still uncertain whether the Pre-Cordilleran basement represents the continental crust formational event of a shield-margin at approximately 1200 Ma, or whether what is to be seen here even is the result of ensialic processes between adjacent more stable cratonic units. At least the geochemical data of the metabasalts and granitoids seem to favor the latter assumption - although one should keep in mind that there are basement outcrops to be found in the Coastal Cordillera further to the west. COBBING (1985) made it quite clear that for the AREQUIPA MASSIF there was nothing to support a concept of an exotic terrane either. All results of different systematic investigations were consistent with a model of ensialic developments related to similar cycles of the Transamazonian nucleus. Therefore this and probably other fragments in the COASTAL CORDILLERA of the Central Andes segment such as the MEJILLONES PENINSULA and the BOLFIN area may represent the outermost portions of a Gondwana continental margin history. Despite these regions having the same recent structural site in front of the modern continental margin, the grounds for connecting them are quite weak. Only the Brazilian orogenic cycle can be identified simultaneously in the AREQUIPA MASSIF and the MEJILLONES PENINSULA, the Transamazonian is restricted to the AREQUIPA MASSIF and all recent data suggest that it is not present in the MEJILLONES PENINSULA basement. Finally the line of reasoning remains unconvincing if the MEJILLONES and BOLFIN basement areas were also connected to the Transamazonian craton in the same position as they are sited today. However, this could not be proved for the AREQUIPA MASSIF as well, where the structural trends cannot be correlated with the structures in the adjacent basement provinces of the Brazilian shield which are even found as mobile belts with Andean trending (LITHERLAND et al., 1985). COBBING et al. (1977) already mentioned that a longstanding regime of lateral shearing along strike-slip faults at the margin of the Transamazonian cratonic units may have dominated the plate boundary kinematics - due to systems of oblique convergence or even nonconvergent transcurrent slip of the continental- against the oceanic plate. Perhaps this type of continental margin activity - caused especially by rotational effects in the continental segments - was of even greater importance for the Pre-Andean development than the very straight subduction of an oceanic plate. Therefore the recent positions of the AREQUIPA MASSIF as well as the MEJILLONES PENINSULA and the BOLFIN area are somewhat parautochthonous segments that could have been variably brittled and twisted, at least from their Precambrian formation until Early Andean stages.

Acknowledgement

This paper is a synthesis of results of several project groups funded by the "Deutsche Forschungsgemeinschaft" (ref. no. Ze 6/37-1/3, Gi 31/51-1/3). We would also like to acknowledge the financial support of the D.A.A.D. (Deutscher Akademischer Austauschdienst). Analytical data were gained at the Institut für Mineraloge and the Institut für Geologie, Freie Universität Berlin (FRG) and also in cooperation with the Max-Planck Institut für Chemie, Mainz (FRG), the Zentrallaboratorium für

Geochronologie, Universität Münster (FRG), the Institut für Mineralogie der Univerität Tübingen (FRG) and the Centre Nacional de la Recherche Scientifique, Nancy (F).

REFERENCES

ANDERSON, D.L. (1983): Chemical Composition of the Mantle.- Journ. Geophys. Res., **88**, B41-B52.

ARABASZ, W.J. (1971): Geological and geophysical studies of the Atacama fault zone in northern Chile.- unpubl. thesis, Calif. Inst. Techn., 264 p., Pasadena/Calif.

BAEZA, L. (1984): Petrography and tectonics of the plutonic and metamorphic complexes of Limón Verde and Mejillones Peninsula, Northern Chile.- Unpubl. thesis, Univ. Tübingen, R.F.A. 205 p.

BAEZA, L. & PICHOWIAK, S. (1988): Complejos Plutónicos Controlados por Estructuras en la Precordillera del Norte de Chile. Geoquimica y Geochronología de Limón Verde y Catorce de Febrero.- V. Congr. Geol. Chileno, Santiago de Chile (submitted).

BAEZA, L. & VENEGAS, R. (1985): Caracterización petrográfica-   estructural de las rocas de basamento de la parte norte de la   Peninsula de Mejillones, Chile.- IV. Congr. Geol. Chileno, 1,.2-   35 - 2-55, Antofagasta.

BAEZA, L. & VENEGAS, R. (1984): El basamento cristalino de la Sierra Limón Verde, II Región Antofagasta: Consideraciones genéticas.- Revista Geol. Chile, **22**, 25-34.

BAEZA, L. & VENEGAS, R. (1988): Geología del Basamento de la Peninsula de Mejillones.- V. Congr. Geol. Chileno, Santiago de Chile (submitted).

BALDWIN, A.J. & PEARCE, J.A. (1982): Discrimination of productive and non-productive porphyritic intrusions in the Chilean Andes.- Econ. Geol. **77**, 664-674.

BERG, K. & BAUMANN, A. (1985): Plutonic and Metasedimentary Rocks from the Coastal Range of Northern Chile: Rb-Sr and U-Pb Isotopic Systematics.- Earth Planet. Sci. Let., **75**, 101-115.

BREITKREUZ, C. (1986): Plutonism in the Central Andes.- Zbl. Geol. Paläont., Teil 1, **9/10**, 1283-1293.

COBBING, E.J. (1985): The tectonic setting of the Peruvian Andes.- in: PITCHER, W.S., ATHERTON, M.P., COBBING, E.J. & BACKINSDALE, R.D. (eds.): Magmatism at a Plate Edge - The Peruvian Andes, 3-12, John Wiley and Sons, New York.

COBBING, E.J., OZARD, J.M. & SNELLING, N.J. (1977): Reconnaissance geochronology of the crystalline basement rocks of the Coastal Cordillera of Southern Peru.- Geol. Soc. Amer. Bull., **88**, 241-246.

COBBING, E.J. & PITCHER, W.S. (1972): The Coastal Batholith of Central Peru.- Journ. Geol. Soc. London, **128**, 412-460.

CONDIE, K.C. (1982): Plate Tectonics & Crustal Evolution.- 310 p., Pergamon Press Inc.

COIRA, B., DAVIDSON, J., MPODOZIS, C. & RAMOS, V. (1982): Tectonic and magmatic evolution of the Andes of Northern Argentina and Chile.- Earth Sci. Rev., **18**, 303-322

DAMM, K.-W. & PICHOWIAK, S. (1981): Geodynamik und Magmengenese in Nordchile.- Geotekt. Forsch., **61**, 1-166.

DAMM, K.-W., PICHOWIAK, S. & TODT, W. (1986): Geochemie, Petrologie und Geochronologie der Plutonite und des metamorphen Grundgebirges in Nordchile.- Berliner geowiss. Abh., **A 66**, 73-146.

DAMM, K.-W., PICHOWIAK, S., HARMON, R.S., TODT, W., OMARINI, R. & NIEMEYER, H. (1988): The "Old Central Andes"- a reflection of an active continental margin? - Geol. Soc. Amer. Bull. (submitted).

DAVIDSON, J., MPODOZIS, C. & RIVANO, S. (1981): El Paleozoico de Sierra de Almeida al oeste de Monturaqui, Alta Cordillera de Antofagasta.- Revista Geol. de Chile, **12**, 3-23, Santiago de Chile.

De La ROCHE, H., LETERRIER, J., GRANDCLAUDE, P. & MARCHAL, M. (1980): A classification of volcanic and plutonic rocks using R_1R_2-diagram and major element analyses - its relationships with current nomenclature.- Chem. Geol., **29**, 185-210.

DIAZ, M., CORDANI, U.G., KAWASHITA, K., BAEZA, L., VENEGAS, R., HERVÉ, F. & MUNIZAGA,
   F. (1985): Preliminary Radiometric Ages from the Mejillones Peninsula, Northern
   Chile.- Comunicaciones, 35, 59-67, Santiago de Chile.
FARRAR, E., CLARK, A.H., HAYNES, S.J., QUIRT, G.S., CONN, H. & ZENTILLI, M. (1970):
   K-Ar evidence for the post-Paleozoic migration of granitic intrusion foci in the
   Andes of northern Chile.- Earth Planet. Sci. Lett., 10, 60-66.
FERRARIS, F. & DI BIASE, F. (1978): Hoja Antofagasta.- I.I.G., Carta Geológica de
   Chile, No 30, Santiago de Chile
GONZALEZ-BONORINO, F. & AGUIRRE, L. (1970): Metamoprphic facies series of the
   Crystalline basement of Chile.- Geol. Rdsch., 59, 979-994.
HALPERN, M. (1978): Geological significance of Rb-Sr isotopic data of northern Chile
   crystalline rocks of the Andean orogen between 23⁰ and 27⁰ S.- Geol. Soc. Am.
   Bull.,89, 522-532.
HARRINGTON, H. (1961): Geology of parts of Antofagasta and Atacama Provinces,
   Northern Chile.- A.A.P.G. Bull., 45, 169-197.
HERVÉ, F., DAVIDSON, J., GODOY, E., MPODOZIS, C. & COVACEVICH, V. (1981): The Late
   Paleozoic in Chile: Stratigraphy, Structure and possible Tectonic Framework.-
   An. Acad. brasil. Ciènc., 53, 361-371.
HERVÉ, F., MUNIZAGA, F., MARINOVIC, N., HERVÉ, M., KAWASHITA, K., BROOK, M. &
   SNELLING, N. (1985): Geocronología Rb-Sr y K-Ar del Basamento Cristalino de
   Sierra Limón Verde, Antofagasta, Chile.- IV. Congr. Geol. Chileno, Antofagasta,
   3, 4-235 - 4-253.
ISACKS, B.L., OLIVER, J. & SYKES, L.R. (1968): Seismology and the new global
   tectonics.- Journ. Geophys. Res., 73, 5855-5899.
JAMES, D.E. (1981): The combined use of oxygen and radiogenic isotopes as indicators
   of crustal contamination.- Ann. Rev. Earth Planet. Sci., 9, 311-344.
JENKS, W.F. (1948): Geology of the Arequipa quadrangle.- Peru Geol. Inst. Bol., 9,
   204 p.
KLERKX, J., DEUTSCH, S., PICHLER, H. & ZEIL, W. (1977): Strontium isotopic
   composition and element data bearing on the origin of Cenozoic volcanic rocks of
   the Central and Southern Andes.-Journ. Volcanol. Geotherm. Res. 2, 49-71.
LEAKE, B.E. (1964): The chemical distinction between Ortho- and Paraamphibolites.-
   Journ. Petrol., 5 (2), 238-254.
LITHERLAND, M., KLINCK, B.A., O'CONNOR, E.A. & PITFIELD, P.E.J. (1985): Andean-
   trending mobile belts in the Brazilian Shield.- Nature, 314, 345-348.
McNUTT, R., CROCKET, J.H., CLARK, A.H., CAELLES, J.C., FARRAR, E., HAYNES, S. &
   ZENTILLI, M. (1975): Initial $^{87}Sr/^{86}Sr$ ratios of plutonic and vocanic rocks of
   the central Andes between latidudes 26⁰ and 29⁰S.- Earth Planet. Sci. Lett., 27,
   305-333.
MÉGARD, F., DALMAYRAC, B., LAUBACHER, G., MAROCCO, R., MARTINEZ, C., PAREDES, J. &
   TOMASI, P. (1971): La chaîne hercyniène au Pérou et an Bolivie, premiers
   résultats.- Cah. Orstom. Sér. Géol., 3, 5-44.
MISRA, S. (1971): Chemical distinction of high-grade ortho- and para-metabasites.-
   Norsk. Geol. Tidsskrift, 51, 311-316, Oslo.
MONTECINOS, F. (1963): Observaciones de Geología en el Cuadrángulo Campanani.- Univ.
   de Chile, Dpto. de Arica, unpubl. thesis. 225 p.
MPODOZIS, C., HERVÉ, F., DAVIDSON, J. & RIVANO, S. (1983): Los granitoides de Cerro
   de Lila, manifestaciones de un episodio intrusivo y termal del Paleozóico
   Inferiór en los Andes del Norte de Chile.- Revista Geol. de Chile, 18, p. 3-14.
PACCI, D., HERVÉ, F., MUNIZAGA, F., KAWASHITA, K. & CORDANI, U. (1980): Acerca de la
   edad Rb-Sr precambrica de rocas de la Formación Esquistos de Belén, Departamento
   de Parinacota, Chile.- Revista Geol. de Chile, 11, 43-50.
PALMA, M.A., PARICA, P.D. & RAMOS, V.A. (1986): El granito archibarca: su edad y
   significado tectónico, provincia de Catamarca.- Asoc. Geol. Argent. Rev., 41,
   414-419, Buenos Aires.
PEARCE, J.A. (1983): Role of the sub-continental lithosphere in magma genesis at
   active continental margins.- in: HAKESWORTH, C.J. & NORRY, M.J. (ed.):
   Continental Basalts and Mantle Xenoliths, 230-249; Shiva publ. ltd., Nantwich,
   UK.
PEARCE, J.A. & CANN, J.R. (1973): Tectonic setting of basic volcanic rocks
   investigated using trace element analyses.- Earth Planet. Sci. Lett., 19, 290-
   300.

PEARCE, J.A., HARRIS, N.B.W. & TINDLE, A.G. (1984): Trace Element Discrimination Diagrams for the Tectonic Interpretation of Granitic Rocks.- Journ. Petrol., **25**, 956-983.

PICHOWIAK, S., BUCHELT, M. & DAMM, K.-W. (1988): Mesozoic magmatic activity and tectonic setting in the N-Chile Central Andes Region: Granitoid magmagenesis and relations to volcanic activity during early stages of the Andean cycle.- Geol. Soc. Amer. Bull. (submitted).

RAMIREZ, C. & GARDEWEG, M. (1982): Hoja Toconao, Región de Antofagasta.- Serv. Nac. Geol. Miner., Carta Geol. de Chile, No 54, Santiago de Chile.

REUTTER, K.J., GIESE, P., GÖTZE, H.-J., SCHEUBER, E., SCHWAB, K., SCHWARZ, G. & WIGGER, P. (1988): Structures and crustal development of the Central Andes between 21^0 and 25^0 S.- (this volume).

ROGERS, G. (1985): Geochemical traverse across the North Chilean Andes.- unpubl. thesis, Dept. Earth Sciences Open Univ., Milton-Keynes, 333 p.

RÖSSLING, R. (1988): Petrologie in einem tiefen Krustenstockwerk des jurassischen magmatischen Bogens in der nordchilenischen Küstenkordillere südlich von Antofagasta.- unpubl. thesis, Inst. Geol. FU, Berlin, 165 p.

SCHEUBER, E. (1987): Geologie der nordchilenischen Küstenkordillere zwischen $24^030^/$ und 25^0 S - unter Berücksichtigung duktiler Scherzonen im Bereich des Atacama-Störungssystems.- unpubl. thesis, Inst. Geol. FU, Berlin, 170 p.

SHACKLETON, R.M., RIES, A.C., COWARD, M.P. & COBBOLD, P.R. (1979): Structure, metamorphism and geochronology of the Arequipa Massif of coastal Peru.- Journ. Geol. Soc. of London, **136**, 195-214.

SHERVAIS, J.W. (1982): Ti-V plots and the petrogenesis of modern and ophiolitic lavas.- Earth Planet. Sci. Lett., **59**, 101-118.

SHIBATA, K., ISHIHARA, S. & ULRIKSEN, C.E. (1984): Rb-Sr Ages and Initial $^{87}Sr/^{86}Sr$ Ratios of Late Paleozoic Granitic Rocks from Northern Chile.- Bull. Geol. Surv. Japan, **35**, 537-545.

SKARMETA, J. & MARINOVIC, N. (1983): Hoja Quillagua, Región de Antofagasta.- I.I.G., Carta Geol. Chile, No. 51, Santiago de Chile.

STEWART, J.W., EVERNDEN, J.F. & SNELLING, N.J. (1974): Age determinations from Andean Peru: A reconnaaissance survey.- Geol. Soc. Amer. Bull., **85**(7), 1107-1116.

TURNER, F. (1981): Metamorphic Petrology. Mineralogical, field and tectonic aspects.- Int. Ser. Earth Planet. Sci., 524 p., McGraw-Hill.

VENEGAS, R. (1979): Rocas metamórphicas plutónicas de la Peninsula de Mejillones al sur de los $23^017^/$ Sur y al oeste de los $70^030^/$ Oeste, II Región, Chile.- II. Congr. Geol. Chileno, E1-E20.

VENEGAS, R. & NIEMEYER, H. (1982): Noticia sobre un probable Precámbrico sedimentario-metamórphico en el borde occidental de la puna, al norte de Chuquicamata.- Congr. Geol. Chileno, 3 , F143-F154, Concepción.

VERGARA, H. & SATO, E. (1978): Cuadrangulo Quehuita y sector occidental del cuadrangulo Volcán Miño.- I.I.G., Carta Geológica de Chile, No. 32, Santiago de Chile.

VILJOEN, M.G. & VILJOEN, R.P. (1969): A collection of 9 papers on many aspects of the Barberton granite-greenstone belt, South Africa.- Geol. Soc. S. Afr. Spec. Publ., **2**, 295 p.

WINKLER, H.G.F. (1979): Petrogenesis of Metamorphic Rocks.- 348 p., Springer, New York-Heidelberg-Berlin.

ZEIL, W. (1983): Das präkambrische Basement der Anden. Ein Überblick.- Zbl. Geol. Paläont. **3/4** 246-254.

THE PUNCOVISCANA FORMATION (LATE PRECAMBRIAN - EARLY CAMBRIAN).- SEDIMENTOLOGY, TECTONOMETAMORPHIC HISTORY AND AGE OF THE OLDEST ROCKS OF NW ARGENTINA

Florencio Gilberto Aceñolaza
Facultad de Ciencias, Universidad Nacional de Tucumán
Miguel Lillo 205, 4000 Tucumán, Argentina

Hubert Miller
Institut für Allgemeine und Angewandte Geologie, Univ. München
Luisenstr. 37, D-8000 München, Federal Republic of Germany

Alejandro José Toselli
Facultad de Ciencias, Universidad Nacional de Tucumán
Miguel Lillo 205, 4000 Tucumán, Argentina

Abstract

The Puncoviscana Formation of NW Argentina consists of clastic, very weakly metamorphic sediments, which south of the province of Salta grade into schists, gneisses and migmatites. Three turbidite types, conglomerates, pelagic clays, carbonate rocks and rare volcanic rocks are found. The age of the unit is partly Late Precambrian, partly Early Cambrian proven by a rich trace fauna found at several sites. Palaeo-currents, mineral composition and geochemistry point to a provenance of the sediments from orogens of the Brazilian shield and to deposition on a passive continental margin. Deformation varies from upright folds in the north to flat, shear-bound structures in the south.

I. Introduction

In NW Argentina, from the Bolivian border in the north to about 28° southern latitude, a thick series of clastic sedimentary rocks is exposed. It gradually changes from not or very low grade metamorphic rocks in the north to high grade metamorphics and migmatites in the south. However, this transition is not clearly visible in many parts, because strong fault tectonic movements have uplifted distinct blocks several times from the Middle Cambrian to Recent. The high grade metamorphic rocks of this series rocks will not be considered here in detail.

Turner (1960) used the name Puncoviscana Formation for weakly meta-morphosed clastic sedimentary rocks of more than 2000 m thickness in

Lecture Notes in Earth Sciences, Vol. 17
H. Bahlburg, Ch. Breitkreuz, P. Giese (Eds.),
The Southern Central Andes
© Springer-Verlag Berlin Heidelberg 1988

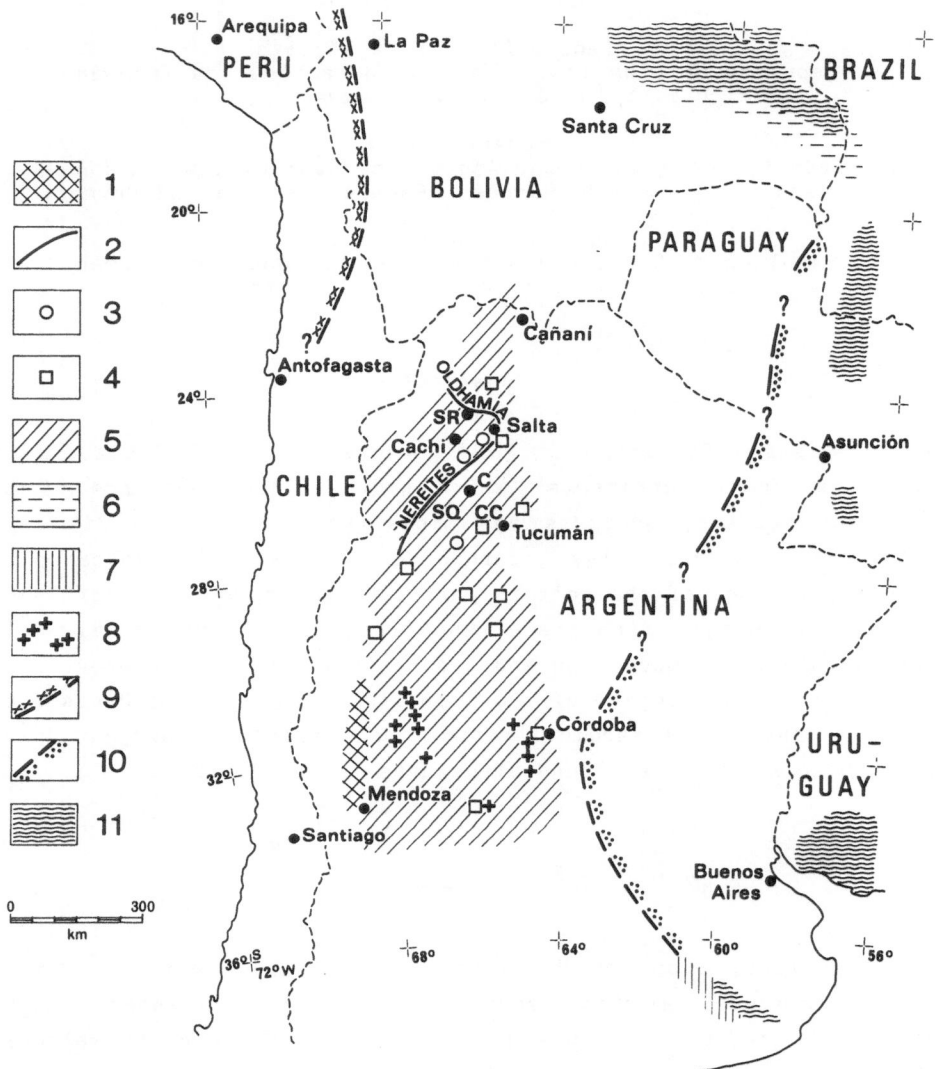

Fig. 1. Central South America during Late Precambrian - Early Cambrian

(1) Lower/Middle Cambrian carbonate rocks of the Precordillera; (2) Limit between distal turbidites of the Nereites facies and more proximal Oldhamia facies sediments in the Puncoviscana Fm.; (3) Important conglomerate occurrences in the Puncoviscana Fm. s.l.; (4) Major carbonate occurrences in the Puncoviscana Fm. and equivalents; (5) Outcrop area of the Puncoviscana Fm. and higher metamorphic equivalents; (6) Outcrop area of the Chiquitos Supergroup (miogeosynclinal sediments, Upper Proterozoic - ? Cambrian); (7) Outcrop area of the Sierras Bayas Fm. (littoral sediments, Upper Precambrian - ? Lower Cambrian); (8) Occurrences of Late Precambrian isotope ages in Central Western Argentina; (9) Edge of the Precambrian Arequipa Massif; (10) Edge of of Pre-Pampean areas; (11) Outcrops of Pre-Pampean cratonized areas. From Jezek et al. (1985). -- C.= Cafayate; CC.= Cumbres Calchaquíes; SQ.= Sierra de Quilmes; SR.= Santa Rosa de Tastil.

the Sierra de Santa Victoria near the Argentinian-Bolivian border (Fig. 1). Later he extended it to rocks as far south as the Quebrada del Toro W of Salta (Turner 1972). For similar rocks further south, several local names have been proposed, e.g. Suncho Formation, Medina Formation or San Javier Formation, but until a clear stratigraphic relationship between these will be known, we prefer to summarize them as "Puncoviscana Formation and equivalents", or "Puncoviscana Formation s.l." (Aceñolaza & Toselli 1981).

Trace fossils (Aceñolaza & Durand 1973) gave reason for an at least partly Cambrian age of the strata. However, preliminar isotopic age determinations (Halpern & Latorre 1973) seemed to prove a Precambrian age of granites that intruded the folded Puncoviscana Fm. Other poorly documented age data (Shell Argentina) were published by Borrello (1971), suggesting an approximate age of 530 Ma for the intrusions.

In the early eighties, Omarini (1983) and Ježek (1986) began to elaborate a facies concept for the sequence and Willner & Miller (1986) were the first to divide the pre-Late Cambrian rocks in several tectonometamorphic units.

II. Sedimentology of the Puncoviscana Formation

The rocks of the Puncoviscana Fm. s.l. are mostly composed of a submarine fan facies association of turbiditic clastic sediments (Baldis & Omarini 1984; Ježek & Miller 1986, 1987). Within this fan complex, six facies types have been distinguished, which overlap each other in a complicated, not yet clearly known system.

1. Facies types

a) Proximal facies

This type of facies is caracterized by a strong predominance of psammitic layers (psammite/pelite ratios > 5), average bed thicknesses above 30 cm, sharp bottom and top contacts of the beds and the nearly complete absence of Bouma-sequences. These sediments may have been deposited within proximal depositional lobes, on inner fan channel margins or on distal parts of mic fan channels. They correspond more to grain flow currents than to turbidites sensu stricto.

b) Intermediate facies

This facies type represents the classic turbidites. Incomplete ACE and ACDE Bouma-sequences are common. Average bed thicknesses are less than 30 cm, the psammite/pelite ratios range from 5 : 1 to 1 : 1. Lower bed contacts are sharp and frequently show load casts, whereas the tops of the beds grade into pelites. The sediments were deposited by high velocity / high density turbidity currents between the mid fan channel margins and the outer fan depositional lobe. They are common between Tucumán and Salta and north of Salta (Fig. 1).

c) Distal facies

Thin to very thin bedded pelite sequences of a maximum grain size of silt to very fine sand prevail in this facies type. Psammite/pelite ratios far below 1 are characteristic. Average thicknesses of the beds are of about a few centimeters. Bouma-sequences are reduced to DE or CDE intervals. The distal facies is common west of Salta and further to the north. It is interpreted as the distal continuation of the inter-mediate turbiditic facies characterizing low velocity / low density turbidity currents off levees or on the outer fan fringe.

d) Channel facies

Conglomerates occur at some localities. They contain pebbles of mostly clastic sedimentary origin. Some basic and acidic magmatic pebbles of unknown origin are also found. They may have been derived from local volcanic events recently described by Omarini & Alonso (1987), Chayle & Coira (1987) and Manca et al. (1987). The most important conglomerate has been named Corralito Fm. by Salfity et al. (1975). The conglome-rates are generally associated with the intermediate facies type; only in Las Tienditas (south of Salta) they are associated with carbonates. Following Ježek & Miller (1986) we interpret them as debris flow depo-sits of inner fan channels or of proximal parts of mid fan channels near submarine canyon mouths.

e) Pelagic facies

Red pelites frequently occur intercalated with the turbidites of the intermediate type. Particularly thick occurrences of the red pelites are found near Ingeniero Maury in the Quebrada del Toro. They are formed by the autochthonous particle rain, party mixed with whirled-up suspension material.

f) Carbonate facies

Various isolated outcrops of carbonate rocks are restricted to the eastern parts of the area. Partly they are associated with conglomerates. We think that at least part of the carbonates are great olistoliths. They derived from carbonate platforms which partly broke down along submarine canyons and the continental slope. To the south, within the higher grade metamorphics, occurrences of marble may have the same origin (e.g. Peñas Azules in the Cumbres Calchaquíes, and within other Pampean Ranges).

g) Volcanic rocks

Various small occurrences of lava flows and dykes have been found recently (Omarini & Alonso 1987, Chayle & Coira 1987, Manca et al. 1987, Toselli & Aceñolaza 1984). These volcanic rocks will help to define the geotectonic environment and the time of deposition of the Puncoviscana Formation.

2. Provenance of Material

The provenance of the sedimentary material from the Brazilian shield has been proven by three methods: analyses of palaeocurrents, petrographic composition and geochemistry.

a) Palaeocurrents

More than 800 palaeocurrent data (ripple marks, flute casts, slump folds and clast imbrication), obtained by Ježek (1986) show a generally unidirectional pattern of transport from east to west (present coordinates). Minor, more meridionally directed currents in the west (see also Sosa Gómez 1984) may be correlated with basin parallel currents that often occur off the virtual sediment fans. In the extreme NW a certain influence of the Arequipa Massif may be possible.

b) Modal analyses of main minerals

The analysis of grain composition using the Dickinson & Suczek (1979) plots, clearly shows that the material derived from recycled orogens, or with other words, from collision orogen sources (Ježek & Miller 1987). Autochthonous material derived from the scarce volcanic rocks within the Puncoviscana Formation itself is not relevant within the general composition of the sediments. Thus, the uplifted orogens of the

Brazilian cycle east of the area of deposition are the most probable source rocks.

c) Geochemistry

Willner et al. (1985) analyzed 120 psammite and pelite samples of different metamorphic grades from outcrops of the Puncoviscana Formation and its higher metamorphic equivalents. There are only little differences between the rocks of various metamorphic grade. All greywacke samples can be defined as being in an intermediate position between the quartz-rich and quartz-intermediate types of Crook (1974). Poorly recycled greywackes of orogenic belts seem to be an adequate source area. Major volcanic influence is only probable in the south. These geochemical analyses support the petrological and sedimentological data, which deny an island arc in the vicinity of the Puncoviscana basin. The area of deposition most probably was a stable continental margin.

III. Metamorphism and deformation

At the type locality and to the south onto a line about 20 km north of Cafayate (Fig. 1), it is not easy to decide from the outcrop whether the Puncoviscana Formation sensu stricto is metamorphosed or not. Toselli & Rossi (1982) and Doherr (1983), however, have shown by illite crystallization studies that the pelites of the Puncoviscana rocks at many places have passed through the diagenesis/metamorphism boundary. Thus, great part of the formation is "very low grade" metamorphic.

To the south, metamorphism increases strongly. From the fossil bearing sediments of the Puncoviscana Formation s.s. to the high grade metamorphic rocks of the northern Sierras Pampeanas a continuous transition is evident (Willner & Miller 1986, Willner et al. 1987). The deposition area of the Puncoviscana Formation, thus, continued far more to the south than the weakly metamorphosed rocks of the Puncoviscana Formation s.s. (Ježek et al. 1985).

This regional dynamometamorphic overprint must not be confounded with the frequent contact metamorphism, which often develops in the Puncoviscana Formation metasediments near early Palaeozoic granitoids.

The increase of metamorphism to the south is not due to an actual increase of metamorphic conditions, but to the exposure of deeper crustal horizons. This is demonstrated by the tectonic zoning described by Willner & Miller (1986) and Willner et al. (1987). On the one hand, a "transposition line" has been defined north of Tucumán, where the primary sedimentary banding as the principal set of s-planes is reemplaced by schistosity (s_1) planes. On the other hand, greater structural units have been exposed by local horst-like uplifts of deeper crustal levels at the Nogalito Ranges area NE of Tucumán and in the Sierra de Quilmes W of Tucumán.

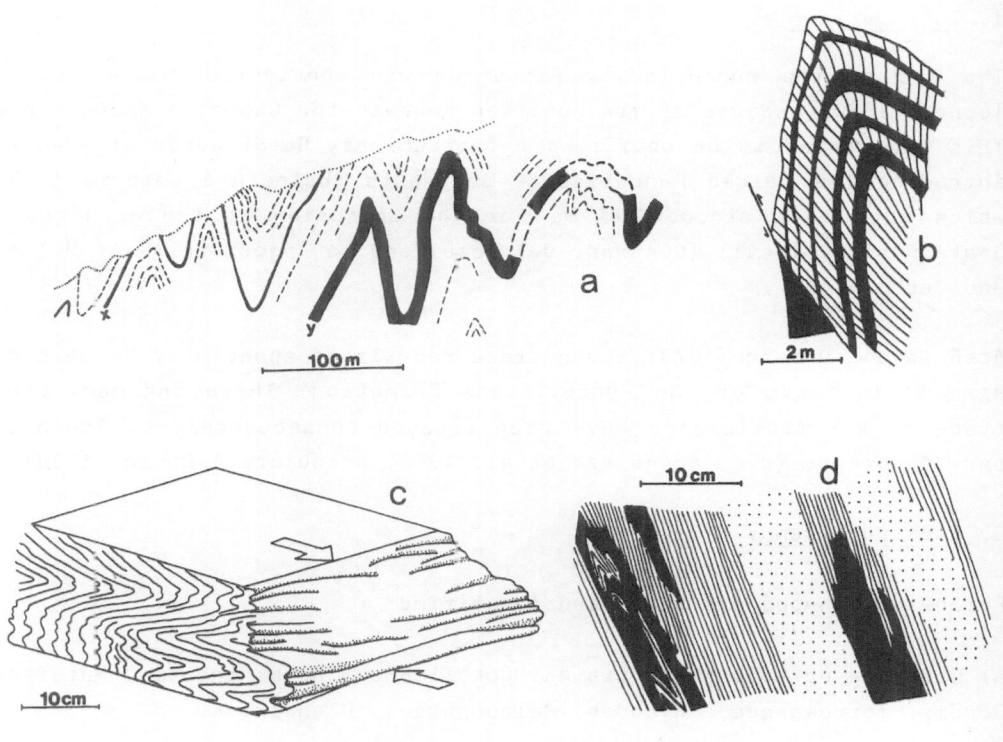

Fig. 2. Mesoscopic structures in various tectono-metamorphic zones.

(a) Zone I b; Río de las Conchas, Quebrada Don Bartolo. - F_1 chevron folds, x = quartzitic layer, y = red pelite layer. (b) Zone II a; Cumbres Calchaquíes, road SW Rearte. - F_1 fold with banding in metapsammitic layers as an axial plane cleavage. (c) Zone II b; Sierra del Campo - Subhorizontal F_2 folds deforming s_1 banding. (d) Zone IV; Cumbres Calchaquíes, Quebrada Los Cuartos. - S_1 banding transposed subparallel to the bedding with destruction of F_1 hinges by slip along banding. Rootless isoclinal folds in pelitic layers; dotted are metapsammites poor in phyllosilicates. (From Willner & Miller 1986).

Folds are generally upright in the Puncoviscana Formation s.s.(Fig. 2). An axial plane cleavage appears in intermediate structural levels, forming a banding produced by pressure solution processes in psammites (e.g. Sa. San Javier; Miller & Willner 1981, Toselli & Rossi 1983). This banding was transposed parallel to bedding in even deeper levels. Horizontal shearing plays an important role in the deformation of deeper crustal zones, leading to thickening of the crust during the Early Palaeozoic (Willner et al. 1987).

IV. The age of deposition, metamorphism and deformation

1. Introduction

The age of the Puncoviscana Formation was considered Precambrian by Turner (1960) because of its position beneath the Cambrian Mesón Group. This age seemed to be confirmed by preliminary Rb-Sr dates of granites intruding the folded Puncoviscana sediments (Halpern & Latorre 1973), which gave ages around 600 Ma for the intrusion of the granites of Santa Rosa de Tastil (Quebrada del Toro) and La Angostura (S of Molinos and Cachi).

Aceñolaza & Durand (1973) found trace fossils of apparently Phanerozoic age within rocks of the Puncoviscana Formation. These and many other trace fossil assemblages have been treated consecutively by Acenolaza and co-workers (e.g. Aceñolaza et al. 1976, Aceñolaza & Durand 1986).

2. Biostratigraphy

Following ichnogenera have been identified:

a) Creeping or tube shape traces: Cochlichnus, Helminthopsis, Nereites, Gordia, Torrowangea, Phycodes, Palaeophycus, Planolites.

b) Appendicular sharp traces: Asaphoidichnus, Diplichnites, Dimorph-ichnus, Monomorphichnus, Oldhamia, Protichnites, Protovirgularia, Tas-manadia.

c) Bilobate trace: Didymaulichnus.

This great amount of trace types, and particularly of forms produced by appendicular structures, enables us to decide that at least part of the Puncoviscana Formation belongs to the Phanerozoic. Traces like Oldhamia, Dimorphichnus or Diplichnites are generally related to trilobites or similar animals (Alpert 1975, Aceñolaza & Durand 1984).

Some recent discoveries of medusoid imprints at several localities (Aceñolaza & Durand 1986) have been tentatively identified as forms similar to Beltanelliformis and Sekwia. They may prove a Late Vendian age for part of the Puncoviscana Formation.

3. Isotope geochronology

The folded and metamorphosed Puncoviscana Formation has been intruded by several granitoid complexes, partly older than Late Cambrian, because they are overlain by the Upper Cambrian Mesón Group (Santa Rosa de Tastil, Cañaní). Various authors recently communicated a Precambrian age of these granites (Omarini et al. 1985, Damm et al. 1986), based on Rb-Sr data of about 720 Ma for samples of the Santa Rosa de Tastil complex. These authors consequently argued for a Pre-Vendian age of the Puncoviscana Formation.

On the other hand, Bachmann et al. (1986, 1987) have published U-Pb and Rb-Sr data of the Santa Rosa de Tastil complex and U-Pb data of the Cananí granitoid, which clearly gave ages between 520 and 535 Ma for this intrusive cycle.

These Middle to Late Cambrian intrusion ages are indirectly confirmed by dating the metamorphism of the Puncoviscana Formation, which as well as the folding occurred before the intrusion. Adams et al. (in prep.) found K-Ar ages of 525 to 530 Ma in very low grade metamorphic rocks of the Puncoviscana Formation and ages of 535 to 570 Ma in the lower structural levels situated NE and NW of Tucumán. Medium grade metamorphic equivalents of the Puncoviscana Formation W and S of Tucumán have yielded similar ages of 540 to 570 Ma for a first metamorphic event (Bachmann et al. 1986).

The granite of La Angostura (Molinos, Salta), considered to be Late Precambrian by Halpern & Latorre (1973) was recently dated as Late Ordovician by the U-Pb method (Lork et al., in prep.).

4. Discussion

The age of metamorphism of the Puncoviscana Formation and the age of the intruded granitoids clearly give an upper limit to its sedimentation age. It must be pre-Late Cambrian in the northern part (Puncoviscana Formation s.s.) and pre-Middle Cambrian to the south, but

not necessarily Precambrian. Biostratigraphic evidence calls for a
partly Vendian, partly early Cambrian age of the deposition of the
sediments.

V. Conclusions

From the Late Vendian to the Early Cambrian a submarine fan system
deposited a hugh pile of clastic sediments on the stable palaeo-Pacific
margin of Gondwana west of the Brazilian Shield. It was accompanied by
a carbonate platform to the east. Volcanic events occurred rarely.

The sediments were metamorphosed and folded from the late Early
Cambrian to the Middle Cambrian; they were intruded at the end of the
Middle Cambrian by several granitoids. These events characterize the
main deformation phase of the Pampean orogenic cycle in the Central
Andes.

VI. Acknowledgements

We thank CONICET, the Deutsche Forschungsgemeinschaft and the Stiftung
Volkswagenwerk for financial support.

VII. References

Aceñolaza, F.G. & Durand, F. (1973): Trazas fósiles del basamento
    cristalino del Noroeste Argentino. - Bol. Asoc. Geol. Córdoba, 2:
    45-55.

Aceñolaza, F.G. & Durand, F. (1984): The trace fossil Oldhamia. Its
    interpretation and occurrence in the Lower Cambrian of Argentina.
    - N. Jb. Geol. Paläont. Mh., 1984: 728-740.

Aceñolaza, F.G. & Durand, F. (1986): Upper Precambrian - Lower Cambrian
    biota from the northwest of Argentina. - Geol. Mag., 123: 367-375.

Aceñolaza, F.G. & Toselli, A.J. (1981): Geología del Noroeste Argen-
    tino. - Publ. Esp. Fac. Ciencias Natur., Univ. Nac. Tucumán: 212
    pp.

Aceñolaza, F.G., Durand, F. & Díaz Taddei, R. (1976): Geología y
    contenido paleontológico del basamento metamórfico de la región de
    Cachi (Provincia de Salta). - Actas VI. Congr. Geol. Argent., 1:
    319-332.

Adams, Ch., Miller, H. & Toselli, A.J. (in prep.): New K-Ar ages on the
    metamorphic history of the oldest sedimentary basin on the Pacific
    margin of Gondwana in NW Argentina (Puncoviscana Formation and
    equivalents, Vendian to Early Cambrian).

Alpert, S.P. (1975): Trace fossils of the Precambrian-Cambrian
    succession. White Inyo Mountains, California. - Reprinted from
    Dissertation Abstracts International, 35 (8).

Bachmann, G., Grauert, B., Kramm, U., Lork, A. & Miller, H. (1986):
    Oberkambrischer Magmatismus im Grundgebirge Nordwest-Argentiniens:
    Isotopengeologische Untersuchungen an Granitoiden der Intrusiv-
    komplexe von Santa Rosa de Tastil und Canani. - Berliner Geowiss.
    Abh., Reihe A, Sonderband 10. Geowiss. Lateinamer.-Koll.: 111-112.

Bachmann, G., Grauert, B., Kramm, U., Lork, A. & Miller, H. (1987): El
    magmatismo del Câmbrico Medio / Câmbrico Superior en el basamento
    del Noroeste Argentino: Investigaciones isotópicas y geochrono-
    lógicas sobre los granitoides de los complejos intrusivos de Santa
    Rosa de Tastil y Canani. - Actas X. Congr. Geol. Argent., 4: 125-
    127.

Baldis, B. & Omarini, R. (1984): El Grupo Lerma (Precámbrico-Câmbrico)
    en la comarca central Saltena y su posición en el borde Pacífico
    Americano. - Actas IX. Congr. Geol. Argent. 1: 64-78.

Borrello, A.V. (1971): The Cambrian of South America. - In: Holland,
    C.H. ed.: Cambrian of the New World: 385-438, London (Wiley-
    Interscience).

Chayle, W. & Coira, B. (1987): Vulcanitas básicas a ultrabásicas y
    mesosilíceas de la Formación Puncoviscanaa en el área del Cerro
    Alto de Minas - Departamento Tilcara - Jujuy, Argentina. - Actas
    X. Congr. Geol. Argent., 4: 296-298.

Crook, K.A.W. (1974): Lithogenesis and geotectonics: the significance
    of compositional variation in flysch arenites (greywackes). <In:>
    Dott, R.H. & Shaver, R.M. (eds.): Modern and ancient geosynclinal
    sedimentation . Soc. econ. Paleontol. Mineralog. Spec. Publ., 19:
    304 - 310.

Damm, K.-W., Pichowiak, S. & Todt, W. (1986): Geochemie, Petrologie und
    Geochronologie der Plutonite und des metamorphen Grundgebirges
    in Nordchile. - Berliner Geowiss. Abh. (A), 66, Teil I: 73-145.

Dickinson, W.R. & Suczek, C.A. (1979): Plate tectonics and sandstone
    composition. - Amer. Assoc. Petr. Geol. Bull., 63: 2164-2182.

Doherr, D. (1983): Strukturelle Untersuchungen und Illitkristallini-
    tätsbestimmungen im schwachmetamorphen Basement NW-Argentiniens. -
    Zentralbl. Geol. Paläont., Teil I, 1983: 375-386.

Halpern, M. & Latorre, C.O. (1973): Estudio geocronológico inicial de
    rocas del Noroeste de la República Argentina. - Rev. Asoc. geol.
    Argent., 28: 195-205.

Ježek, P. (1986): Petrographie und Fazies der Puncoviscana Formation,
    einer turbiditischen Folge im Jungpräkambrium und Unterkambrium
    Nordwest-Argentiniens. - Diss. Univ. Münster.

Ježek, P. & Miller, H. (1986): Deposition and facies distribution of
    turbiditic sediments of the Puncoviscana Formation (Upper Pre-
    cambrian - Lower Cambrian) within the basement of the NW Argentine
    Andes. - Zentralbl. Geol. Paläont. Teil I, 1985: 1235-1244.

Ježek, P. & Miller, H. (1987): Petrology and facies analysis of
    turbiditic rocks of the Puncoviscana trough (Upper Precambrian -
    Lower Cambrian) in the basement of the NW Argentine Andes. - In:
    Gondwana Six: Structure, tectonics, and geophysics, Geophys.
    Monogr., 40: 287-293.

Ježek, P., Willner, A.P., Aceñolaza, F.G. & Miller, H. (1985): The Puncoviscana trough - a large basin of Late Precambrian to Early Cambrian age on the Pacific edge of the Brazilian shield. - Geol. Rundschau, 74: 573-584.

Lork, A., Miller, H. & Kramm, U. (in prep.): U-Pb zircon age of the La Angostura granite, an approach to the dating of the orogenic history of the NW Argentine basement. -

Manca, N., Coira, B., Barber, E. & Pérez, A. (1987): Episodios magmáticos de los ciclos Pampeano y Famatiniano en el Río Yacoraite, Jujuy. - Actas X. Congr. Geol. Argent., 4: 299-301.

Miller, H. & Willner, A.P. (1981): Del sedimento al esquisto: desarrollo de la esquistosidad en el ejemplo del Cerro San Javier, Tucumán. - Actas VIII. Congr. geol. Arrgent., 4: 979-986.

Omarini, R. (1983): Caracterización litológica, diferenciación y génesis de la Formación Puncoviscana entre el Valle de Lerma y la Faja Eruptiva de la Puna. - Tesis doctoral, Univ. Nac. Salta, 202 pp.

Omarini, R. & Alonso, R. (1987): Lavas en la Formación Puncoviscana, Río Blanco, Salta, Argentina. - Actas X. Congr. geol. Argent., 4: 292-295.

Omarini, R., Aparicio, A., Párica, C., Pichowiak, S., García, L., Damm, K., Viramonte, J., Salfity, J. & Alonso, R. (1985): Nuevos datos geocronológicos acerca de la edad precámbrica de la Formación Puncoviscana, Noroeste Argentino. - Comunicaciones Dept. Geol. U. Chile, 35: 181-183.

Salfity, J., Omarini, R., Baldis, B. & Gutiérrez, W. (1975): Consideraciones sobre la evolución geológica del Precámbrico y Paleozoico del Norte Argentino. - Actas II. Congr. Iberoamer. Geol. Econ., 4: 341-361.

Sosa Gómez, J. (1984): Zur Geologie im oberen Abschnitt der Quebrada del Toro und am SW-Rande der Salinas Grandes de Jujuy, unter besonderer Berücksichtigung der Puncoviscana-Formation/Argentinien. -- Diss. Univ. Clausthal, 140 pp.

Toselli, A.J. & Aceñolaza, F.G. (1984): Presencia de eruptivas basálticas en afloramientos de la Fm. Puncoviscana en Coraya, Dpto. Humahuaca - Jujuy. - Rev. Asoc. Geol. Argent., 39: 158-159.

Toselli, A.J. & Rossi, J.N. (1982): Metamorfismo de la Formación Puncoviscana en las Provincias de Salta y Tucumán, Argentina. - Actas V. Congr. Latinoamer. Geol., 2: 37-52.

Toselli. A.J. & Rossi, J.N. (1983): Controles del metamorfismo y deformación en las parametamorfitas de las Cumbres de San Javier, Tucumán. - Rev. Asoc. Geol. Argent., 38: 137-147.

Turner, J.C. (1960): Estratigrafía de la Sierra de Santa Victoria y adyacencias. - Bol. Acad. Nac. Cienc. Córdoba, 42: 163-206.

Turner, J.C. (1972): Cordillera oriental y Puna. - In: Leanza, A. ed.: Geología Regional Argentina: 91-144, Córdoba (Acad. Nac. Cienc.).

Willner, A.P. & Miller, H. (1986): Structural division and evolution of Lower Paleozoic basement in the NW-Argentine Andes. - Zentralbl. Geol. Paläont. Teil I, 1985: 1257-1268.

Willner, A.P., Miller, H. & Ježek, P.: Geochemical features of an Upper
    Precambrian - Lower Cambrian greywacke/pelite sequence (Puncovis-
    cana trough) from the basement of the NW-Argentine Andes. - N.Jb.
    Geol. Paläont. Mh., 1985: 493-512.

Willner, A.P., Lottner, U.S. & Miller, H. (1987): Early Paleozoic
    structural development in the NW Argentine basement of the Andes
    and its implication for geodynamic reconstructions. - In: Gondwana
    Six: Structure, tectonics, and geophysics, Geophys. Monogr. 40:
    229-239.

GEOLOGY AND SEDIMENTOLOGY OF THE CAMBRIAN GRUPO MESON
(NW ARGENTINA)

M. Kumpa
Institut für Geologie Freie Universität Berlin,
Altensteinstr. 34a, 1000 Berlin 33, Federal Republic Germany

M. C. Sanchez
Universidad Nacional de Salta
Buenos Aires 177, 4400 Salta, Argentina

Abstract

For the region studied in the eastern range of the Andes between 22º S
and 25º S (northwestern Argentina), the Cambrian transgression produced thick quartz arenite series and is described as a coastal, tidal
and subtidal influenced process. Lithology, sedimentology and the
spatial distribution of the facies types made a model of enclosed extended platform or basin areas under shallow marine conditions plau-
sible. The position of a postulated feeder channel is assumed to be
located in present-day south Bolivia at the northern border of the
Cratógeno Central.

Introduction

As well as giving a stratigraphic review of the Cambrian Grupo Mesón,
the spatial correlation of thick quartz arenite series and intercalated laminated silty mudstones shall be discussed for the area studied
between 22º S and 25º S in the Eastern Cordillera of the Argentine
Andes (fig. 1). These series unconformably overlie the sedimentary
rocks of the Precambrian and Lower Cambrian Formación Puncoviscana. In
order to obtain more detailed information on the paleogeographic
situation of this important period in Andean geology, sedimentary
structure and associated trace fossils were examined along selected
sections.

Lecture Notes in Earth Sciences, Vol. 17
H. Bahlburg, Ch. Breitkreuz, P. Giese (Eds.),
The Southern Central Andes
© Springer-Verlag Berlin Heidelberg 1988

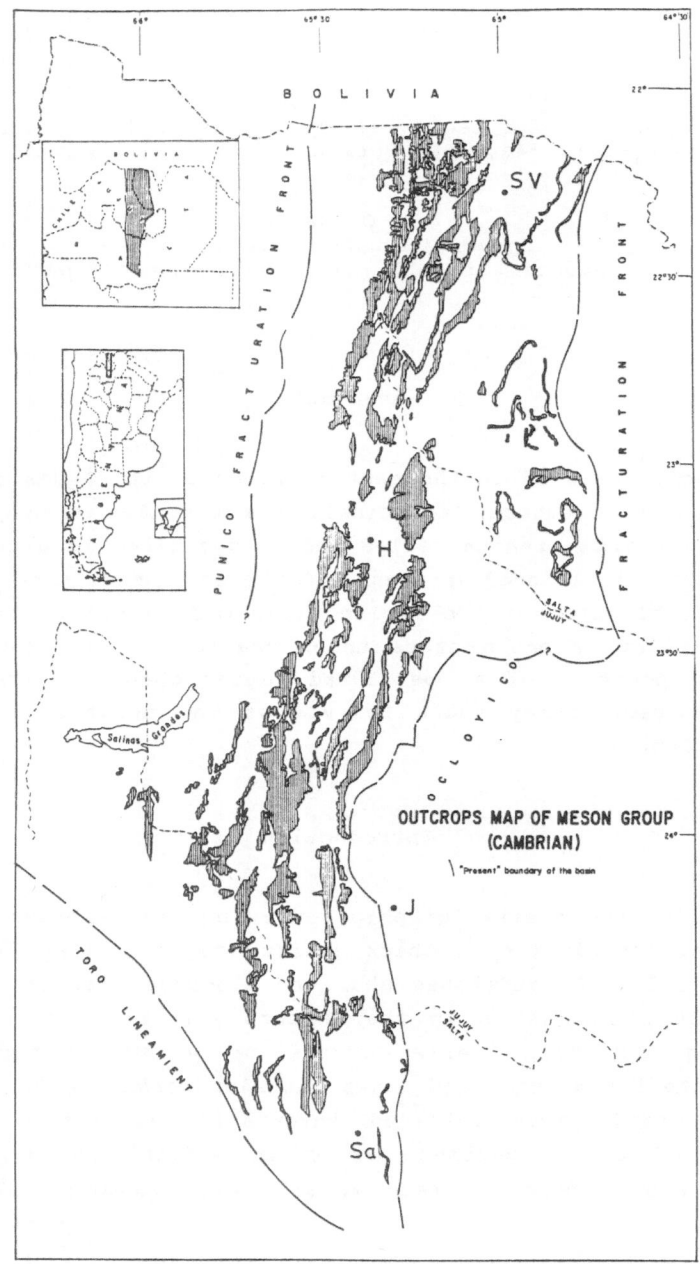

Fig. 1. Map of the region studied showing Cambrian outcrops.
H - Humahuaca, J - San Salvador de Jujuy,
Sa - Salta, SV - Santa Victoria.

Five different types of facies could be distinguished belonging to the
fluvially influenced coastal, tidal and to the subtidal environment.
The spatial thickness distribution of the sections studied suggests,
at least for the region studied, a gradually increasing rate of subsidence in a northern direction.

Stratigraphy

The Grupo Mesón has already been described several times at selected
localities in the Argentine Puna and in the eastern range of the Andes
(TURNER, 1960, 1970, 1972; SCHWAB, 1973; MENDEZ, 1974; TURNER &
MENDEZ, 1979). The basal conglomerates overlie unconformably the Formación Puncoviscana, a sedimentary sequence of low grade metamorphic
sandstones, greywackes, shales, phyllites and few carbonate horizons.
The 70-800 m thick exposed sediments of the Grupo Mesón are lithologically composed of conglomerates, quartz arenites and alternating
layers of fine grained quartzose sandstones and laminated silty
mudstones.

TURNER (1960) divided the Grupo Mesón into three lithological units:
Formación Lizoite, Formación Campanario, Formación Chalhualmayoc. Regarding the surroundings of Santa Victoria (Fig. 1) in the northern
part of the Argentine Eastern Cordillera, TURNER (1960) assessed the
thickness of the Grupo Mesón as 3000 m. However, a complicated tectonic structure led to E-W orientated overthrusts and layer re-
petitions. While the basal conglomerates are sometimes accumulated to
horizons of 30-40 m thick-ness or laterally substituted by coarse
grained quartz arenites (MENDEZ etc. al., 1979), the sedimentary top
of the Grupo Mesón is unknown. The fine grained quartz arenite sequences of the Formación Chalhualmayoc unconformably - Fase Irúyica -
underlie Ordovician strata (TURNER & MENDEZ, 1975; see also MOYA, this
volume), which are characterized by a well documented trilobite fauna
(HARRINGTON & LEANZA, 1957).

Due to the unconformity in relation to the underlying Precambri-
an/Cambrian Formación Puncoviscana series and the overlying Ordovician strata, the Grupo Méson layers are likely to be of an Cambrian
age; the basal unconformity must have resulted from the Pampean Orogeny.

ACENOLAZA (1973) and ACENOLAZA & DURAND (1973) reported on the presence of *Oldhamia* sp. in several Formación Puncoviscana horizons. They
therefore concluded that at least the upper part of the Formación Puncoviscana was Lower Cambrian. With regard to this the basal Grupo Mesón unconformity is of an Intracambrian age.

Sedimentology

On the whole, the facial classification corresponds to the lithological subdivision of the Grupo Mesón by TURNER (1960). Further
lithotypes of local importance were distinguished, but were not considered as individual types of facies due to their spatially inconstant
appearence.

Facies Type I (Formación Lizoite) - 1. Description: Basal conglomerates form sedimentary sequences of up to 30 m in thickness. Following DOTTs (1964) sedimentary petrography classification for sandstones and greywackes, most of the conglomerates can be said to have
the mineral composition of lithic arenites. Only few of them can be
classified as lithic greywackes. The pebbles in the lithic greywackes
are usually sandstone and greywacke clasts from the underlying Formación Punciviscana. Only few elongated mudstone clasts can be found.
The pebble composition of the lithic arenites is reduced to quartz.
The diameters of the usually well rounded quartz pebbles are about 2-4
cm, while the sandstone and greywacke clasts can reach up to 8 cm. The
poorly sorted lithic arenites have a grain size distribution reaching
from coarse silt to coarse sand including portions of fine gravel. In
the Santa Victoria area, well rounded feldspars (orthoclase), diameters 2-4 mm, complete the clast spectrum.
The conglomerate beds range from 20 cm to 200 cm. The thin beds especially wedge out laterally. The vertical facial development of the basal conglomerates takes its course from thick unstratified lithic arenites first to trough cross-stratified beds (fig. 2). This is accompanied by increased clast sorting. At the top of these sequences clast
diameters decrease and become less frequent; laminated coarse

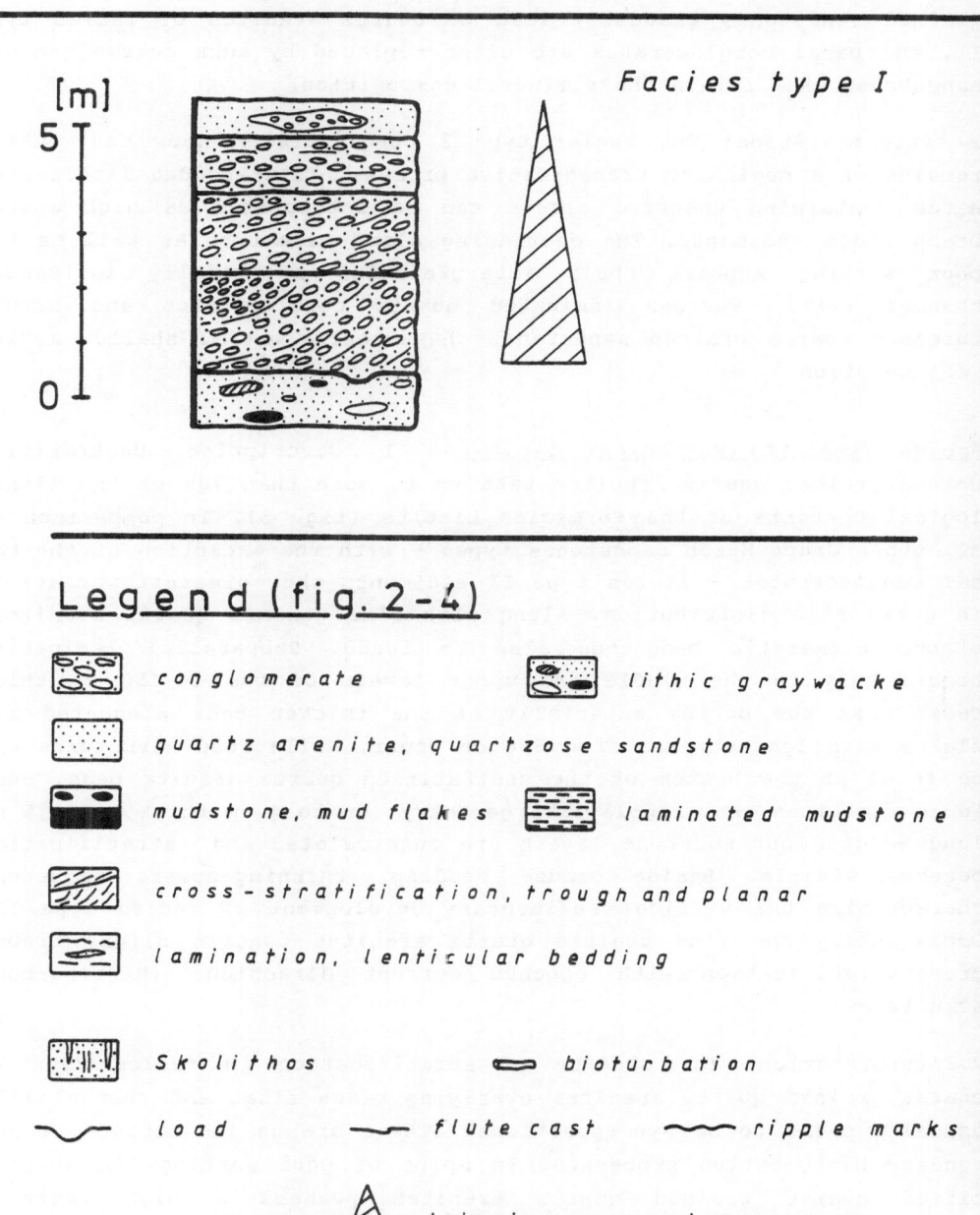

Facies type I

Legend (fig. 2-4)

conglomerate

lithic graywacke

quartz arenite, quartzose sandstone

mudstone, mud flakes

laminated mudstone

cross-stratification, trough and planar

lamination, lenticular bedding

Skolithos

bioturbation

load

flute cast

ripple marks

thinning-upward sequence

Fig. 2. Characteristic section of the basal conglomerates and coarse grained sandstones (facies type I); also including legend for fig. 2-4.

grained sandstones gradually lead to quartz arenites of facies type II. The basal conglomerates are often replaced by such coarse grained sandstones that retain their mineral composition.

2. Interpretation: The facies type I conglomerates show sedimentary remains of a beginning transgressive process. Thick-bedded lithic arenites containing unsorted clasts can be interpreted as high energy beach ridge sediments. The decreasing clast diameters as well as the poor sorting support their interpretation as tidally influenced channel fills, whereas laminated quartz conglomerates and structureless coarse grained sandstones characterize normal shallow marine sedimentation.

Facies Type II (Formación Lizoite) - 1. Description: Unstratified coarse grained quartz arenites take up to more than 70% of the lithological contents of the Formación Lizoite (fig. 3). In comparison to all other Grupo Mesón sandstones types - with the exception of the basal conglomerates - facies type II sediments show greatest variations in grain size distribution. Along with fine grained quartz arenites, microconglomeratic beds can also be found. Subparallel lamination occurs only in the middle and upper levels of the 10-150 cm thick beds. Near the bottom especially of the thicker beds elongated mud flakes with longest axes from 3-6 cm occur. While load structures can be found at the bottom of the unstratified quartz arenite beds, surfaces sometimes show ripple-like geometry. Up to 4 cm thick and 1,5 cm long wedging out mudstone layers are intercalated and stratification becomes visible. Beside graded bedding, thinning-upward sequences characterize the vertical sedimentary development of facies type II. Occasionally the fine grained quartz arenites contain slight trough cross-stratification with opposed current directions (herring-bone structures).

2. Interpretation: The absence of stratification in facies type II coarse grained quartz arenites overlying beach ridge and channel fill deposits seems to be syndepositional. There are no indications of secondary bioturbation processes. In spite of poor sorting the unstratified coarse grained quartz arenites possess a high textural maturity, acquired by intensive littoral sediment movement.

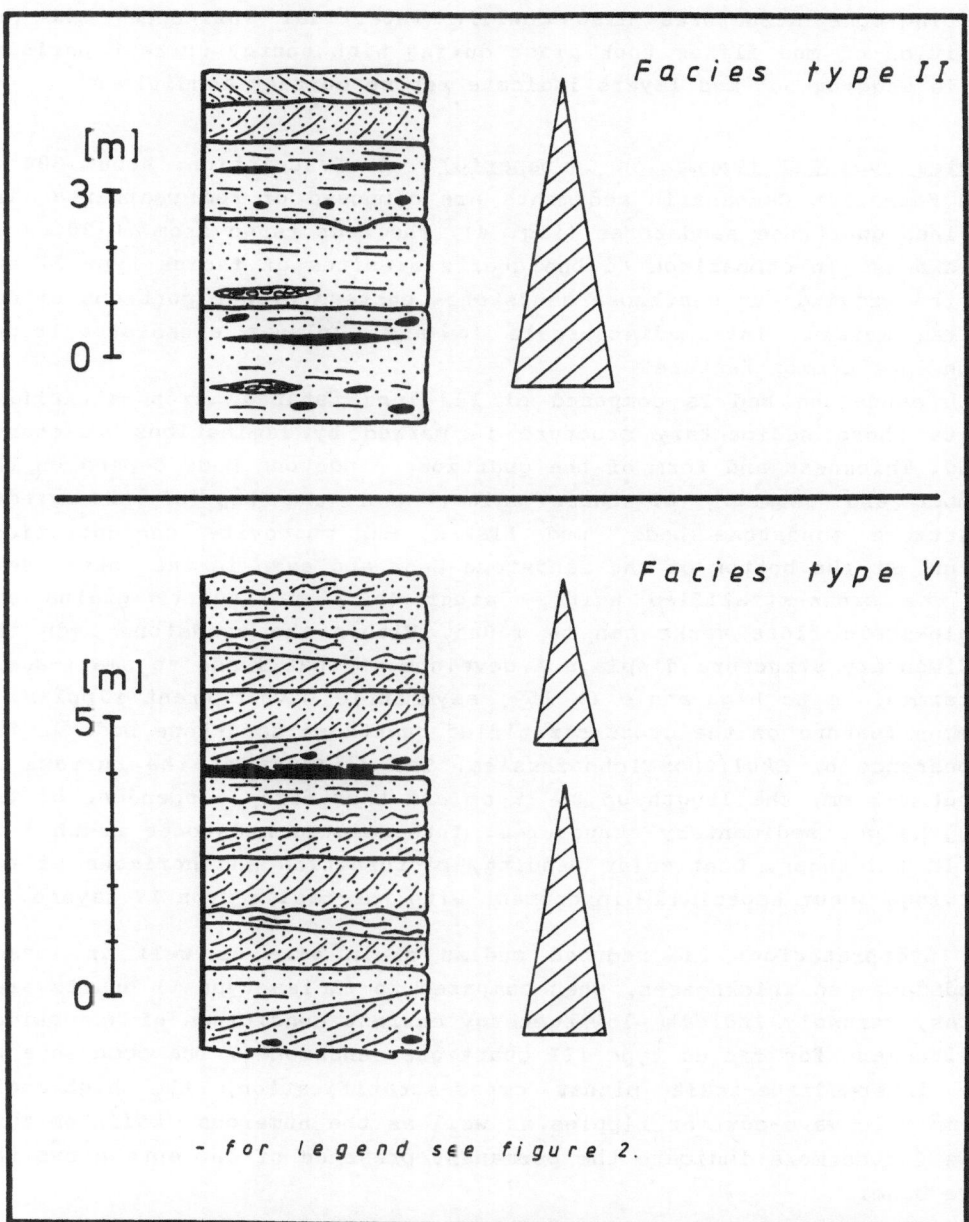

Fig. 3. Characteristic sections of the quartz arenite sequences
of facies type II and facies type V.

Herring-bone structures also confirm intertidal influence. The deposition of mud flakes took place during high energy current periods, while wedging out mud layers indicate repose current conditions.

Facies Type III (Formación Campanario) - 1. Description: About 60% of the Formación Campanario sediments are composed of red weathered fine grained quartzose sandstones (fig. 4). The beds range from 20-90 cm in thickness. In comparison to the quartz arenites of facies type II and V, the wedging out quartzose sandstones possess higher portions of detrital matrix. Intermediate-scale low-angle planar cross-stratification is a common feature.

Each sandstone bed is composed of 2-3 lense shaped, cross-stratified units whose sedimentary stucture is marked by laminations of coarse sand. Thickness and form of the quartzose sandstone beds depend on the number and geometry of these units. Near the top of the graded quartzose sandstone beds, mud flakes and muscovite concentrations occur. At the bottom of the sandstone beds and even in the bottom sets of the cross-stratified units, larger (> 2 mm) quartz grains and smale-scale flute marks can be found. Within the sandstone beds the sedimentary structure displays a development that leads to small-scale intermediate to high-angle (> 15°) asymmetric wave-current ripples. A common feature of the cross-stratified quartzose sandstone beds is the appearence of *Skolithos* ichnofossils. The diameter of the burrows is about 2-3 mm, the length up to 35 cm and therefore independent of the wedging out sedimentary structures. Intercalated mudstones reach 5-15 cm in thickness. Lenticular bedding, bioturbation and horizons of reworkings occur especially in contact with the facies type IV layers.

2. Interpretation: The reduced median grain size as well as lesser sandstone bed thicknesses, when compared to facies type II quartz arenites, strongly indicate lower energy current conditions of a subtidal environment for facies type III quartzose sandstones. The occurence of the intermediate-scale planar cross-stratification, the high-angle asymmetric wave-current ripples as well as the numerous *Skolithos* burrows furthermore indicate the permanent presence of currents above the wave base.

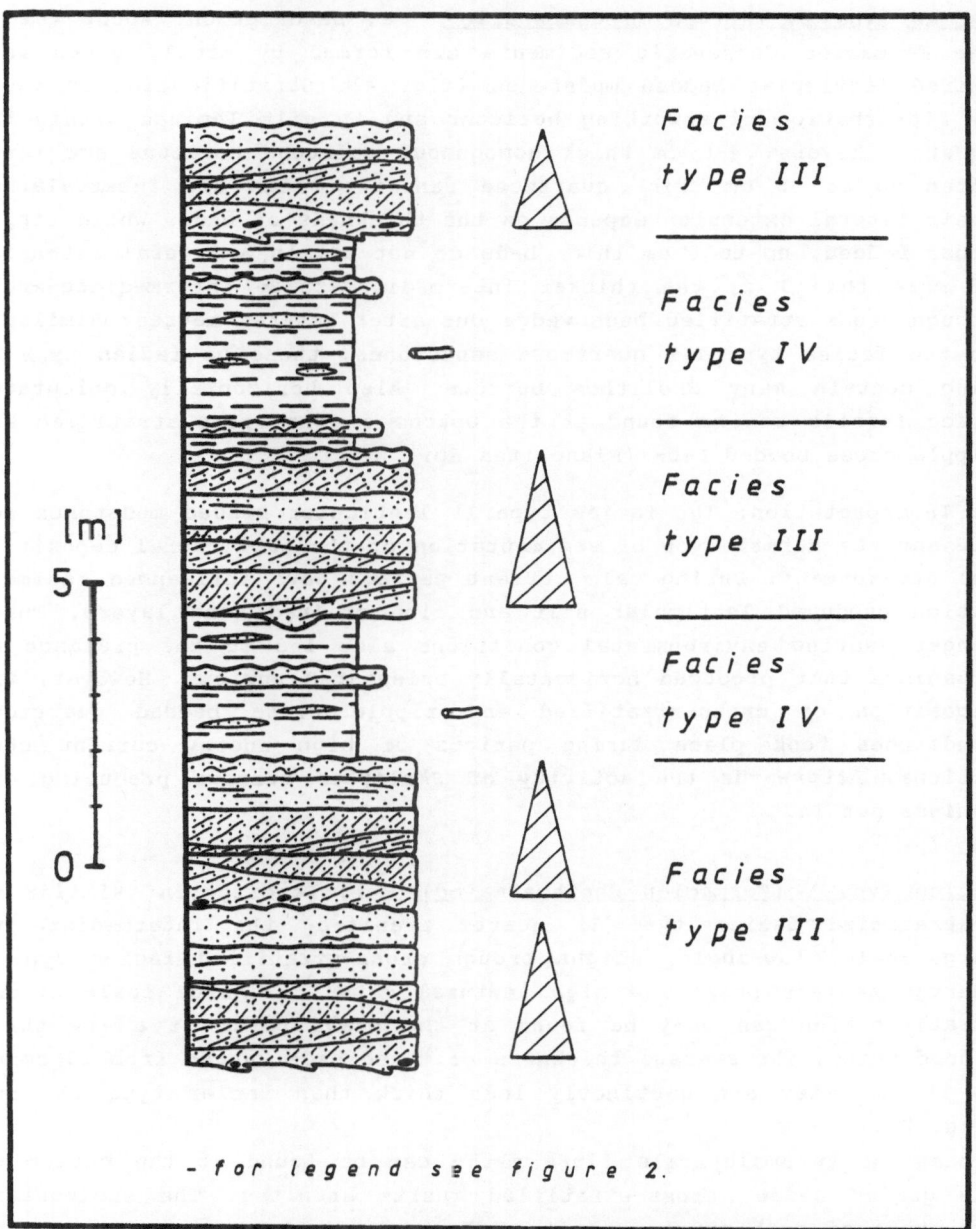

Fig. 4. Charecteristic section of a mixed layer from facies
type III quartzose sandstones and facies type IV
lenticularly bedded silty mudstones.

Facies Type IV (Formación Campanario) - 1. Description: About 40% of the Formación Campanario sediments are formed by mostly green weathered lenticular bedded mudstones (fig. 4). Stratification is shown by bioturbation of reworking horizons and by silt laminae within the mudstone layers. 4-6 cm thick homogenous mudstone horizons are rare. Often up to 35 cm thick quartzose sandstone beds are intercalated. Their lateral extension depends on the thickness of beds. While ripple cross bedded, up to 8 cm thick beds do not attain a lateral extension of more than 1 m, the thicker intermediate-scale intermediate-angle trough cross-stratified beds wedge out after several metres. Similarly to the facies type III quartzose sandstones, those of facies type IV also contain many *Skolithos* burrows. Also horizontally orientated trace fossils can be found at the bottom of the cross-stratified and ripple cross bedded beds (Planolites sp.).

2. Interpretation: The facies type IV lenticular bedded mudstones represent the normal type of sedimentation of the sublittoral depositional environment. During calm current periods wave influenced sedimentation produced lenticular silt and fine grained sand layers. These longer lasting environmental conditions also led to the presence of organisms that produced horizontally orientated burrows. However, the deposition of cross-stratified and ripple cross bedded quartzose sandstones took place during periods of high energy current conditions. Afterwards the activity of *Skolithos* burrows producing organisms set in.

Facies Type V (Formación Chalhualmayoc) - 1. Description: Similar to unstratified facies type II quartz arenites, the intermediate to large-scale, low-angle, slight trough cross-stratified facies type V quartz arenites possess a high textural maturity. Large scale cross-stratification can only be found at the basal parts of a few thick bedded units. The average thickness of these beds ranges from 15 cm up to 55 cm. They are distinctly less thick than facies type II ones (fig. 2).
Occassionally smale-scale flute marks can be found at the bottom of the graded bedded cross-stratified quartz arenites. The sedimentary structure of each quartz arenite bed can be defined as a system composed of 2-4 laterally wedging out cross-stratified units. All cross-stratification sets incline with different angles but with the same

orientation. The geometrical appearance of several cross-stratified beds resembles megaripples with a ripple index (RI; TANNER, 1967) <2. Up to 12 cm thick horizontal laminated partly bioturbate reworked silt horizons are intercalated in the facies type V quartz arenite sequences. They usually reach only a few millimetres in thickness.

2. Interpretation: The development of facies type V large- and intermediate-scale low-angle trough cross-statification needed comparable current conditions to the littoral facies type II quartz arenites. TANNER (1967) and REINECK & WUNDERLICH (1968) interpreted the calculated ripple index and ripple-symmetry index values as characteristic of large-scale asymmetric wave-current ripples. Considering the fine grained detritus ALLEN (1970) declared the necessary current velocity producing such ripples to be 10-50 cm/s. Graded bedding and the change from large-scale to intermediate-scale cross bedding are indications of continuously abating currents. Seen dimensionally the facies type V quartz arenite sequences build extended megaripple systems of migrating subtidal bars.

Sediment Petrography

Most of the Grupo Mesón sediments can be described as quartz arenite and quartzose sandstone sequences. Silty mudstones occur occasionally as do conglomerates and greywackes restricted to the base of Formación Lizoite. The sandstones gererally show more than 90% detrital quartz grains. Infrequently occur microcline, plagioclase, composite quartz rock fragments, volcanic rock fragments, muscovite and heavy minerals (fig. 5; 1). The matrix clay minerals are altered into illite. Authigenetic chlorite is a common feature. Secondary quartz overgrowths (fig. 5; 2) have produced partly suturic mosaic structures which can be seen in thin sections (fig. 5; 3).

The Grupo Mesón basin

The Pampean orogen delivered high quantities of erosional detritus to the transgressive Cambrian Grupo Mesón basin. Under shallow marine depositional conditions, detritus was accumulated to the Grupo Mesón

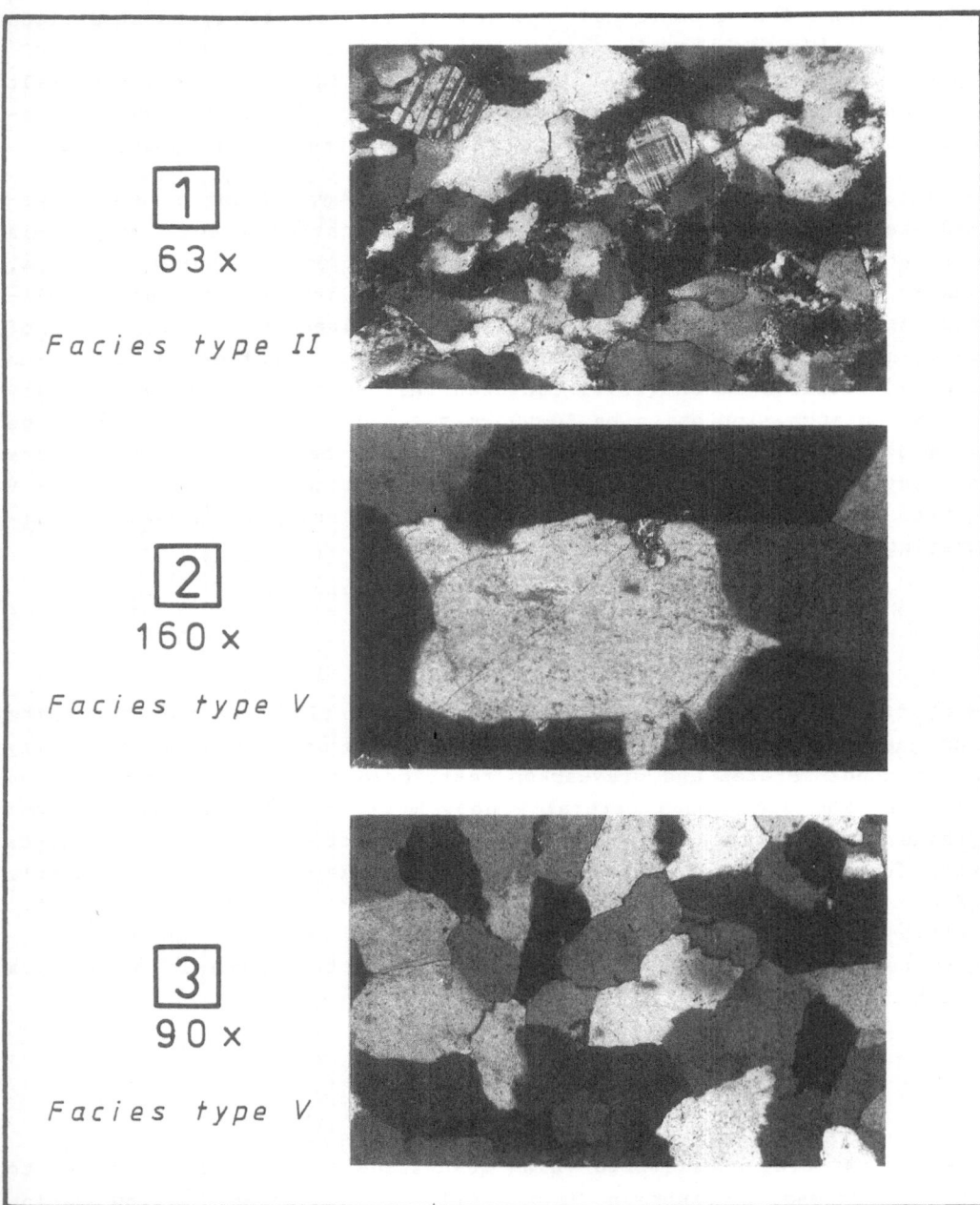

1
63 x

Facies type II

2
160 x

Facies type V

3
90 x

Facies type V

Fig. 5. Sedimentary petrography. 1 – mineral composition,
2 – secondary quartz overgrowths, 3 – partly suturic
mosaic structure.

quartz arenite sequences. The facies types characterize the transgression from beach ridge horizons to migrating subtidal bars. The vertical sequences could be traced throughout the whole NW-Argentine Eastern Cordillera. Important changes in lateral facies could not be discerned. The continuously increasing thickness towards the north of all Grupo Mesón formations (Lizoite, Campanario, Chalhualmayoc) also indicate increasing rates of subsidence for the Grupo Mesón basin in this direction. The position of the Cambrian deeper marine basin is unknown. According to CUERDA (1973), the Ordovician transgression progressed southeastward from Bolivia towards NW-Argentina along the northern border of the Cratógeno Central. This could also have been the connection between the Grupo Mesón basin and the open sea in Cambrian times. ACENOLAZA et al. (1982) also interpreted the Grupo Mesón basin as an enclosed area influenced by a NW-SE orientated marine inlet situated in present day south Bolivia.

Acknowledgement

This project was funded by the "Deutsche Forschungsgemeinschaft" (ref. no. Gi 31/51-3) and was realised in close cooperation with the members of the Geological Institute of the "Universidad Nacional de Salta", Argentina, in particular with J. A. Salfity.

References

ACENOLAZA, F.G. (1973): Sobre la presencia de Oldhamia Sp. en la
  Formación Puncoviscana de Cuesta Munano, Provincia de Salta,
  República Argentina. - Rev. Asoc. Geol. Arg., 28: 56-60;
  Buenos Aires.

ACENOLAZA, F.G. & DURAND, F. (1973): Trazas fósiles del basamento
  cristalino del Noroeste Argentino. - Bd. A soc. Geol. Córdoba,
  2: 45-55; Córdoba.

ACENOLAZA, F.G., FERNANDEZ, R. & MANCA, N. (1982): Caracteres
  bioestratigraficos y paleoambientales del Grupo Mesón
  (Cambrico medio-superio), centro-oeste de America del Sur. -
  Estud. geol., 38: 385-392; Madrid.

ALLEN, J.R.L. (1970): Physical processes of sedimentation. An introduction. - 248 S., George Allen & Unwin LTD.; London.

CUERDA, A.J. (1973): Resena del Ordovicico Argentino. - Ameghiniana, 3: 272-312; Buenos Aires.

DOTT, R.H. (1964): Wacke, graywacke and matrix - what approach to immature sandstone classification? - J. sed. Petrol., 34 (3): 625-632; Tulsa.

HARRINGTON, H.J. & LEANZA, A. (1957): Ordovician trilobites of Argentina. - Spec. Publ., I. Dept. of Geol. Univ. of Kansas; Lawrence.

MENDEZ, V. (1974): Estructuras de las Provincias de Salta y Jujuy a partir del Meridiano 65°30' Oeste, hasta el limite con las Républicas de Bolivia y Chile. - Rev. Asoc. Geol. Arg., 29: 391-424; Buenos Aires.

MENDEZ, V. et al. (1979): Geología de la region noroeste, provincias de Salta y Jujuy, Rep. Arg. - 118 S., Direccion General de Fabricaciones Militares; Buenos Aires.

MOYA, M.C. (this volume): Lower Ordovician in the southern part of the Argentine Eastern Cordillera.

REINECK, H.-E. & WUNDERLICH, F. (1968): Zur Unterscheidung von asymetrischen Oszillationsrippeln und Strömungsrippeln. - Senckenb. Lethaea, 49: 321-345; Frankfurt/M.

SCHWAB, K. (1973): Die Stratigraphie in der Umgebung des Salar de Cauchari (NW Argentinien). Ein Beitrag zur erdgeschichtlichen Entwicklung der Puna. - Geotekt. Forsch., 43 (1-2): 1-168; Stuttgart.

TANNER, W.F. (1967): Ripple mark indices and their uses. - Sedimentology, 9: 89-104; Amsterdam.

TURNER, J.C.M. (1960): Estratigrafía de la Sierra de Santa Victoria y adyacencias. - Bol. Acad. Nac. de Ciencias Cordoba, 41 (2): 163-196; Córdoba.

TURNER, J.C.M. (1970): The Andes of Northwestern Argentina. - Geol. Rundschau, 59 (3): 1028-1063; Stuttgart.

TURNER, J.C.M. (1972): Cordillera Oriental. - in: LEANZA, A.F.
(Dir. & Hrsg.): Geología Regional Argentina. - Acad. Nac. de
Ciencias Córdoba: 117-142; Cordoba.

TURNER, J.C.M. MENDEZ, V. (1975): Geología del Sector Oriental
de los Departamentos de Santa Victoria e Iruya, Provincia de
Salta, Republica Argentina. - Acad. Nac. Cs. Córd., LI (1-2):
11-24; Córdoba.

TURNER, J.C.M. & MENDEZ, V. (1979): Puna. Geología Regional
Argentina. - Acad. Nac. Cs. Córd., 1: 13-56; Córdoba.

LOWER ORDOVICIAN IN THE SOUTHERN PART OF THE ARGENTINE EASTERN CORDILLERA

Maria Cristina Moya
Univ. Nacional de Salta and CONICET
Buenos Aires 177, 4400 Salta, Rep. Argentina

Summary

In this paper, the stratigraphic evolution of the Santa Victoria Group (Latest Cam-
brian-Lower Ordovician) in the southern half of the Argentine Eastern Cordillera is
analysed and the sequence of the main sedimentary episodes that define its depositional cycle is determined. The equivalence of formational units identified thus
far by means of varied nomenclature is established. The main structural elements
that may have regulated the geometry of the Lower Ordovician basin are outlined and
a probable paleoenvironmental model is presented.

Introduction

Several authors have analysed different aspects of the Ordovician basin of Northwestern Argentina, the most important ones being: ACENOLAZA (1973, 1976), ACENOLAZA
& TOSELLI (1984), BALDIS (1978), COIRA et al. (1982), CUERDA (1973), FURQUE & CUERDA
(1979), HARRINGTON (in HARRINGTON & LEANZA 1957), KEIDEL (1943), RAMOS (1973), RUIZ
HUIDOBRO (1975), SALFITY et al. (1975, 1984a), TURNER (1972). There is a wealth of
paleontologic research though much of it deals with findings restricted to isolated
areas, except for some synthesis papers such as the one by HARRINGTON & LEANZA
(1957). This paper analyses the main sedimentary episodes occurring in the southern
half of the Argentine Eastern Cordillera during the Lower Ordovician, which define
the depositional cycle of the Santa Victoria Group.

The Eastern Cordillera was selected not only because the Lower Ordovician is well
recorded there but also because this region is the only one in the whole Andean
Cordillera where the basal relation between the Ordovician and the Cambrian may be
analysed clearly. There are no Upper Ordovician records in the Eastern Cordillera.
They outcrop only in the neighbouring regions of the Puna to the west and the Subandean Ranges to the east, though in these areas the exposures of Ordovician basal
units (Tremadocian) are either buried or very few thus preventing an adequate analysis of the initial evolutionary stges of the basin.

The limits of the area under study can be seen in Figures 1a and 1c; the latter
shows the twenty-nine Ordovician sections that provided the necessary information to

Lecture Notes in Earth Sciences, Vol. 17
H. Bahlburg, Ch. Breitkreuz, P. Giese (Eds.),
The Southern Central Andes
© Springer-Verlag Berlin Heidelberg 1988

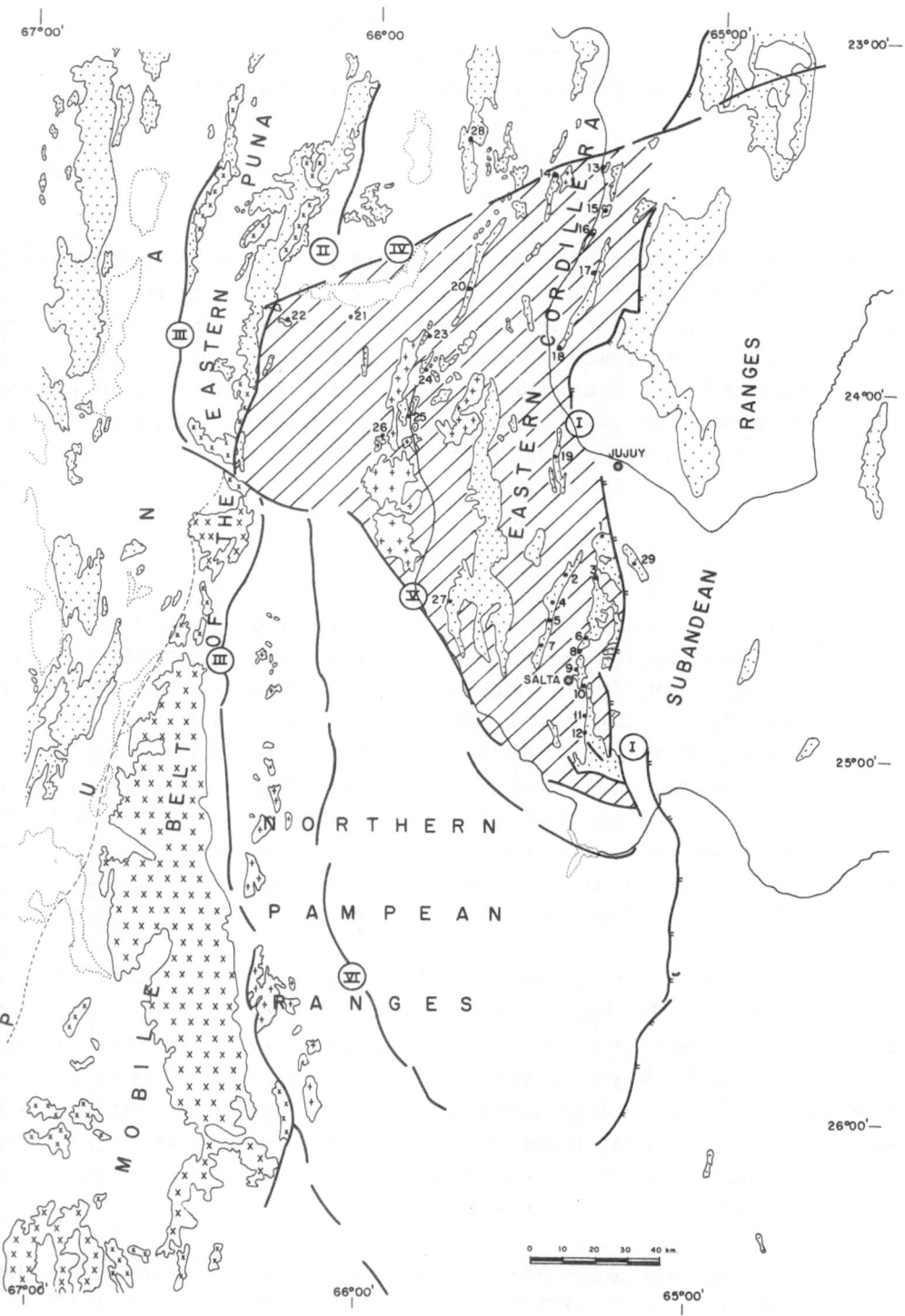

Fig. 1a: Map of Ordovician outcrops.

References

Lacunes and saline deposits

Ordovician intrusions into Lower Ordovician and Precambrian basement (eruptive belt)

Ordovician sediments. Contemporaneous volcanism in Puna included

Upper Precambrian - Middle Cambrian intrusions

Survey area

Border of Precambrian basement with Ordovician intrusions

I   Eastern-Ocloyico Fracturation Front
II   Púnico Fracturation Front
III   Western-Ocloyico Fracturation Front
IV   Salinas Grandes Lineament
V   El Toro Lineament
VI   Calchaquí Lineament
•   Ordovician sections surveyed
a. Mojotoro range - Lesser range
  1   San Antonio
  2   San Alejo ravine
  3   Abra de la Sierra
  4   Yacones river
  5   Lesser ravine
  6   Gallinato ravine
  7   Castellanos ravine
  8   Mojotoro river
  9   Tres Cerritos
 10   La Candelaria - San Bernardo hills
 11   La Pedrera
 12   Pingüiyal ravine
b. Quebrada de Humahuaca area
 13   Chucalezna narrows
 14   Yacoraite ravine (between Iriques and Gallicapor ravines)
 15   El Perchel area (Tres Cienagas, Campo Grande, Arenal and Umacha ravines)
 16   Northern border of Alfarcito range (Conglomerado ravine)
 17   Rupasca ravine
 18   Punta Corral ravine
 19   Yala - Reyes road
c. Toro river - Salinas Grandes area
 20   Agua Blanca
 21   Cangrejillos
 22   Taique
 23   El Moreno
 24   El Angosto (of Moreno river)
 25   Pueblo Viejo
 26   La Quesera narrows
 27   Sococha ravine (Pascha hills)
d. Other sections surveyed
 28   Despensa river (Aguilar range)
 29   Las Maderas river (Las Maderas range)

elaborate the column presented in Figure 2. Figure 3 indicates the geographical features referred to in the text. Figure 4 shows the structural elements that, according to interpretation, may have regulated the sedimentation of the Santa Victoria Group. In Figure 5 a probable paleoenvironmental model is designed.

Research was carried out thanks to the support provided by Universidad Nacional de Salta and CONICET, under the direction of Professor Dr. José Antonio Salfity.

Stratigraphy

The Santa Victoria Group is a great sedimentary cycle limited by erosional surfaces at the top and bottom. Such surfaces originated during the movements of the Irúyica and Guandacol diastrophic phases respectively. Its deposition interval of time ranges from the latest Cambrian to the Upper Arenigian (Lower Llanvirnian?), i. e. the Cambrian-Ordovician time limit is placed within this cycle (SALFITY et al. 1984a).

I. Stratigraphic Relations (see Figs. 1a, 2, 3 and 4)

The Santa Victoria Group lies on different stratigraphic levels of the Mesón Group (Upper Cambrian) and, when it transgresses the limits of the Mesón Group basin, it lies on Precambrian-Cambrian basement (the Puncoviscana Formation and the granites that intrude it). The Puncoviscana Formation (TURNER 1960) is represented by a considerably thick turbiditic sequence, affected by regional metamorphism of very low degree, folded and afterwards intruded by batholithic bodies, La Quesera or Santa Rosa de Tastil being located in the area under study.

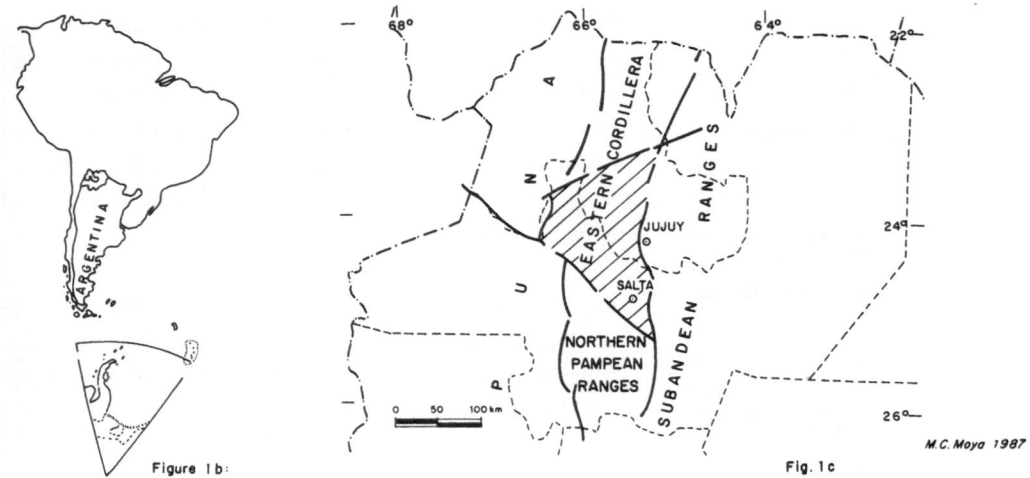

Figure 1b:

Fig. 1c

M.C.Moya 1987

Fig. 1b: Relative geographical location.
Fig. 1c: Geological Provinces of northwestern Argentina.

The Mesón Group lies on the Puncoviscana Formation and its intrusives through an erosion surface of clear angularity. The formational entities tht constitute it are, from bottom to top, the Lizoite, Campanario and Chalhualmayoc Formations (TURNER 1960, see also KUMPA & SANCHEZ, this volume); they represent a sedimentation cycle defined by a transgressive episode followed by a regression (Fig. 2).

The Lizoite and Chalhualmayoc Formations consist of light pink, royal purple-pink and whitish-grey quartzites; they present thick to medium stratification of tabular to cuneiform type with pronounced cross-bedding. Only ichnofossils have been found in these formations thus far. The fining upward Lizoite Formation starts with conglomerates almost entirely composed of well-rounded, well-sorted quartz grains followed by coarse and medium-grained quartzites; even though basal psephites represent a typical feature of the Lizoite Formation, sometimes they may be absent; this can be observed in the western part of the area under consideration (RAMOS 1973, MOYA 1986) (Fig. 3). On the contrary, the Chalhualmayoc Formation is coarsening upward with mainly fine quartzites at the bottom that range to medium-size at the top, sometimes with scattered clasts that occasionally form thin conglomeratic levels.

The Campanario Formation is the intermediate unit; it presents transitional passages with the underlying and overlying formations. It is formed by shales and sandstone intercalations with fine to medium stratification of tabular type and parallel lamination; cross-bedding is scarce. The Campanario Formation contains abundant sedimentary structures and fossil traces. In almost all the area under consideration two members are present: the lower green one, more micaceous, with a greater percentage of sandstones than shales; shales are sandy and sandy-silty. The lower green member transitionally passes to the royal purple one of finger granulometry and with mainly silty shales on fine-grained sandstones. The Chalhualmayoc Formation does not present the features described above all over the area under study. In three profiles of the western region (Sections 20, 21 and 24, Fig. 1a) the regressive process becomes evident in the upper parts of the Campanario formation where frequent sabulite levels intercalate with royal purple shales and quaartzose medium grained royal purple and grey sandstones.

The Mesón Group shows tidal and current sedimentary structures. Even though it is considered as the marine record of a shallow platform, it is likely that the Mesón Group partially corresponds to the records of a paleodelta. The textural and mineralogic maturity of the Mesón Group quartzites evidences great stability in the Cambrian basin.

The Santa Victoria Group represents the records of a series of sedimentary episodes in which an increasing instability can be observed from an initial interval of time in which sedimentation conditions may have varied. This period included erosion and smooth uplift hat correspond to the movements of the Irúyica phase (TURNER & MENDEZ 1975). The final expression of this diastrophism is the transgression with which the Santa Victoria Group is initiated. The Santa Victoria Group lies on different formations of the Mesón Group without evident angularity through a well-defined erosional surface. A basal psephitic-psammitic unit with a pronounced fining upward tendency lies on this erosional surface. Similar observations were performed in the northern part of the Argentine Eastern Cordillera by RUSSO (1948, 1949), MAURI (1949), NESOSSI (1950), TURNER (1960) and TURNER & MENDEZ (1975). Based on these observations, TURNER (1960) defines the Mesón and Santa Victoria Groups as sequences separated by a diastrophic episode. The Irúyica Phase movements probably had little intensity and duration and they were not likely to be orogenic ones; this can be deduced from the good conservation degree that the Mesón Group entities have as well as from the scarce angularity of the discordance that separates the Mesón Group from the Santa Victoria Group. The deposition of the Santa Victoria Group ends in the Upper Arenigian-Lower Llanvirnian? with a generalized regression and a probable emergence of some parts of the Eastern Cordillera as a result of the Guandacol diastrophic phase (SALFITY et al. 1984b).

In the Argentine Eastern Cordillera, the Santa Victoria Group is covered by Cretaceous and Cenozoic units. In most cases, the overlying entity in the region under consideration is the Yacoraite Formation (Maastrichtian-Paleocene) of the Salta Group. When analysing this contact and taking into account the great hiatus that separates the Paleozoic sequences from the Mesozoic ones, it is surprising to observe the scarce angularity of this unconformity. On the other hand, in neighbouring

60

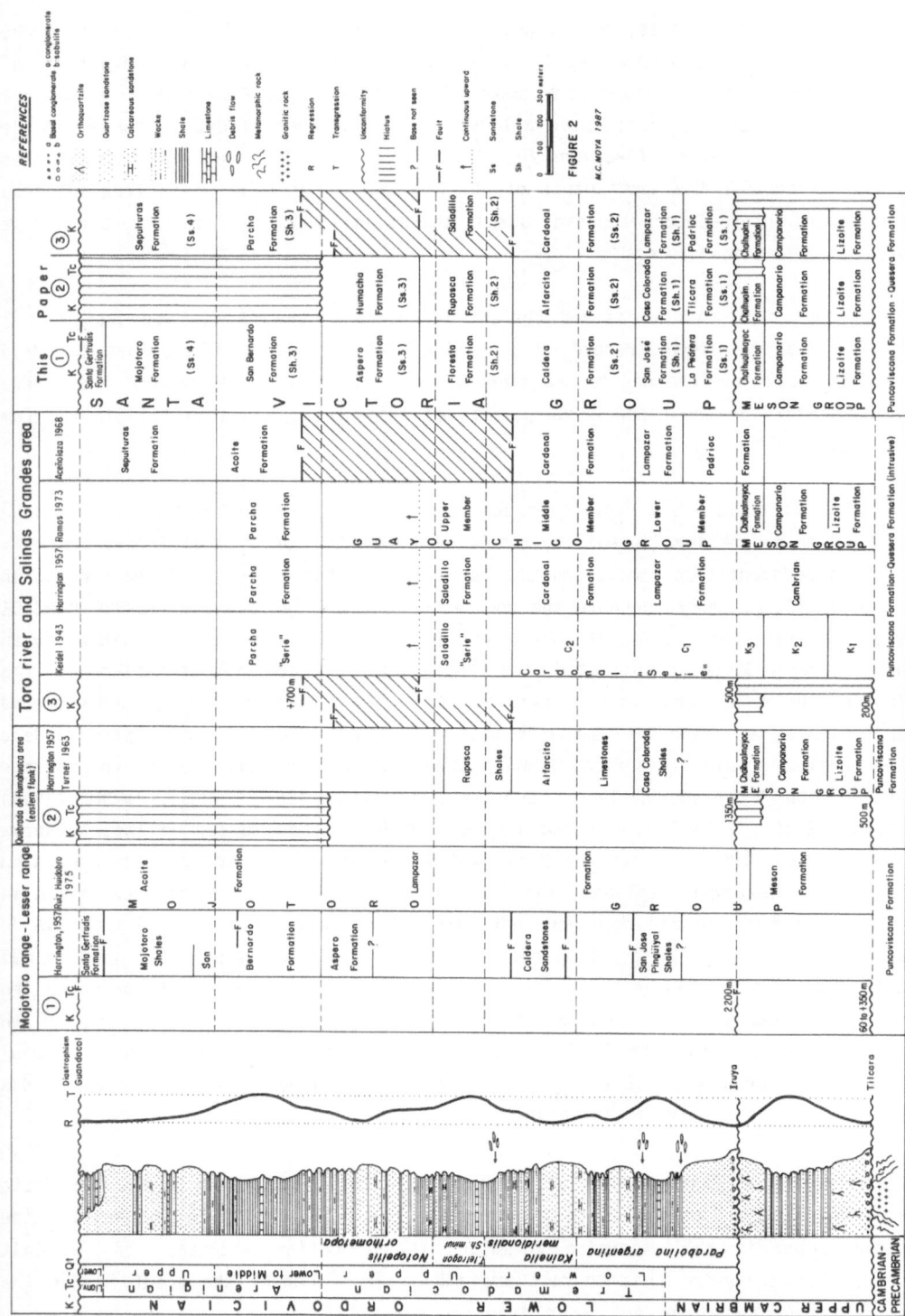

Fig. 2: Stratigraphic column.

areas the effects of the Oclóyica (Upper Ordovician) and Chánica (Upper Devonian-Lower Carboniferous) phases, among the most important Paleozoic diastrophisms, are evident. When the Lower Ordovician is covered by Upper Tertiary-Quaternary units, the unconformity presents manifest angularity. There are frequent tectonic contacts between the Santa Victoria Group and tne Tertiary units (Lesser range and Mojotoro range in its eastern flanks, Figs. 1a, 2).

II. Stratigraphic Evolution of the Santa Victoria Group
   (see Figs. 1a, 2, 3, and 4)

The Santa Victoria Group corresponds to a sequence of marine platform clastics with a rich fauna of trilobites, graptolites, gastropods, pelecypods, cephalopods and echinoderms. In the stratigraphic evolution of the Santa Victoria Group it is possible to recognize successive transgressive and regressive episodes represented in Figure 2 by the most characteristic lithologic records (sandstone and shales) whose depositional interval was determined by means of paleontologic data. These rock bodies were partially or totally recognized by other authors; the parts of the sequence not identified before are named according to the geographic location of the outcrops.

Figure 2 shows that the thicknesses exposed in the eastern region are greater than the ones registered in the western region (Fig. 3). In the latter, shales bearing graptolites of uncertain Upper Tremadocian age are scarce and present little thickness; section 27 (Fig. 1a) is the exception. Sandy units of that age were not identified in this region. Suggestively, in profiles 20, 22, 24, 25 and 27 the Tremadocian-Arenigian contact is tectonic as well as in the Cajas range (ACENOLAZA 1968) and Aguilar range (MARTIN et al. 1986). When the Arenigian is not present, the Tremadocian sequence either is truncated due to faulting (section 23), or is covered by Cretaceous deposits (26) or modern sediments (fault?)(21). On the other hand, a pronounced regressive episode (Fig. 2) developed in the eastern region during the Upper Tremadocian (*Notopeltis orthometopa* zone). It is likely that part of the Upper Tremadocian was not deposited in the western region, a shallower platform; this region was probably subjected to erosion. This plane of sedimentary discontinuity may thus constitute an adequate slide surface for tectonic action.

Considering the observations performed, it is deduced that the faults of the Tremadocian-Arenigian contact are pre-Cretaceous since they are interrupted and shifted by a younger system of fracturation that affects both Ordovician and Cretaceous records. Due to the reasons mentioned above, the thicknesses plotted in the column presented (Fig. 2) corrspond to the maxima of each unit outcropping in the eastern region. The interpretation of the stratigraphic evolution of the Santa Victoria

Group initiates in the final stages of the Irúyica phase (Uppermost Cambrian) with a generalized transgression (Fig. 2).

a. Sandstones 1 unit (La Pedrera, Tilcara and Padrioc Formations, Fig. 2)

The first records of the initial transgressive episode are constituted by a basal conglomerate, well represented in the eastern region of the area under study and, according to bibliographic references, also present in the northern part of the Argentine Eastern Cordillera (RUSSO 1948, 1949, MAURI 1949, NESOSSI 1950, TURNER 1960). In the western region, basal psephites were observed in Sections 22, 25 and 27. In the other profiles, they are replaced by thin banks of quartz sabulites and quartzose coarse grained sandstones with scattered quartz clasts and quartzites that lie through net contact, without transition, on the La Quesera Formation (granite) (26, Fig. 1a) and on the Campanario Formation (Sections 20, 21 and 24). It is worth remembering that in this western region no psephites were observed at the bottom of the Mesón Group but only scarce thin sabulitic quartz banks.

The basal psephite member of the Santa Victoria Group is not a totally conglomeratic unit except for rare cases (Sections 10 and 16); it generally consists of intercalations of conglomerates or sabulites, with conglomeratic sandstones and sandstones. Conglomerates are polymictic, medium and fine grained (5-2 cm) and clast-supported. Skeletons of quartz, quartzites, green and royal purple pelites as well as dark psammo-pelites with small quartz veins occur; clasts of plutonites are rare. Quartz percentage ranges from 40 to 85 % and the proportion of other components is variable though quartzites are more frequent. The matrix is sandy-quartzose. Drab and dark-drab alteration colours are frequent, sometimes with impregnation of Fe and Mn oxides. Sorting is good, roundness good to very good. Thicknesses range from 3 to 22 m. A member of quartzose sandstones with pronounced fining upward grain tendendy lies on the psephitic sabulitic member. These psammites present yellowish white, whitish gray and yellowish drab colours, thick tabular and lenticular stratification, a drab and yellowish drab alteration colour. In some sections these sandstones present crumbs of heavy minerals (Sections 24 and 26).

When examined under the microscope, they are quartzose (80-90 %) with subordinate feldspars (mainly microcline), muscovite, biotite and zircon. Clasts are rounded to sub-rounded, sorting is good. The amount of quartzose-micaceous matrix (chloritic) ranges from 10 to 15 %. The rocks are partially cemented by a siliceous and ferruginous cement. Shales in increasing percentage intercalate at the top of the unit.

b. Shale 1 unit (San José, Casa Colorada and Lampazar Formations, Fig. 2)

The passage from the underlying unit is gradual with slow shale increase until this lithology prevails. In the eastern region, shales are generally silty and muddy; in the western region, they are silty and silty-sandy with lateral facies-changes that become more sandy in marginal profiles (Sections 22, 25 and 26). Colours are variable: green, yellowish drab, greenish drab and dark greenish gray (Sections 21, 22, 25 and 26 as well as nearly all the profiles of the Mojotoro range and the Quebrada de Humahuaca area), dark gray to black (Sections 2, 19, 20, 23, 24, 27 and 28) and reddish drab (Section 1). This unit bears a fauna from the *Parabolina argentina* zone. *Parabolina argentina* is the most common fossil in the basal parts of the entity.

c. Sandstone 2 unit (Caldera, Alfarcito and Cardonal Formations, Fig. 2)

The first regressive pulse is gradually evidenced at the top of the *Parabolina argentina* zone. Its sedimentary records, just as the ones from the units described above, are present in almost all the sections surveyed. This unit is not a continuous sequence of sandstones; shales and wackes intercalate in different proportions in its middle part whereas there is a pronounced predominance of medium-grained, quartzose, quartzose-micaceous and calcareous sandstones at the bottom and top. They present tabular and lenticular stratification with different types of cross-bedding and yellowish white, drab, whitish gray, pink and royal purple-pink colours. In all the unit, frequent calcareous coquinas and thin fossiliferous levels of limestone can be observed. Important bioturbation, deformational and storm structures, ripple cross-lamination and flaser lamination are important.

In the eastern region, thin banks formed by glauconitic pellets intercalate whereas in the western region pyrite crystals can be observed within the psammitic banks. This unit is paleontologically rich not only in variety of forms but also in quantity of organisms. This entity and the one to be described below contain the highest percentage of carbonatic material in the Santa Victoria Group.

d. Shale 2 unit (Floresta, Rupasca and Saladillo Formations, Fig 2)

In a transitional way, with intercalations of sandstones, wackes and gray, green and greenish drab shales, pelitic material increases until shales prevail (at the top of the *Kainella meridionalis* zone, bottom of the *T. tetragonalis-Sh. minutula* zone, Fig. 2). Shales are generally dark coloured: black, dark greenish-gray, bluish gray

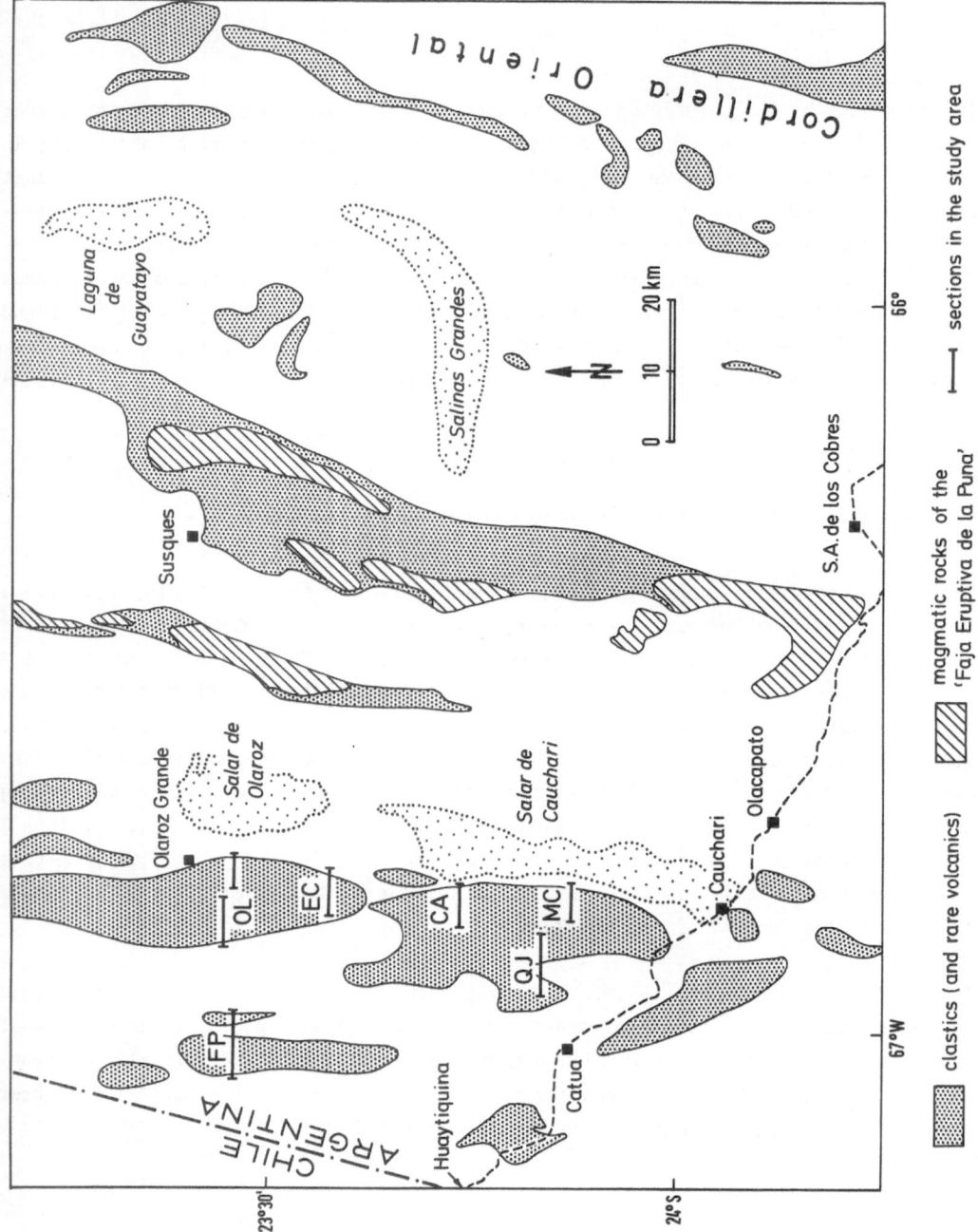

clastics (and rare volcanics)

magmatic rocks of the
'Faja Eruptiva de la Puna'

sections in the study area

and greenish drab (Sections 2, 4, 5, 7, 10, 11, 15, 17, 20? and 27). The fauna is restricted to few forms of trilobites, lingulids, straight nautiloids, pelecypods and predominant graptolites (*Bryograptus* sp.) (Sections 10, 11, 27).

e. Sandstone 3 unit (El Aspero and Humacha Formations, Fig. 2)

This unit was not recognized in the western region. In the eastern region, towards the top of the *T. tetragonalis-Sh. minutula* zone and the bottom of the *Notopeltis orthometopa* zone, the sequences become gradually richer in gray and green wackes followed by quartzose sandstones. The latter present gray (Section 7), pink (Section 10) and royal purple-pink colours with *Skolithos* (Sections 2, 15 and 29). Levels of coquinas, trilobites and brachicpods are frequent.

f. Shale 3 unit (San Bernardo and Parcha Formations, Fig. 2)

In a transitional way, the sequence again becomes richer in pelites (Uppermost Tremadocian-Lowermost Arenigian). During the Lower and Middle Arenigian, the most extensive transgression of the Lower Ordovician sea takes place. Its lithologic records are clearly shaly with mainly graptolite fauna. Shales present greenish drab and dark greenish-gray colours in Sections 2, 5 and 10 and bluish gray ones in all of the western region. Here turbiditic intercalations and plenty of levels of fossiliferous concretions with diameters ranging from 3 to 50 cm can be observed. Equivalent units are the Acoite Formation in the Puna and northern part of the Cordillera Oriental and the Zanjón Formation in the Subandean Ranges (MONALDI et al. 1986).

g. Sandstone 4 unit (Mojotoro and Sepulturas Formations, Fig. 2)

A generalized regression takes place in the Argentine Eastern Cordillera during the Upper Arenigian-Lower Llanvirnian?. The sedimentary record is formed by clearly coarsening upward sequences well represented in the western region and consisting of fine wackes and green shales at the base and yellowish white, pink and royal purple micaceous sandstones at the top (Sections 20, 25 and eastern part of Section 24). According to bibliographic references, they are also present in the Cajas range (ACENOLAZA 1968) and Aguilar range (MARTIN et al. 1987).

In the eastern region, the regressive process is represented by the pink quartzites with *Cruziana* sp. and *Skolithos* sp. of the Mojotoro Formation; in the Subandean Ranges, by the redish and royal purple sandstone with *Skolithos* sp. of the Labrado

Formation (MONALDI et al. 1986). As for the Santa Gertrudis Formation, only a single reference point provided by MONALDI (1976) is known. He indicates net contact between this unit and the Mojotoro quartzites; therefore its location in the column is tentative.

Main Structural Elements of the Lower Ordovician Basin
(Fig. 4)

The El Toro lineament puts a sudden end to Ordovician outcrops and may have constituted the southern border of the Ordovician basin in the area of influence of the La Quesera batholith, which is an element considered to be partially emerged. However, outcropping Ordovician units to the East of the Batholith expose important thicknesses and high shale percentage (Sections 7, 12 and 27). Shales are also recognized as tectonic scales in the Castillejo range. That is why the El Toro lineament, to the East of the La Quesera batholith, is considered to correspond to an erosional border; the Calchaquí lineament, eastern limit of the Argentine Central Cratogene (BRACACCINI 1960), is tentatively proposed as a probable border of the Lower Ordovician basin. A long prolongation of the Central Cratogene - the Cobres High - now shifted westwards and formed part of the Mobile Belt of the Eastern Puna. It would seem to have behaved as an emerged to subemerged element during the Tremadocian. During the Lower and Middle Arenigian, the sea may have invaded this area; both in it and in neibouring western regions, volcanic episodes appear to have taken place (MOYA & SALFITY 1982).

The Lipán swell is another emerged to subemerged element whose axis was traced by following the strike of present Precambrian basement outcrops. Its existence was deduced by analysing the pronounced differences in thickness, sedimentary structures, paleocurrent directions, fauna and geochemical features exhibited by the units in the eastern and western region. This has to be added to the fact that, even though debris flows are locally frequent in different levels of the Lower Ordovician sequence, their occurrence is continuous along tens of kilometres within shaly entities to both sides of the southern area of the axis pointed out. In all cases, debris flows keep very similar textural and compositional features specially with regard to percentage and maximum size of components. On the other hand, frequent coquina blocks of articulated brachiopods indicate provenance from shallower areas. Paleocurrents coming from E-SE were measured in Section 27; in the sections of the Lesser range, paleocurrents are bidirectional (E-W with eastern predominance).

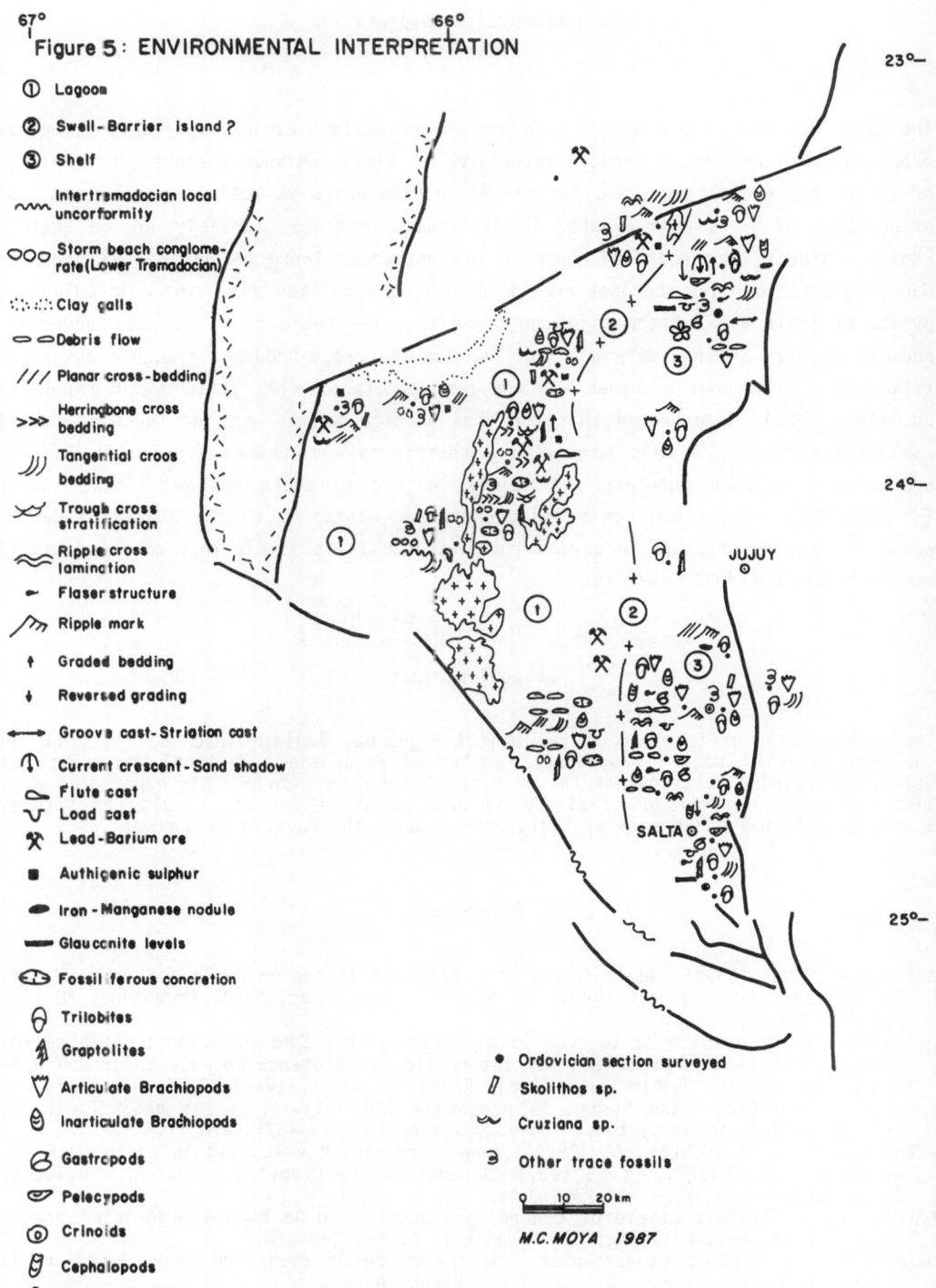

Figure 5: ENVIRONMENTAL INTERPRETATION

① Lagoon

② Swell-Barrier Island?

③ Shelf

〰 Intertremadocian local unconformity

∞∞ Storm beach conglomerate (Lower Tremadocian)

⋯⋯ Clay galls

⌒⌒ Debris flow

/// Planar cross-bedding

>>> Herringbone cross bedding

))) Tangential cross bedding

⤳ Trough cross stratification

〰 Ripple cross lamination

⌐ Flaser structure

/⋀⋀ Ripple mark

† Graded bedding

↓ Reversed grading

↔ Groove cast-Striation cast

�↑ Current crescent-Sand shadow

⌒ Flute cast

⊍ Load cast

✗ Lead-Barium ore

■ Authigenic sulphur

⬮ Iron-Manganese nodule

▬ Glauconite levels

⬭ Fossiliferous concretion

Ө Trilobites

⚡ Graptolites

▽ Articulate Brachiopods

Ө Inarticulate Brachiopods

⌀ Gastropods

⌒ Pelecypods

◎ Crinoids

Ө Cephalopods

❀ Medusoid remains

• Ordovician section surveyed

ǁ Skolithos sp.

〜 Cruziana sp.

∋ Other trace fossils

0 10 20 km

M.C.MOYA 1987

67°   66°   23°—

24°—

25°—

JUJUY

SALTA

Environmental Interpretation
(Fig. 5)

The Lower Ordovician sequence is complete in the eastern region. Sedimentary records expose greater thickness, exhibit structures of tidal and wave currents, bear plenty of varied benthonic fauna, and include glauconite banks as well as nodules and impregnations of Fe and Mn oxides. The thickness decreases sensibly in the western region, without the record of part of the sequence. There is good development of dark shale facies with abundant organic matter, graptolites and authigenic sulphurs; pyrite crystals are scattered in the psammites. Benthonic fauna is less abundant, reduced in size and not very diverse. Herringbone cross-bedding, trough cross-stratification, calcareous coquinas and abundant levels of clay galls are frequent in psammites. It is interpreted that the area located to the west of the Lipán swell would correspond to a calm zone with a restricted environment (Lagoon in Fig. 5) protected from wave currents and mainly subject to tides and occasional storms (Section 26). The eastern region that continues eastwards to the Subandean Ranges would correspond to a better oxygenated platform. It is likely that the Lipán swell may have been partially emerged.

Acknowledgements

The authoress is grateful to Víctor Daniel Figueroa, English Teaching Assistant in the Department of Modern Languages, Faculty of Humanities, National University of Salta, for helping to translate the original Spanish version of this paper into English. Likewise, a debt of gratitude is owed to Jorge Flores, Faculty of Natural Sciences, National University of Salta, for drawing the maps in this report.

References

ACENOLAZA, F. G. (1968): Geología estratigráfica de la región de la Sierra de Cajas, Departamento de Homahuaca (Prov. de Jujuy). - Rev. Asoc. Geol. Argentina, 28 (3): 207-224; Buenos Aires.
-- (1973): El Ordovícico de la Puna Salto-Catamarquena. Consideraciones sobre su importancia en la interpretación del desarrollo de là cuenca Eo-Paleozoica del Noroeste Argentino. - Quinto Congr. Geol. Argentina Actas, IV: 3-18; Buenos Aires.
-- (1976): The Ordovician System in Argentina and Bolivia. - In: BASSETT, M. G. (ed.): The Ordovician System. - Univ. Wales Press: 479-487; Cardiff.
ACENOLAZA, F. G. & TOSELLI, A. (1984): Lower Ordovician volcanism in North West Argentina. - In: BRUTON, D. L. (ed.): Aspects of the Ordovician System. - Paleont. Contr. Univ. Oslo, 295: 203-209; Oslo.
BALDIS, B. A. (1978): Líneas de control estructural en la cuenca ordovícica argentina. - Acta Geol. Lilloana, XIV (Suplem.): 11-14; Tucumán.
BRACACCINI, O. (1960): Lineamientos principales de la evolución estructural de la Argentina. - Petrotecnia, X (6): 57-69; Buenos Aires.
COIRA, B., DAVIDSON, J., MPODOZIS, C. & RAMOS, V. (1982): Tectonic and magmatic evolution of the Andes of northern Argentina and Chile. - Earth Sci. Rev., 18: 303-332; Amsterdam.

CUERDA, A. J. (1973): Resena del Ordovícico argentino. - Ameghiniana, 10 (3): 272-312; Buenos Aires.

FURQUE, G. & CUERDA, A. J. (1979): Ordovícico argentino. - Asoc. Geol. Argentina, Publ. Esp. Serie B, 7: 56 p.; Buenos Aires.

HARRINGTON, H. J. & LEANZA, A. F. (1957): Ordovician Trilobites of Argentina. -Dept. Geol. Lniv. Kansas, Spec. Publ., 1: 276 p.; Lawrence.

KEIDEL, J. (1943): El Ordovícico Inferior en los Andes del norte argentino y sus depósitos marino-glaciales. - Bol. Acad. Nac. Cienc., 36: 140-229; Córdoba.

KUMPA, M. & SANCHEZ, M. C. (this volume): Geology and sedimentology of the Cambrian Grupo Mesón (NW Argentina).

MARTIN, J. L., MALANCA, S. & SUREDA, R. J. (1987): La fauna graptolitica de la sierra de Augilar, Jujuy, Argentina. Algunos comentarios sobre las formaciones ordovícicas. - Cuarto Congr. Latinoam. Paleont. Actas, II: 599-619; Santa Cruz de la Sierra, Bolivia.

MARTIN, J. L., MALANCA, S. & SUREDA, R. J. (in press): Nuevos hallazgos paleontológicos en el Ordovícico de la sierra de Aguilar, Jujuy, Argentina. Implicancias estratigráficas y consideraciones geológicas. - Octavo Congr. Geol. Boliviano; La Paz.

MAURI, E. (1949): Comentario al trabajo del doctor Baez sobre su "Informe geológico de las cuencas de los ríos Canas, Piedras y San Andrés". - Informe interno Y.P.F., inédito, Buenos Aires.

MONALDI, C. R. (1976): Paleontología de las unidades aflorantes a lo largo de la Qda. del Gallinato-Dpto. La Caldera-Pcia. de Salta. - Univ. Nac. de Salta, Dpto de Cienc. Nat., Seminario I, inédito; Salta.

MONALDI, C. R., BOSSO, M. A. & FERNANDEZ, J. C. (1986): Estratigrafía del Ordovicico de la sierra de Zapla, provincia de Jujuy. - Asoc. Geol. Argentina Rev., 41 (1-2): 62-69; Buenos Aires.

MOYA, M. C. (1986): Estratigrafía del Tremadociano en el tramo austral de la Cordillera Criental argentina. Parte II: sierra de Mojotoro, cordón de Lesser, área de Salinas Grandes y curso superior río Toro. - Informe Final beca de perfeccionamiento, CONICET, inédito, Buenos Aires.

MOYA, M. C. & SALFITY, J. A. (1982): Los ciclos magmáticos en el noroeste argentino. - Quinto Congr. Latinoam. de Geol. Actas, III: 523-536; Buenos Aires.

NESOSSI, E. (1950): Estudio Geológico de los departamentos de Santa Victoria (Salta) y Yavi (Jujuy). - Informe interno Y.P.,F., inédito, Buenos Aires.

RAMOS, V. A. (1973): Estructuras de los primeros contrafuertes de la Puna salto-jujena y sus manifestaciones volcánicas asociadas. - Quinto Congr. Geol. Argentina Actas, IV: 159-202; Córdoba.

RUIZ HUIDOBRO, O. J. (1975): El Paleozoico Inferior del centro y sur de Salta y su correlación con el Grupo Mesón. - Primer Congr. Argentino de Paleont. y Bioestr. Actas, I: 91-107; Tucumán.

RUSSO, A. (1948): Levantamiento geológico de la cuenca del río Santa Cruz. - Informe interno Y.P.F., inédito; Buenos Aires.

-- (1949): Levantamiento geológico de la parte del río Iruya, entre sus cabeceras y el Río Astilleros. - Informe interno Y.P.F., inédito,; Buenos Aires.

SALFITY, J. A., OMARINI, R., BALDIS, B. & GUTIERREZ, W. J. (1975): Consideraciones sobre la evolución geológica del Precámbrico y Paleozoico del norte argentino. - Segundo Congr. Iberoam. Geol. Econom. Actas, IV: 341-361; Buenos Aires.

SALFITY, J. A., MALANCA, S., MOYA, M. C., MONALDI, C. R. & BRANDAN, E. M. (1984a): El límite Cámbrico-Ordovícico en el norte de la Argentina. - Noveno Congr. Geol. Argentina Actas, I: 568-575; Bariloche.

SALFITY, J. A., MALANCA, S., BRANDAN, E. M., MONALDI, C. R. & MOYA, M. C. (1984b): La fase Guandacol (Ordovícico) en el norte de la Argentina. - Noveno Congr. Geol. Argentina Actas, I: 555-567; Bariloche.

TURNER, J. C. M. (1960): Estratigrafía de la sierra de Santa Victoria y adyacencias. - Acad. Nac. Cienc., XLI (2): 163-196; Córdoba.

-- (1963): The Cambrian of Northern Argentina. - Tulsa Geol. Soc. Digest, 31: 193-211; Tulsa.

-- (1972): Ordovícico. - Cuartas Jorn. Geol. Argentina Actas, III: 191-208; Buenos Aires.

TURNER, J. C. M. & MENDEZ, V. (1975): Geología del sector oriental de los departamentos de Santa Victoria e Iruya, Provincia de Salta, República Argentina. -Acad. Nac. Cienc., 51 (1-2): 11-24; Córdoba.

GEOLOGY OF THE COQUENA FORMATION (ARENIGIAN-LLANVIRNIAN) IN THE NW ARGENTINE PUNA: CONSTRAINTS ON GEODYNAMIC INTERPRETATION

H. Bahlburg, C. Breitkreuz & W. Zeil
Institut für Geologie und Paläontologie, Technische Universität,
Ernst Reuter Platz 1, 1000 Berlin 10, Federal Republic Germany.

Abstract

The turbidite series of the Coquena Formation (upper Arenigian-Llanvirnian) are exposed in the northwest Argentine Puna to the west of the salars of Cauchari and
Olaroz between 23°30'S and 24°S. Together with the volcaniclastic Aguada de la Perdiz Formation (middle Arenigian) to the west of the Chilean border, it constitutes a
series of turbidites and intercalated volcanics of at least 6000 m thickness. Synsedimentary volcanism occurs only in the presumably oldest parts of the Coquena Formation. The turbidite greywackes consist mainly of detritus of intermediate and acidic
magmatic rocks. The turbidites were deposited by northward directed paleocurrents in
an at least 2700 m thick submarine fan system. In the upper 2000 m the fan system
displays a marked fining upward trend from mid fan channel environments to outer fan
depositional lobe associations and basin plain pelites. Basin subsidence increased
when active volcanism represented by the Aguada de la Perdiz Formation faded away
after the middle Arenigian and uplift of the Proto-Cordillera Oriental was initiated
in the Late Arenigian (Guandacol diastrophic phase). Subsequently the thick submarine fan system developed, fed mainly by erosional debris of Proto-Cordillera
Oriental and Aguada de la Perdiz Formation. The alleged volcanics constituting the
'Faja Eruptiva de la Puna Oriental' are in fact foliated, silicic porphyric intrusives. For structural and stratigraphic reasons the intrusives have to be younger
than Late Ordovician in age and are not the magmatic arc of the alleged east-dipping
subduction zone contemporaneous to the Coquena Formation.

Introduction

In the Puna highlands of the northwest Argentine provinces Salta and Jujuy, Ordovician sediments occur in elongated NNE-SSW striking horst structures. The horsts sur-
pass the almost 4000 m high Puna plain by at least a further 1000 m. In our studies
we concentrated on the Ordovician strata to the west of the salars of Cauchari and
Olaroz between 23°30'S and 24°S (Fig. 1). These rocks are closely linked by stratigraphy and facies to the middle Arenigian Aguada de la Perdiz Formation (GARCIA et
al. 1962, BREITKREUZ 1986)(Fig. 7). The geotectonic situation that controlled the
development of this and further Ordovician formations in northern Chile is poorly
understood (BREITKREUZ et al., this volume). In this paper we portray the evolution
of the Ordovician basin in the western part of the northern Puna.

Lecture Notes in Earth Sciences, Vol. 17
H. Bahlburg, Ch. Breitkreuz, P. Giese (Eds.),
The Southern Central Andes
© Springer-Verlag Berlin Heidelberg 1988

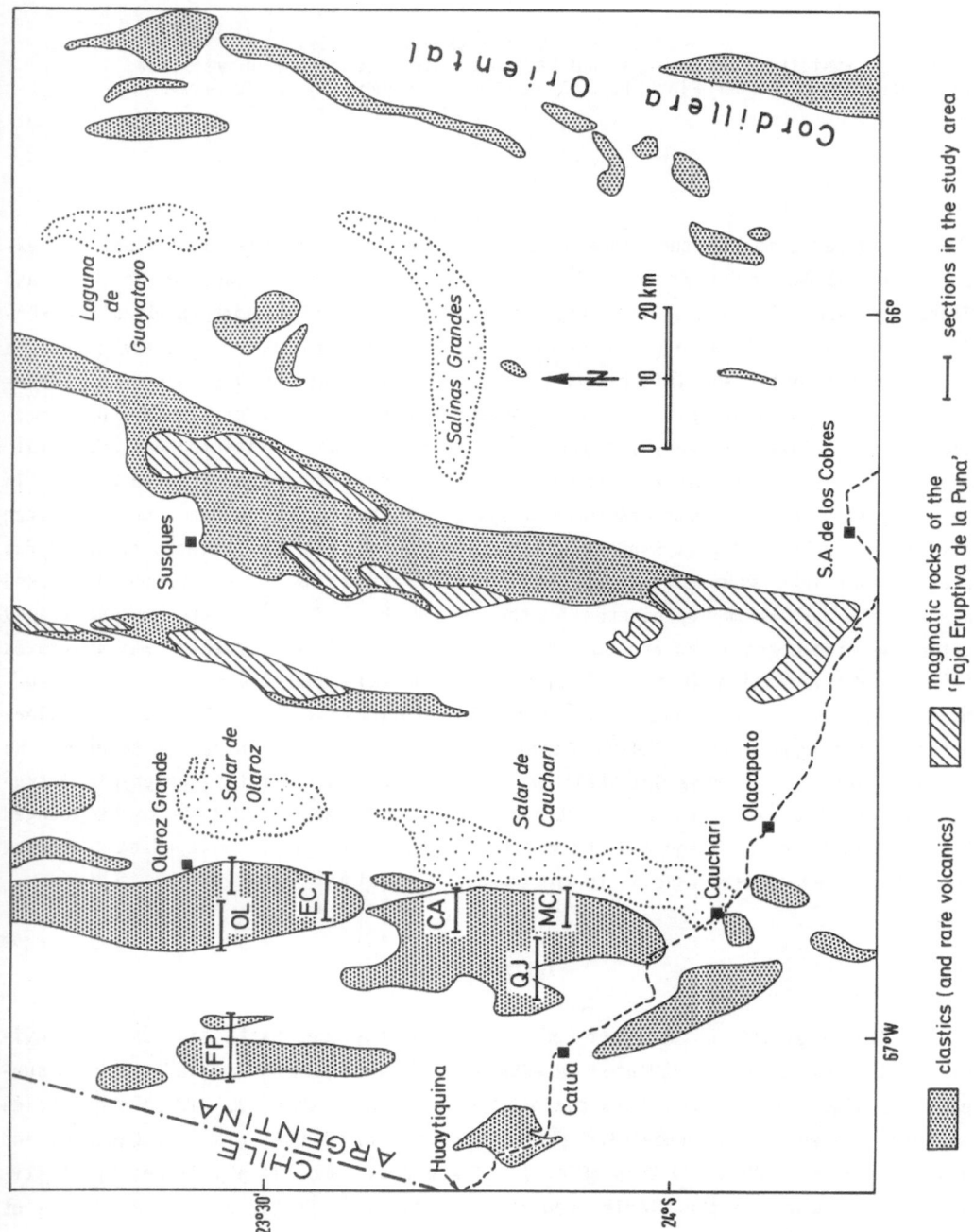

Fig. 1: Distribution of Ordovician rocks in the northwestern Argentine Puna and the
sections studied in the Coquena Formation: Olaroz Grande (OL), Quebrada Esquina
Colorada (EC), Filon Pircas (FP), Cerro Agua Chica (CA), Moro Corralito (MC) and
Quebrada Juntas (QJ).

Stratigraphy

The clastic series of the Puna were considered Precambrian or Ordovician until graptolite finds of ROLLERI & MINGRAMM (1968), ACEÑOLAZA & TOSELLI (1971), RAMOS (1972)
and SCHWAB (1973) established a Tremadocian to ?Caradiocian age for the rocks. The
evolution of the northwestern Puna through geological time was first described by
SCHWAB (1973). The base of the Ordovician is not exposed in the area studied. However, in the eastern Puna and the Eastern Cordillera Tremadocian beds lie unconformably on top of the Cambrian 'Grupo Mesón' (TURNER 1960, MENDEZ et al. 1979, KUMPA &
SANCHEZ, this volume). The oldest known fossils of the northwestern Puna belong to
the upper Arenigian (SCHWAB 1973), whereas the Aguada de la Perdiz Formation on the
western side of the Chilean border is of middle Arenigian age (BREITKREUZ 1986). The
series under consideration in this paper was named Coquena Formation by SCHWAB
(1973) who assigned it to the upper Arenigian-Llanvirnian. He estimated the thickness of the formation to be 3000 m. For this study only preliminary determinations
of those specimen of the graptolite material that we collected are available that
pertain to the Quebrada Esquina Colorada (EC) and Olaroz (OL, Fig. 1) sections (B.
ERDTMANN, Berlin, written com.):

sample EC 26: *Rectograptus (= Orthograptus) cf. R. rugosus* (Emmons 1955)
  (Llanvirnian/Llandeilian)

  EC 23: *Diplograptus cf. ellesi* BULMAN 1963 or
  Glyptograptus dentatus
  (base of Llanvirnian)

  EC 20: *Glyptograptus (Undulograptus) cf. austrodentatus* (HARRIS & KEBLE 1932)
  (top of Arenigian) or
  G. intersitus
  (base of Llanvirnian)

  EC 11a: *cf. Didymograptus artus* ELLES & WOOD 1901
  (base of Llanvirnian)

  EC 28: *Glyptograptus (Undulograptus) cf. austrodentatus* (HARRIS & KEBLE 1932)
  (top of Arenigian)

  OL 10: *Glyptograptus (Undulograptus) dentatus*
  Pseudoclimatograptus cf. P. cumbrensis BULMAN 1963
  (Arenigian/Llanvirnian boundary zone)

  OL 28: *Didymograptus cf. D. artus* ELLES & WOOD 1901
  (or *D. pluto* JENKINS 1983)
  (Lower Llanvirnian)

The stratigraphic results document that the largest part of the turbidites of the
Coquena Formation have been deposited from the Late Arenigian to the Early Llanvirnian. The preliminary determinations confirm the stratigraphic position of the formation as determined by TURNER in (SCHWAB 1973). The strata were folded during the
Oclóyic phase at the Ordovician/Silurian transition (TURNER & MENDEZ 1979). They are
discordantly overlain by Cenozoic clastics and volcanics.

Tectonics

We studied the Coquena Formation along 6 sections (Fig. 1). The rocks are well exposed to the south of Olaroz Grande (OL) and in the Quebrada Esquina Colorada (EC,
Fig. 1). In these areas the strata form large scale open folds around NNW to N
trending fold axes (Figs. 2 and 3). The folded strata were uplifted as NNE trending
elongated horsts along reverse faults during the Tertiary (SCHWAB 1985). In the
immediate vicinity of the faults the uplift led to secondary inclination of the
folds. In the western part of the Olaroz Grande section (OL, Figs. 1 and 2), we were
in a position to measure a sedimentological section with a thickness of 2700 m in
the eastern limb of an open fold (Figs. 2 and 4). In the other sections mentioned in
Figure 1, effects of compressive tectonics are more pronounced and led in parts to
isoclinal folds and locally to east or west vergencies. Cleavage at an angle to
bedding planes is common in pelites and may occur in greywackes within the center of
folds and in isoclinally folded areas.

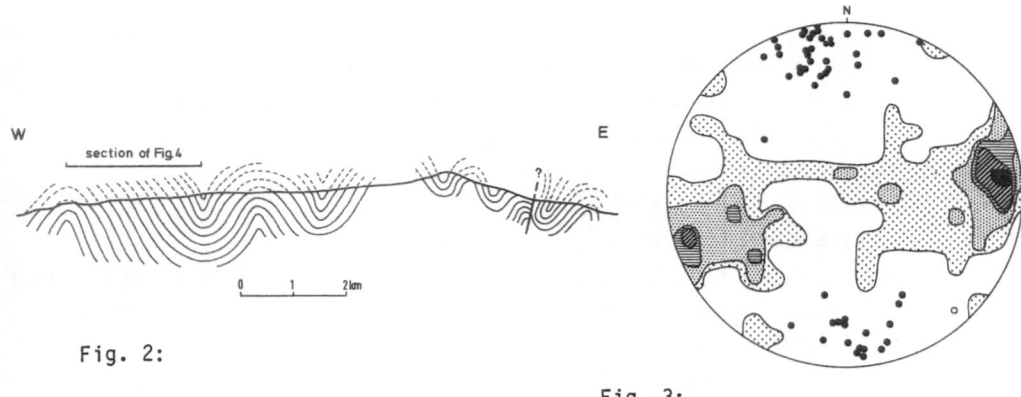

Fig. 2:

Fig. 3:

Fig. 2: Cross section through the Coquena Formation to the south of Olaroz Grande
(OL-section, Fig. 1).

Fig. 3: Poles to bedding planes and fold axes (dots) of the Coquena Formation.

Sedimentology

The sediments of the Coquena Formation consist predominantly of coarse and medium
grained greywackes, pelites and less frequent pebbly sandstones and pebbly conglomerates (see SCHWAB 1973). They were deposited by high and low density turbidity
currents (LOWE 1982). Apart from the dominant turbidites, fine grained, siliceous
rocks of up to 10 cm thickness occur, especially in the 'Filon Pircas' area (FP,
Fig. 1). On the basis of thin sections we interpret them as vitric tuffs and tuffi-
tes originally rich in glass of feldspatic composition. On the crest of 'Cerro

Coquena' (at the western end of the 'Quebrada Juntas' section (QJ), Fig. 1), black basaltic tuffs of a few meters thickness occur. SCHWAB (1973) mentions the presence of diabasic lavas and diabasic tuffs in this area.

The turbidites of the OL-section represent the thickest, undisturbed series (2700 m) that we studied. We thus chose it for detailed representation. The strata of the EC-section are exposed with a thickness of approx. 1300 m. Due to stratigraphical results, photogeologic interpretation and its similar cyclicity it can be correlated with the upper part of the OL-section. The sediments of both profiles can be divided into 5 facies types which, however, display gradual transitions among them (Fig. 4):

Facies type 0: This group of sediments is composed of pebbly sandstones and pebbly conglomerates deposited by high density turbidity currents. They correspond largely to the S-divisions of the LOWE-sequence (LOWE 1982), equivalents of R-divisions are less common. The deposits are usually channelled and amalgamated to thick packets and display normal and inverse grading. They represent channel fill sediments of the distal inner fan or the proximal mid fan (facies A of MUTTI & RICCI LUCCHI 1978).

Facies type 1: This subdivision deals with successions of relatively level bedded coarse and medium sandy greywackes and pebbly sandstones (S-divisions of the LOWE-sequence). Bed thicknesses are in the range of 10-15 cm and 1 m. The rocks are equivalent to the massive sandstones of WALKER (1984) and resemble facies B of MUTTI & RICCI LUCCHI (1978). We assign them to mid fan channels or channeled portions of supra fan lobes (WALKER 1984).

Facies type 2: Type 2 comprises (i) turbidites that consist of successions of 15-60 cm thick T_{a-c} of the BOUMA-sequence (facies type 2a) and (ii) sequences of approx. 50 cm thick $T_{a-d(e)}$ and $T_{a,cd(e)}$ with thin T_a and T_b (facies type 2b). Due to dia-genetic and tectonic changes it is usually not possible to distinguish between T_d and T_e. Facies type 2 corresponds in parts to facies C of the MUTTI & RICCI LUCCHI-model. Type 2a belongs to depositional lobes of the middle and outer fan but may also occur as the upper part of channel fill. Type 2b is more distal in character and also contains sediments of middle to outer fan depositional lobes.

Facies type 3: Here fine grained, "distal turbidites" consisting of $T_{cd(e)}$-sequences are grouped together. Single turbidites are usually not thicker than 20 cm. They closely resemble facies D of MUTTI & RICCI LUCCHI (1978) and could have been deposited in outer fan depositional lobe, interchannel or inner and mid fan levee environments.

Facies type 4: Pelites of up to 60 m thickness repeatedly occur. They are interpreted as pelagic pelites belonging to the basin plain.

On the basis of the introduced facies types we can describe the evolution of the series (Fig. 4). Base and top are, however, not exposed in the area studied: The first 700 m of the section are characterized by 2 coarsening upward megacycles (fa-

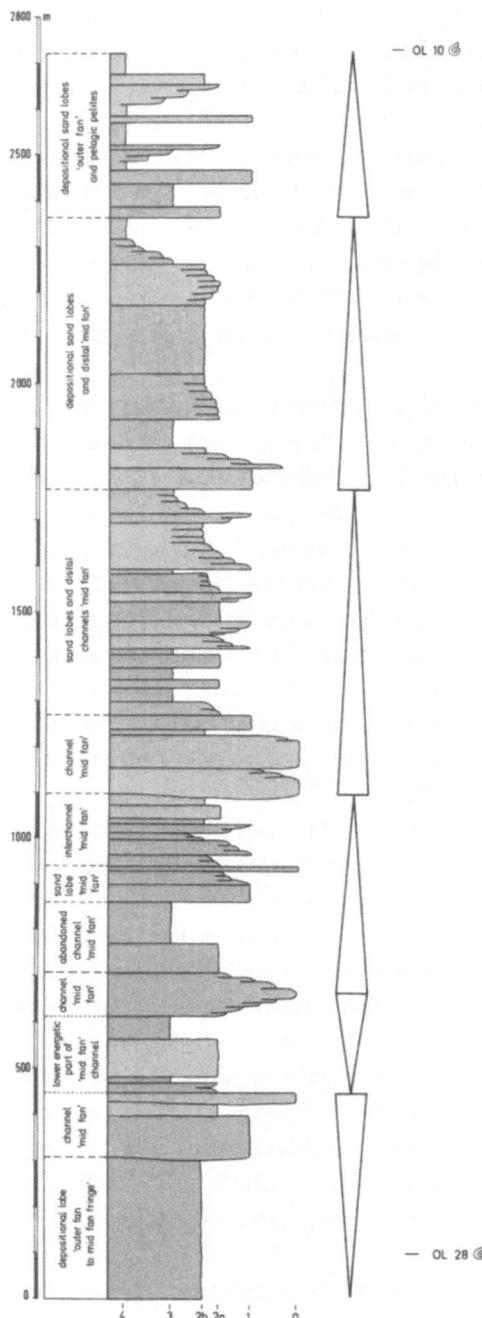

Fig. 4: Schematical sedimentological section of the Coquena Formation to the south of Olaroz Grande (OL-section, Figs. 1 and 2). See text for explanation of facies-types **0**, **1**, **2a**, **2b**, **3** and **4**.

cies type **2** to **O**). They were produced by the migration of a mid fan channel into outer fan to mid fan fringe depositional lobes. This development is reversed in the upper 2000 m by 4 marked fining upward megacycles (max. **O** to **4**). They represent an episodic retreat of the submarine fan system. The depositional environment moved from mid fan channel deposits to mid fan depositional lobes and outer fan sand lobes. Thick pelites intercalated near the top belong to the basin plain. Figure 4 also demonstrates that the megacycles consist of oscillating subcycles. Deposition of the series was very rapid as it took place during the Late Arenigian and the Early Llanvirnian.

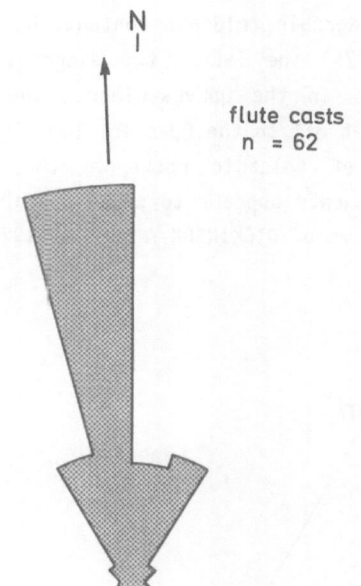

flute casts
n = 62

In all measured sections of Figure 1 we were able to determine paleocurrent directions with the help of flute casts (Fig. 5). The currents were directed northward and have a pronounced maximum to the NNW. Currents directed to the SW occur in the Chiquero Formation to the east of the study area in the region of the 'Faja Eruptiva de la Puna' (SCHWAB 1973, MENDEZ et al. 1973)(Figs. 1 and 7). The Chiquero Formation has a Tremadocian to early Arenigian age and underlies the Coquena Formation (SCHWAB 1973). Turbidite deposition developed parallel to NNW trending fold axes (Fig. 2). This leads to the interpretation that the turbidites filled an elongate trough which suffered synsedimentary compressive tectonics.

Fig. 5: Paleocurrent directions in the Coquena Formation as indicated by flute casts.

Petrofacies

The turbidite greywackes have a modal matrix content of up to 12%. The detritus is usually angular to subrounded and the rocks are badly sorted. The greywackes have a quartz content of 25-60%. Monocrystaline, mainly non-undulatory and crescent shaped quartz grains predominate and are probably of volcanic origin. Polycrystaline quartz is usually very fine grained and 'chert-like'. At least parts of it may correspond to groundmass fragments of rhyolites. Feldspar at 10-20% is less abundant whereas plagioclase is slightly enriched against K-feldspar. The percentage of rock fragments can be as high as 40%. Fragments of intermediate volcanics are more abundant than those of basalt. Detritus of feldspar rich, crystalized vitric tuffs is conspicuous, as is their similarity to those of the Aguada de la Perdiz Formation as des-

cribed by BREITKREUZ (1986). Related rocks also occur in the 'Filon Pircas' area
(Fig. 1). Sedimentary and metasedimentary clasts are rare and consist of fine grained arenites, pelites and phyllites. Occasional gneissic fragments constitute the
detritus of the highest metamorphic grade. Muscovite is present in variable but low
amounts, biotite is rare. Carbonate clasts do not occur in the studied rocks.

Plots of selected frame work components (Fig. 6) emphasize the intermediate quartz
content of the greywackes (QFL-diagram) and the abundance of rock fragments (QmFLt-
diagram). The rock fragments are dominated by volcanic clasts and polycrystaline
quartz, the latter of which could also be of volcanic origin (QpLvLs-diagram). The
QFL- and QmFLt-diagrams also underline the considerable feldspar content. Using the
discriminant schemes of DICKINSON & SUCZEK (1979) the latter two diagrams would
class the rocks as 'recycled orogenic', whereas in the QpLvLs-diagram the rocks
would plot close to the 'arc orogen' field but set off to the Qp-pole. The petrology
of the greywackes demonstrates the dominance of volcanic rocks in the source
area(s). A pronounced influence of rhyolitic fragments appears to cause difficulties
in the interpretation of provenance using the scheme of DICKINSON & SUCZEK (1979).

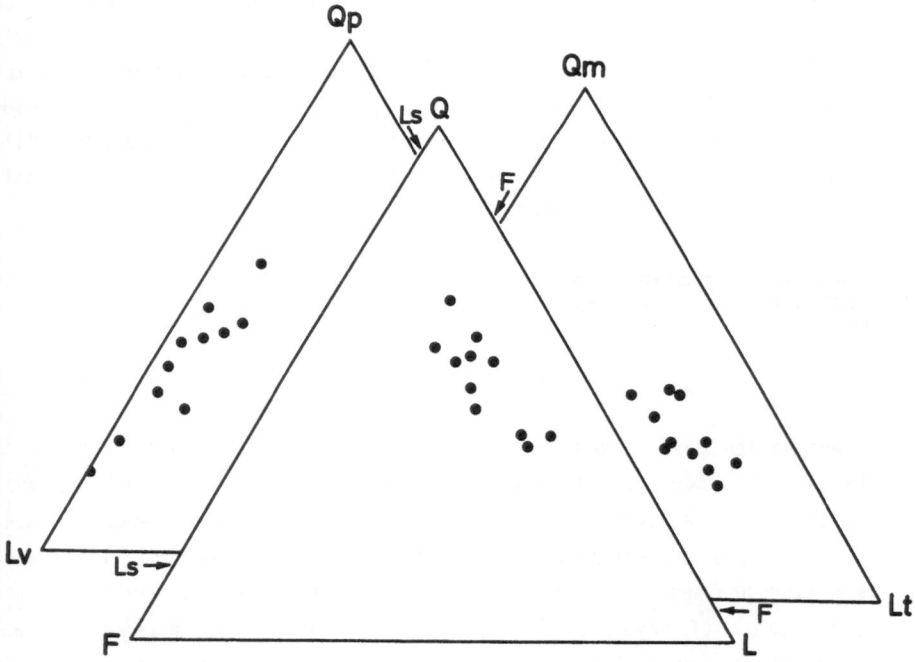

Fig. 6: Framework modes of turbidite greywackes of the Coquena Formation.

The source area of the volcanic detritus could have been the Aguada de la Perdiz Formation to the west and its southsoutheastern continuations (Fig 7) (SCHWAB 1973, COIRA & BARBER 1987, KOUKHARSKY et al. 1987). In this formation intermediate and silicic volcaniclastics dominate over basic rocks (BAHLBURG et al. 1987). The (meta-)sedimentary detritus was probably derived from the rising Proto-Cordillera Oriental to the east ('Cratógeno Central', SALFITY et al. 1975).

Basin evolution and problems of geodynamic interpretation

The turbidite series of the Coquena Formation is characterized by a dominance of volcanic detritus. Lavas and tuffs document active volcanism only in the western and presumably oldest outcrops of the study area ('Filon Pircas' and 'Quebrada Juntas', Fig.1). These lie at a short distance from the middle Arenigian Aguada de la Perdiz Formation (Fig. 7). We presume that at 'Filon Pircas' (and 'Cerro Coquena'?) the oldest strata (upper Arenigian) of the Coquena Formation are exposed and that in these the waning of Aguada de la Perdiz volcanism is documented. The series of the OL- and EC-sections lack this volcanic influence and therefore might constitute the parts of the formation that lie above the 'Filon Pircas' strata. The Aguada de la Perdiz Formation was formed in shallow water conditions in a basin controlled by volcanism (BREITKREUZ 1986). Concomitant to the waning of volcanic activity, subsidence of the basin increased and the turbidites of the Coquena Formation were deposited in a submarine fan system by northward directed paleocurrents. The thicknesses of the Aguada de la Perdiz Formation (2700 m) and the Coquena Formation (2700 m plus a few hundred meters at 'Filon Pircas') add to approx. 6000 m of volcanics and sediments that accumulated in the northwestern Puna from the middle Arenigian to the Llanvirnian.

Another source area was most probably the Proto-Cordillera Oriental to the east (Fig. 1). Uplift in this area was initiated in the Late Arenigian (Guandacol diastrophic phase, SALFITY et al. 1984) and probably also contributed to increased basin subsidence in the western part of the basin that was apparently situated west of the Proto-Cordillera Oriental. As a result of subsidence, the depositional site of the Coquena Formation moved to a more distal position relative to the source area. This is expressed by the longstanding fining upward trend in the upper part of the formation. The turbiditic system was fed by increased erosion of the source regions.

Whereas the volcanic influence in the isolated outcrop of the Aguada de la Perdiz Formation became known only very recently (BREITKREUZ 1986) and is attributed to a volcanic arc regime (PICHOWIAK et al. 1987), the geotectonic role of the 'Faja Eruptiva de la Puna Oriental' (MENDEZ et al. 1973)(Fig. 7) has not been agreed upon.

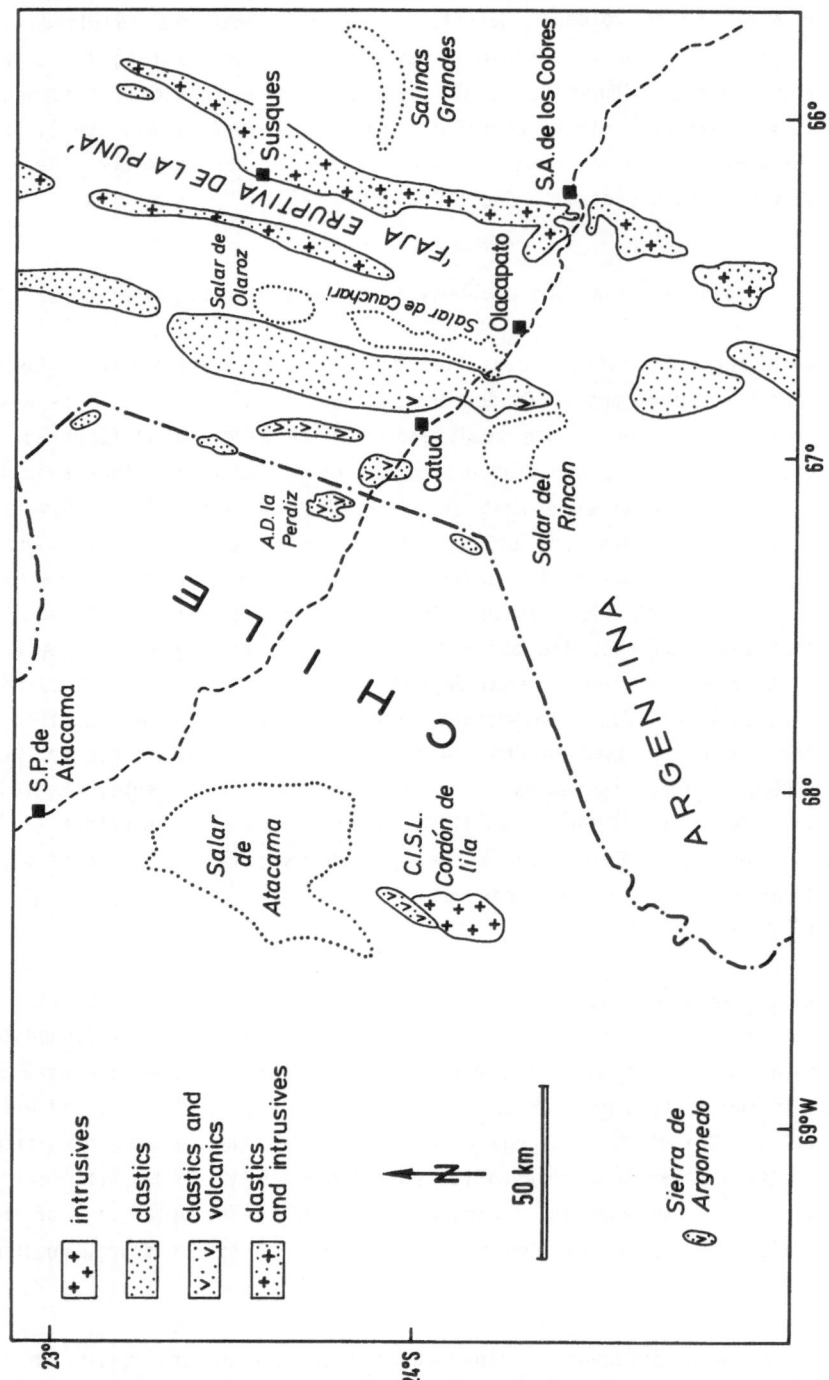

Fig. 7: Distribution of Ordovician rocks in the northwestern Argentine Puna and northern Chile.

Radiometric dating (Rb-Sr) of granitoids in the 'Faja Eruptiva' (471 +/- 12 Ma, OMARINI et al. 1984) in the vicinity of San Antonio de los Cobres (Fig. 7) indicates that magmatism might have been active in Middle Ordovician (Llanvirnian) times. A radiometric age of 374 +/- 7 Ma (Rb-Sr) determined further north by OMARINI et al. (1979) is seen as the result of a thermal reactivation (COIRA et al. 1982). COIRA & RAMOS in COIRA et al. (1982) and ALLMENDINGER et al. (1983) interpret the 'Faja Eruptiva' as the magmatic arc of an Early Ordovician east dipping subduction zone which became inactive after the Arecuipa-Massif docked to Gondwana during the Oclóyic phase (RAMOS 1986). In this context the Coquena Formation is seen as a fore arc flysch. To the south of 25°S, isolated ultrabasite associations (ARGANARAZ et al. 1973, ALLMENDINGER et al. 1983) are correlated with occurrences of similar rocks in the Argentine Precordillera. They are assumed to be ophiolite fragments of the Early Ordovician subduction zone (ALLMENDINGER et al. 1983, RAMOS et al. 1986). As opposed to this DAVIDSON & MPODOZIS in COIRA et al. (1982) and ACENOLAZA & TOSELLI (1984) consider the magmatism of the 'Faja Eruptiva' as a product of ensialic distension. The Early and Middle Ordovician sedimentary basin of the Coquena Formation correspondingly forms part of an ensialic marginal basin between a southern extension of the Arequipa-Massif and the Paraguay craton. A further hypothesis is offered by DALZIEL & FORSYTHE (1985). They classify the Coquena basin as oceanic. It was closed during the Oclóyic phase as a result of the collision of the Arequipa-Massif with the Paraguay craton. Lithospheric shortening is compensated for by two subduction zones of opposing polarity.

All subduction models mentioned postulate pronounced and uniform west vergencies (or vergencies to the east, RAMOS 1986) as observed in southern Puna and Precordillera (ALLMENDINGER et al. 1983, RAMOS et al. 1986) for the folded strata of the Coquena Formation. However, in the study area vergencies to the east and west are equally represented and are only of local importance. Pronounced and uniform vergencies do not occur.

During our visits to various outcrops of the 'Faja Eruptiva' between San Antonio de los Cobres (Fig. 7) and the Argentine-Bolivian border to the north we exclusively observed silicic porphyric rocks intruding the already folded Ordovician sediments. The intrusives therefore have a maximum age of Late Ordovician to Silurian. Our results coincide with the original observations presented by MENDEZ et al. (1973). The rocks of the 'Faja Eruptiva' in the northern Puna are not ignimbrites as described by COIRA (e.g. 1979) and they do not constitute the magmatic arc of a subduction zone contemporaneous to deposition of the mentioned Ordovician sediments. However, the porphyric intrusives are foliated to varying degrees by a north trending subvertical foliation.

As in the southern Puna the 'Faja Eruptiva' is locally accompanied by metamorphic rocks displaying augengneissic and migmatic structures (see also SCHWAB 1973). The field relations suggest that they are metamorphosed Ordovician clastics. In the southern Puna, the sediments were subject to prograde regional metamorphism of medium pressure (VIRAMONTE et al. 1976). These authors assume a genetic connection with the intrusions of the 'Faja Eruptiva' whose granitoids have high Sr-initials of 0,7183 and 0,7100 (OMARINI et al. 1979, 1984). They seem to be either of anatectic origin or seem to have suffered considerable crustal contamination. SALFITY et al. (1975) interpret the 'Faja Eruptiva' as an intracontinental magmatic and migmatic zone for which they coined the term 'Faja Movil de la Puna Oriental'.

The regional review receives further complication if the north Chilean sediments and volcanics of the 'Sierra de Argomedo' (Argomedo beds, BREITKREUZ 1986) and of the 'Cordón de Lila' (C.I.S.L., NIEMEYER et al. 1985)(Fig. 7) are included. The Argomedo beds were deposited in a shallow marine, coastal environment during the Early Ordovician (BREITKREUZ 1986). Few altered intermediate and acidic volcanics are intercalated. In the 'Cordón de Lila' turbiditic sediments occur which are probably of Ordovician or even pre-Ordovician age (DAMM et al. 1986). Intercalated tholeiitic pillow basalts either extruded in a back or fore arc rift regime or bear witness to an initial island arc magmatism (PICHOWIAK et al. 1987). As long as the exact stratigraphic position of C.I.S.L. and Argomedo beds is not determined, any discussion of their connection with the Aguada de la Perdiz Formation and the Ordovician strata of the Argentine Puna is highly speculative (see also BREITKREUZ et al., this volume).

Conclusion

In spite of the presented variety of models, it is still almost impossible to advance an evolutionary interpretation of the Ordovician sedimentary basin of the northern Argentine Puna. It appears to be debatable whether the relatively well established geotectonic concepts proposed for the southern Puna (ALLMENDINGER et al. 1983) and Precordillera (RAMOS et al. 1986) can be easily applied to the northern Puna. The basic problem may lie in variations in the evolution of the Puna to the north and south of the NW-SE striking Calama-Olacapato-Toro lineament as outlined by ALONSO et al. (1984). This megafault had apparently directed the structural development since the Precambrian (SALFITY 1985). As long as differences in the history of northern and southern Puna, including the 'Faja Eruptiva', are not explained, one can only speculate which geodynamic regime can be ascribed to the Early and Middle Ordovician basin of the northern Puna. The accumulation of sediments within the basin had been strongly influenced by the rising Proto-Cordillera Oriental and the intermediate and silicic volcanism of the Aguada de la Perdiz Formation at least since the middle Arenigian as well as by subsequent compressive tectonics. There are no outcrops of oceanic crust known in the northern Puna. It is therefore possible that the basin formed on continental crust (see DAVIDSON & MPODOZIS in COIRA et al. 1982 and ACENOLAZA & TOSELLI 1984). The questions of origin and duration of magmatism in Aguada de la Perdiz Formation and 'Faja Eruptiva' and a possible relationship between the two are vital for a solution of the problem. Only when these problems are solved, can a geodynamic concept for the evolution of the Early and Middle Ordovician basin of the northern Puna be realistically proposed.

However, on the basis of our results we can describe the depositional history of the Coquena Formation in the NW Argentine Puna: The Coquena Formation is at least 3000 m thick and consists of turbidites which interfinger with volcanics in the lower part of the formation. The sediments were deposited very rapidly in a submarine fan system during Late Arenigian and Early Llanvirnian times by northward directed, longitudinal paleocurrents. To the east the basin was limited by the Proto-Cordillera Oriental. In the western part of the ensialic basin the Coquena Formation overlies the shallow marine sediments and volcanics of the middle Arenigian Aguada de la Perdiz Formation. The cessation of volcanism in this formation after the middle Arenigian and the onset of uplift of Proto-Cordillera Oriental during the Guandacol diastrophic phase in the Late Arenigian led to increased subsidence mainly in the western part of the basin. As a result of subsidence the depositional site of the Coquena Formation moved to a more distal position relative to the source areas. This is expressed in the longstanding fining upward trend in the upper part of the turbidite series. The turbidites are characterized by a dominance of intermediate and silicic volcanic detritus. It was derived from the Aguada de la Perdiz Formation and its equivalents as a product of syn- and postsedimentary erosion of the volcanic zone.

According to SCHWAB (1973), deposition of the Coquena Formation took place until the Late Llanvirnian. The occurrences of ?Llandeilian graptolites in the area of Mucar to the north of Filon Picas (Fig. 1)(GARDEWEG & RAMIREZ 1985) and of ?Lower Caradocian graptolites in the greywacke and pelite series of the 'Sierra de Lina' (RAMOS 1972) to the north of the area studied at approx. 23°S point to the possibility that sedimentation in the basin continued into Caradocian times. If this assumption is correct, the total thickness of the sediments must be considerably higher than the added thicknesses of Aguada de la Perdiz Formation and Coquena Formation (approx. 6000 m).

The Coquena Formation was deposited by paleocurrents that developed parallel to the tectonic axes of the subsequent Oclóyic tectonic phase. Thus we assume that contemporaneous to sedimentation the basin was already subject to compressive movements. These led to the formation of NNW to N trending folds during the Oclóyic phase at the Ordovician/Silurian transition.

Acknowledgement

We would like to thank B. Erdtmann, Berlin, for the determination of the graptolites. The graptolite material will be deposited in the fossil collection of the Universidad Nacional de Salta, Argentine, under the catalogue nos. CNS-I 088/625-8, CNS-I 091/625-6 and CNS-I 091/629. This project is funded by the 'Deutsche Forschungsgemeinschaft' (ref. no. Gi 31/51-3) and was realized in close cooperation with the members of the Geological Institute of the 'Universidad Nacional de Salta', Argentine, in particular with J.A. Salfity, C. Moya and C. Monaldi. J.A. Salfity, Salta, and K. Şchwab, Clausthal-Zellerfeld, reviewed the manuscript.

References

ACENOLAZA, F.G. & TOSELLI, A.J. (1971): Hallazgo de graptolites en el supuesto precambrico de la Puna de Catamarca.- Rev. Asoc. Geol. Arg., 21(2): 274; Buenos Aires.

ACENOLAZA, F.G. & TOSELLI, A.J. (1984): Lower Ordovician volcanism in North West Argentina.- In: BRUTON, D.L. (ed.): Aspects of the Ordovician System.- Palaeont. Contr. Univ. Oslo, 295: 203-209; Oslo.

ALLMENDINGER, R.W., RAMOS, V.A., JORDAN, T.E., PALMA, M. & ISACKS, B.L. (1983): Paleogeography and Andean structural geometry, northwest Argentina.- Tectonics, 2(1): 1-16; Washington.

ALONSO, R., VIRAMONTE, J. & GUTIERREZ, R. (1984): Puna Austral - Bases para el sub-provincialismo geológico de la Puna Argentina.- 9. Congr. Geol. Arg. Actas, 1: 43-63; Bariloche.

ARGANARAZ et al. (1973): Sobre el hallazgo de serpentinitas en la Puna Argentina.- 5. Congr. Geol. Arg. Actas, 1: 23-32; Córdoba

BAHLBURG, H., BREITKREUZ, C. & ZEIL, W. (1987): Paleozoic basin development in northern Chile (21°-27°S).- Geol. Rdsch., 76(2): 633-646; Stuttgart.

BREITKREUZ, C. (1986): Das Paläozoikum in den Kordilleren Nordchiles (21°-25°S).- Geotekt. Forsch., 70: 1-88; Stuttgart.

BREITKREUZ, C., BAHLBURG, H. & ZEIL, W. (this volume): The Paleozoic strata of Northern Chile: Geotectonic implications.-

COIRA, B. (1979): Descipción geológica de la Hoja 3c, Abra Pampa.- Carta Geológico-Económica de la Republica Argentina 1:200.000, Servicio Geológico Nacional, Buenos Aires.

COIRA, B. & BARBER, E. (1987): Vulcanismo submarino Ordovícico (Arenigiano-Llanvirniano) del Rio Huaytiquina, Provincia de Salta, Argentina.- 10. Congr. Geol. Arg. Actas, 4: 305-307; Tucuman.

COIRA, B., DAVIDSON, J., MPODOZIS, C. & RAMOS, V. (1982): Tectonic and magmatic evolution of the Andes of northern Argentina and Chile.- Earth Sci. Rev., 18: 303-332; Amsterdam.

DALZIEL, I.W.D. & FORSYTHE, R.D. (1985): Andean evolution and the terrane concept.- In: HOWELL, D.G. (ed.): Tectonostratigraphic terranes of the Circum-Pacific-Region.- Circum-Pacific council for energy and mineral resources earth science series, 1: 565-581; Houston.

DAMM, K.-W., PICHOWIAK, S. & TODT, W. (1986): Geochemie, Petrologie und Geochronologie der Plutonite und des metamorphen Grundgebirges in Nordchile.- Berliner geowiss. Abh., A, 66: 73-146; Berlin.

DICKINSON, W.R. & SUCZEK, C.A. (1979): Plate tectonics and sandstone composition.- Am. Ass. Petrol. Geol. Bull., 63: 2164-2182; Tulsa.

GARCIA, A.F., PEREZ D'ANGELO, E. & CEBALLOS, S.E. (1962): El Ordovícico de Aguada de la Perdiz, Puna de Atacama, Provincia de Antofagasta.- Rev. Miner., 77: 52-61; IIG, Santiago.

GARDEWEG, M. & RAMIREZ, C.F. (1985): Hoja Zapaleri, II. Región de Antofagasta. Carta geol. Chile, 1 : 250.000, Sernageomin, 66; Santiago.

KOUKHARSKY, M., COIRA, B. & MORELLO, O. (1987): Vulcanismo Ordovícico de la Sierra de Guayaos, Provincia de Salta, Argentina. Caracteristicas petrológicas e implicancias tectonicas.- 10. Congr. Geol. Arg. Actas, 4: 316-318; Tucuman.

KUMPA, M. & SANCHEZ, C. (this volume): Geology and sedimentology of the Cambrian Grupo Mesón (NW Argentina).-

LOWE, D.R. (1982): Sediment gravity flows, II: Depositional models with special reference to the deposits of high density turbidity currents.- Jour. Sed. Petrol., 52: 279-297; Tulsa.

MENDEZ, V., NAVARINI, A., PLAZA, D. & VIERA, V. (1973): Faja Eruptiva de la Puna oriental.- 5. Congr. Geol. Arg. Actas, 4: 89-100; Córdoba.

MENDEZ, V., TURNER, J.C.M., NAVARINI, A., AMENGUAL, R. & VIERA, V. (1979): Geología de la región noroeste, Provincias Salta y Jujuy, Republica Argentina.- Dir. Gral. Fab. Militares: 1-118; Buenos Aires.

MUTTI, E. & RICCI LUCCHI, F. (1978): Turbidites of the northern Apennines; introduction zo facies analyses.- Int. Geol. Rev., 20(2): 125-166; Church Falls, Va.

NIEMEYER, H., URZUA, F., ACENOLAZA, F. & GONZALEZ, C.R. (1985): Progresos recientes en el conocimiento del Paleozoico de la región de Antofagasta.- 4. Congr. Geol. Chileno Actas, 1: 1/410-439; Antofagasta.

OMARINI, R.H., CORDANI, U.G., VIRAMONTE, J.G., SALFITY, J. & KAWASHITA, K. (1979): Estudio isotópico Rb-Sr de la 'Faja Eruptiva de la Puna' a los 22°35'LS, Argentina.- 2. Congr. Geol. Chileno Actas: E258-269; Arica.

OMARINI, R.H., VIRAMONTE, J.G., CORDANI, U.G., SALFITY, J.A. & KAWASHITA, K. (1984): Estudio geochronológico Rb-Sr de la Fja Eruptiva de la Puna en el sector de San Antonio de los Cobres, Provincia de Salta.- 9. Congr. Geol. Arg. Actas, 3: 146-158; Bariloche.

PICHOWIAK, S., BAHLBURG, H. & BREITKREUZ, C. (1987): Paleozoic volcanic and geotectonic evolution in northern Chile.- 10. Congr. Geol. Arg. Actas, 4: 302-304; Tucuman.

RAMOS, V. (1972): El Ordovícico fosilífero de la Sierra de Lina, Departamento Susques, Provincia de Jujuy, Republica Argentina.- Rev. Asoc. Geol. Arg., 27: 84-94; Buenos Aires.

RAMOS, V. (1986): El diastrofismo Oclóyico: Un ejemplo de tectónica de collisión durante el Eopaleozoico en el noroeste Argentino.- Rev. Inst. Geol. Miner., 6: 13-28; Jujuy.

RAMOS, V.A., JORDAN, T.E., ALLMENDINGER, R.W., MPODOZIS, C., KAY, S.M., CORTES, J.M. & PALMA, M. (1986): Paleozoic terranes of the Central Argentine-Chilean Andes.- Tectonics, 5(6): 855-880; Washington.

ROLLERI, E.O. & MINGRAMM, A. (1968): Sobre el hallazgo del Ordovícico inferior al oeste de San Antonio de los Cobres (Provincia de Salta).- Rev. Asoc. Geol. Arg., 23(2): 101-103; Buenos Aires.

SALFITY, J.A. (1985): Lineamentos transversales al rumbo andino en el noroeste Argentino.- 4. Congr. Geol. Chileno Actas, 2: 2/119-137; Antofagasta.

SALFITY, J.A., MALANCA, S., BRANDAN, M.E., MONALDI, C.R. & MOYA, C. (1984): La Fase Guandacol en el norte de la Argentina.- 9. Congr. Geol. Arg. Actas, 1: 555-567; Bariloche.

SALFITY, J.A., OMARINI, R.H., BALDIS, B. & GUTIERREZ, W.J. (1975): Consideraciones sobre la evolución geológica del Precambrico y Paleozoico del norte argentino.- 2. Congr. Iberoam. Geol. Econ., 4: 341-361; Buenos Aires.

SCHWAB, K. (1973): Die Stratigraphie in der Umgebung des Salars de Cauchari (NW-Argentinien). Ein Beitrag zur erdgeschichtlichen Entwicklung der Puna.- Geotekt. Forsch., 43:1-168; Stuttgart.

SCHWAB, K. (1985): Basin formation in a thickening crust - the intermontane basins in the Puna and the Eastern Cordillera of NW-Argentina (Central Andes).- 4. Congr. Geol. Chileno Actas, 2: 2/139-158; Antofagasta.

TURNER, J.C.M. (1960): Estratigrafía de la Sierra de Santa Victoria y adyaciencias.- Bol. Acad. Nac. Cienc. Córdoba, 41(2): 163-196; Córdoba.

TURNER, J.C.M. & MENDEZ, V. (1979): Puna.- 2. Simp. Geol. Regional Arg., Acad. Nac. Cs. Córdoba, 1: 13-56; Córdoba.

VIRAMONTE, J., SUREDA, R. & RASKOVSKY, M. (1976): Rocas metamórficas de alto grado al oeste del Salar Centenario, Puna Saltena.- 6. Congr. Geol. Arg. Actas, 2: 191-206; Buenos Aires.

WALKER, R.G. (1984): Turbidites and associated coarse clastic deposits.- In: WALKER, R.G. (ed.): Facies models. 2nd edition.- Geoscience Canada, reprint series 1: 171-188; Ottawa.

THE PALEOZOIC EVOLUTION OF NORTHERN CHILE: GEOTECTONIC IMPLICATIONS

Chr. Breitkreuz, H. Bahlburg & W. Zeil
Institut für Geologie und Paläontologie,
Technische Universität Berlin, West Germany

Abstract

The geological record of Paleozoic strata in Northern Chile (21°-27°S) comprises Ordovician and Devonian to Permian sediments and volcanics. An evaluation of geotectonic concepts concerning its formation is given.

The few results available from the scarce Ordovician outcrops are ambiguous with respect to geotectonic implications. The relation to the Ordovician series in Bolivia and Argentina is still unclear.

We presume that intracontinental processes led to the formation of the Devonian to Permian series rather than processes related to a subduction regime. Tectonic control of basin subsidence, deposition and closure was possibly linked to the collision of Chilenia with Gondwana south of 29°S which started in the Middle Devonian. It might be possible that the collision induced a dextral strike-slip cycle in the north Chilean area as a kind of escape movement according to the reverse indenter model of EISBACHER (1985).

We consider the north Chilean Late Carboniferous-Triassic volcanic series in the Pre- and High Cordillera to have been formed in an intracontinental tensional regime as it is presumed for the formation of the Mitu Group in the Eastern Cordillera of Peru and Bolivia.

Introduction

Knowledge of Paleozoic volcanosedimentary development has increased significantly since GARCIA's (1967) classic survey of north Chilean geology. A wealth of newly discovered outcrops of Paleozoic strata have been reported, existing ones have been reinterpreted with the help of modern methods (see e.g. DAVIDSON et al. 1981a,b, 1985, NIEMEYER et al. 1985). Our research group has contributed detailed studies and surveys (see references).

Lecture Notes in Earth Sciences, Vol. 17
H. Bahlburg, Ch. Breitkreuz, P. Giese (Eds.),
The Southern Central Andes
© Springer-Verlag Berlin Heidelberg 1988

Simultaneous to these studies, geotectonic concepts have been developed on a large scale for the Paleozoic of the Central and Southern Andes (HERVE et al. 1981, BELL 1982, COIRA et al. 1982, FORSYTHE 1982, DALZIEL & FOSYTHE 1985, RAMOS et al. 1986 and others.) Here, models of the north Chilean Paleozoic development have been frequently obtained by transfering implications based on the geology of the Southern Andes. The following is an interpretation of the north Chilean Paleozoic evolution from a north Chilean viewpoint.

Geological features

We begin with a short resumé of the north Chilean Paleozoic - detailed descriptions and a summary of results are given in BREITKREUZ (1986), BAHLBURG (1987a) and BAHL-BURG, BREITKREUZ & ZEIL (1987). Para- and orthometamorphic rocks of probable and certain **Precambrian** age occur in the north Chilean Precordillera and in the Mejillones Peninsula near Antofagasta (Fig. 1). Detailed examinations have just been started (PACCI et al. 1980, ZEIL 1983, BAEZA 1984, DAMM et al. 1986, see also BAEZA & PICHOWIAK this vol.). Contemporaneous equivalents of the Argentinian Precambrian/ Cambrian Puncoviscana Fm. and the subsequent Cambrian Meson Group (see ACENOLAZA et al. and KUMPA & SANCHEZ this vol.) have not been documented in Northern Chile.

Early Ordovician marine strata occur in the north Chilean Puna (Aguada de la Perdiz Fm. and adjacent outcrops), in the northern Sierra de Almeida (CISL), and in the Precordillera (Argomedo Beds) (Fig. 1). The Aguada de la Perdiz Fm. (GARCIA et al. 1962) is composed of at least 2700 m thick, mainly silicic volcaniclastic rocks and sand-/siltstones that display a low diversity middle Arenigian graptolite fauna (ERDTMANN in BREITKREUZ 1986).

Probably Early Ordovician tholeiitic pillow lavas and associated hypabyssal stocks are reported in a hemipelagic-turbiditic facies from the 'Complejo Igneo y Sedimentario del Cordón de Lila (= CISL; NIEMEYER 1984, et al. 1985, DAMM et al. 1986). Silicic volcaniclastic rocks also occur. The depositional features of the volcanic rocks indicate fairly shallow water formation (BREITKREUZ 1986).

Fig. 1: Distribution of pre-Mesozoic strata in Northern Chile (21°-27°S). The numbers indicate the formations and localities mentioned in the text: 1) Quebrada Arcas, 2) El Toco, 3) Sierra del Tigre and Cerros de Cuevitas Fms. (Salar de Navidad), 4) Cerro Palestina, 5) Cerro 1584, 6) Aguada de la Perdiz, 7) Cordon de Lila (including CISL), 8) southern Sierra de Almeida, 9) Sierra de Argomedo, 10) Estratos Cerro del Medio, 11) Las Tórtolas, 12) Chinches Fm.

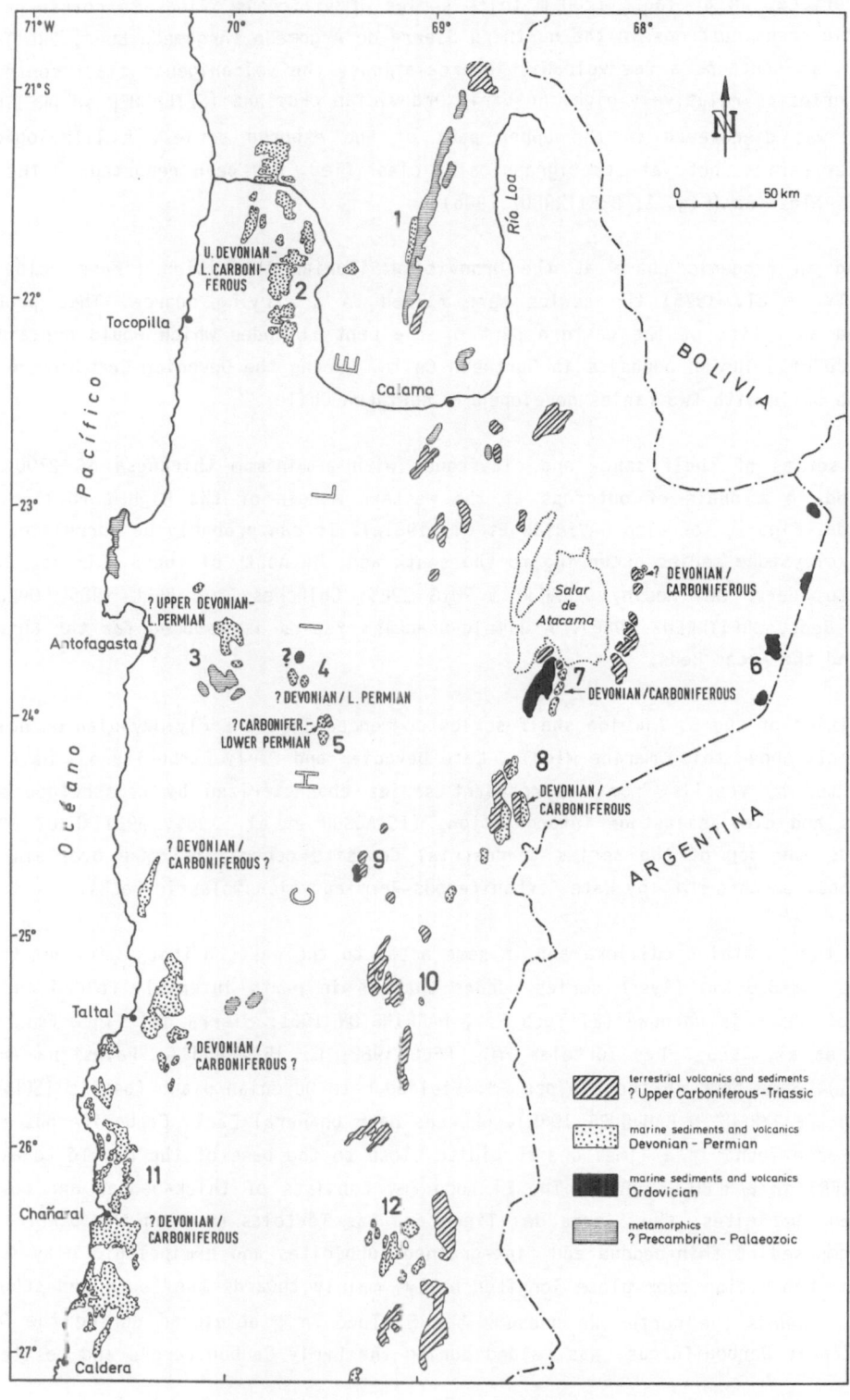

Furthermore, an at least 1200 m thick series of siliceous, fine- to coarse-grained clastic rocks outcrops in the northern Sierra de Argomedo (Argomedo Beds, BREITKREUZ 1985). It contains a few volcanic intercalations, the volcanigenic clast content in the series is relatively high. An Early Ordovician cruziana (SEILACHER in BREITKREUZ 1986) was discovered in the upper part of the exposed series. A lithologically similar series, not yet stratigraphically classified, has been reported in the area of C. Palestina (Fig. 1; BREITKREUZ 1986).

During an orogenic phase at the Ordovician/Silurian transition ('Fase Oclóyica', SALFITY et al. 1975) the series were folded to a varying degree. This probably caused an uplift of the western part of the Central Andes which would explain the absence of Silurian deposits in Northern Chile. During the **Devonian-Carboniferous**, a marine basin with two facies developed in Northern Chile:

1. A series of shelf sand- and claystones with a minimum thickness of 2700 m is exposed in a chain of outcrops at the western margin of the High Cordillera (S. Almeida, Fig. 1, see also DAVIDSON et al. 1981a). It can probably be correlated with sand-/claystone series occurring to the south and the north of the S. Almeida (i. e. Estratos Cerro del Medio, NARANJO & PUIG 1985; Chinches Fm., BELL 1985; Quebrada Arcas Beds, BREITKREUZ 1986). A limnic-brackish facies is assumed for the Chinches Fm. and the Arcas Beds.

Deposition of the S. Almeida shelf series commenced in the Early Devonian on oclóyic basement. Above this, marine Middle, Late Devonian and Early Carboniferous have been confirmed by fossils from a concordant series characterized by crossbedded sand-stones and clay-/siltstone intercalations (ISAACSON et al. 1985, BREITKREUZ 1986). Towards the top of the series terrestrial deposits occur. They are overlain with erosional unconformity by Late Carboniferous-Permian volcaniclastic rocks.

2. In the Coastal Cordillera and in some areas to the east, a thick (minimum thick-nesses 2,3-3,6 km) flysch series occurs that is in parts intensely folded and the base of which is unknown (El Toco Fm., HARRINGTON 1961; Sierra del Tigre Fm., NIE-MEYER et al. 1985; Las Tórtolas Fm., BELL 1982; C. 1584 and C. Palestina areas, Fig. 1). Some beds in the El Toco Fm. yielded Late Devonian plant fossils (SCHWEIT-ZER in BREITKREUZ & BAHLBURG 1985), whereas near Chañaral Early Carboniferous cono-donts were found in a limestone turbidite close to the base of the Las Tórtolas Fm. (OLIVIERI in BAHLBURG 1987a). The El Toco Fm. consists of thick-bedded and coarse-grained turbidites; the Sierra del Tigre and Las Tórtolas Fm., on the other hand, are composed of thin-bedded and fine-grained turbidites and hemipelagic clay-/silt-stones. Deposition took place longitudinally, mainly towards the south and subordi-nately towards the north. We presume the El Toco Fm., developed during the Devo-nian/?Early Carboniferous, was folded during the Early Carboniferous and posttectonically intruded during the middle Carboniferous (320 Ma, SKARMETA & MARINOVIC 1981). As previously mentioned, sedimentation still took place during the Early Carboniferous near Chañaral, where posttectonic intrusions, in parts with S Type affinity, did not occur until the very late Carboniferous (DAMM & PICHOWIAK 1981, BERG et al. 1983).

The southwardly progressing folding led to a shallowing of the marine basin, as confirmed by a transition of the lithofacies from turbidites to mud flows to spiculite sandstones to shallow-water limestones. There are no indications of an Early Carboniferous angular unconformity as postulated by DAVIDSON et al. (1981b) for the Cerro 1584 area (see also NIEMEYER et al. 1985). The fossiliferous shallow-water limestones (C. 1584-, C. Palestina areas, Cuevitas Fm., Fig. 1) with **Early Permian** brachiopods (HOOVER in BREITKREUZ 1986) can probably be correlated to the Copacabana Fm. that is widely exposed in the Central Andes (Fig. 2, see also BARTH 1972). Some basic lavas and acidic pyroclastic and epiclastic rocks are intercalated in this platform series.

These Early Permian volcanic rocks can be seen in connection with parts of the thick Late Carboniferous-Triassic volcanic series which occur in the Chilean Pre- and High Cordillera (Fig. 2, RAMIREZ & GARDEWEG 1982, DAVIDSON et al. 1985). The latter, predominantly acidic, volcanic rocks, referred to in Argentina as Choiyoi Fm. (Fig. 2, see also ZEIL 1981), are associated with limnic-brackish epiclastic rocks. The epiclastic rocks have until now been confined to the Carboniferous-Permian (RAMIREZ & GARDEWEG 1982, OSORIO & RIVANO 1985). Thus their synchronicity with the Early Permian limestones, as assumed in Fig. 2, is speculative.

Geotectonic implications

The scarcity of pre-Devonian outcrops in Northern Chile and the lack of data allows only vague geotectonic conclusions concerning the Early Paleozoic evolution. The relation of the Early Ordovician, predominantly volcanic Aguada de la Perdiz Fm. to the magmatic rocks in the Argentinian Eastern Puna ('Faja Eruptiva') is outlined by BAHLBURG et al. (this volume).

The geochemical characteristics of the tholeiitic magmatic rocks in the CISL, documented by DAMM et al. (1986), indicate the following: The shallow extrusion depth and the occurrence of acidic volcaniclastic rocks render an interpretation of CISL as a remnant of a normal ocean floor improbable. However, one cannot exclude the possibility of CISL having been formed in an ocean island setting. The geochemical data also allows one to presume a subduction-related formation: either as magmatism during subduction initiation or in a back- or forearc rift setting. CISL might

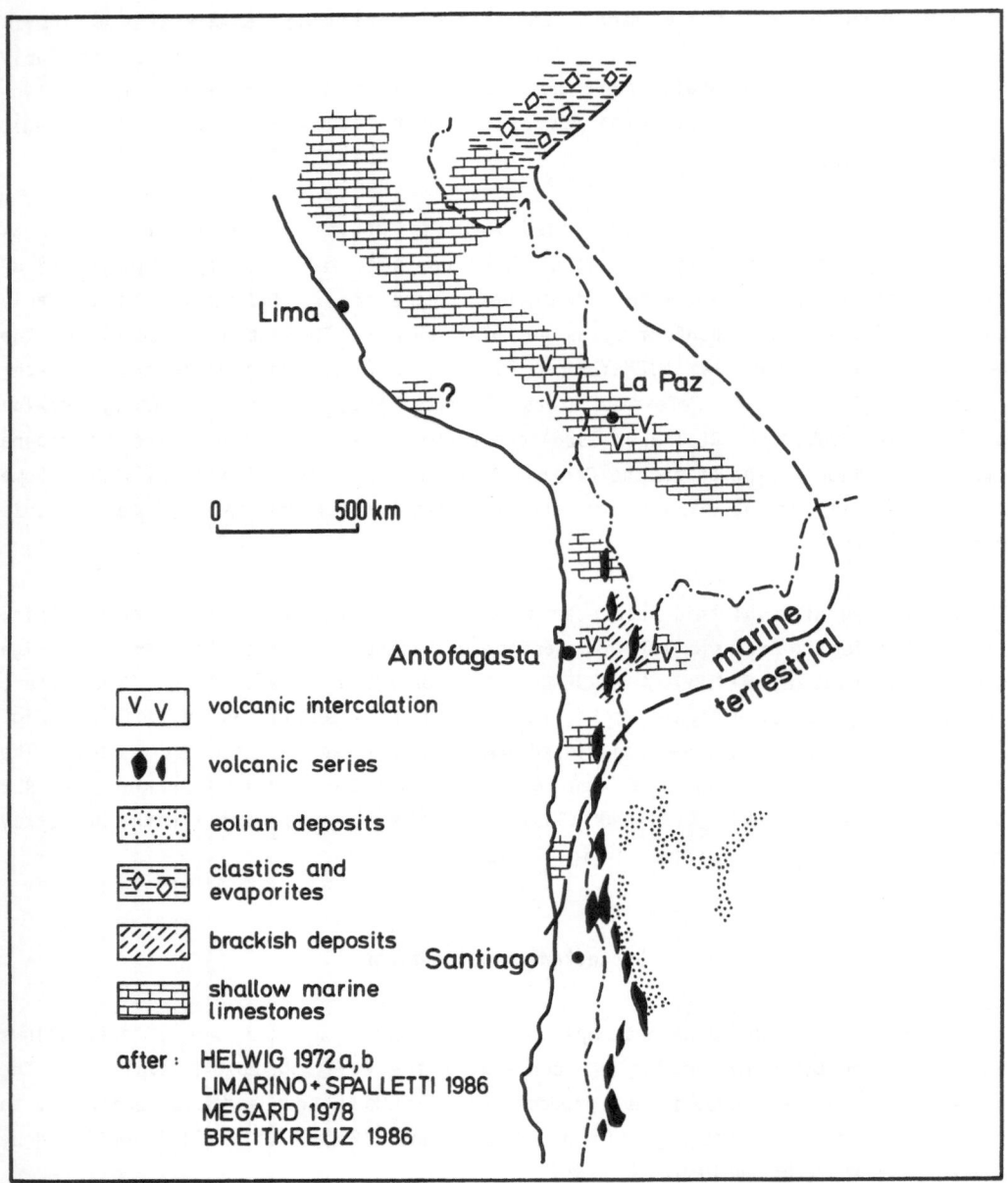

Fig. 2: Facies distribution in the Central and Southern Andes during the Lower Permian.

possibly be considered as the northern continuation of the chain of Ordovician ophiolite outcrops as described by DALZIEL & FORSYTHE (1985) and RAMOS et al. (1986) from the Argentinian Precordillera and Cordillera Frontal. They interpreted it as a relict suture of the Middle Devonian collision of Chilenia with Gondwana (see below). However, closure and folding of the CISL basin must already have taken place during the ?Late Ordovician because the CISL is posttectonically intruded by plutons at the Ordovician/Silurian transition (MPODOZIS et al. 1983, NIEMEYER et al. 1985).

The existence of a Late Paleozoic subduction zone in the area of Central and Southern Chile is seen as confirmed (see e.g. HERVE et al. 1974, 1982, FORSYTHE 1982). Some authors postulate a Late Paleozoic active continental margin also for the north of Chile: **a)** The flysch series of the Coastal Cordillera which in parts display clear SW- to W-vergent folding are seen as forearc deposits (HERVE et al. 1981). **b)** The dismembered formation on the coast near Chañaral is interpreted as a remnant of the accretion prism ('Chañaral mélange', BELL 1982, 1984, 1987). **c)** The basic and ultrabasic volcanic rocks that also occur in the vicinity of Chañaral are presumed to have developed in an ocean island setting, implying that the turbidites were deposited on oceanic crust (BELL 1984). **d)** Extensive Late Carboniferous to Triassic magmatism caused the formation of the thick calcalkaline volcanic series and associated high level intrusions in the Chilean Pre- and High Cordillera. This is presumed with reservation by some authors as to have been formed in a subduction-related setting (COIRA et al. 1982, DAVIDSON et al. 1985, HERVE et al. 1985, NIEMEYER et al. 1985).

We would like to submit the following concerning points a to d:
a) The petrographic and geochemical composition of the Coastal Cordilleran turbiditic rocks corresponds to that of a collisional orogenic source and not to that of an arc system (BAHLBURG 1987a,b).

Sedimentation and folding of the flysch series had allready been completed by the time magmatism started in the Pre- and High Cordillera (so far oldest ignimbrite age: 290 Ma, DAVIDSON et al. 1985). This is certainly true of the EL Toco Fm. that was posttectonically intruded as early as during the middle Carboniferous. The series near Chañaral were also folded during the Carboniferous as they were posttectonically intruded in the very Late Carboniferous and Permian (DAMM & PICHOWIAK 1981, BERG et al. 1983, PANKHURST & BROOK 1987). This time schedule renders any interpretation of the flysch series as a forearc deposit of the assumed magmatic arc in the Pre- and High Cordillera impossible.

b) BELL (1982, 1984, 1987) reports impressive phenomena from the 'Chañaral Mélange' which are also characteristic of subduction-related mélanges. However, because of the paucity of exotic blocks this complex should correctly be classified as a dismembered formation (sensu RAYMOND 1984)(BAHLBURG 1987a). The basic volcanic rocks also occur as concordant lava flows in the less tectonized turbidites to the east of the 'Chañaral Mélange' and should thus be considered as authochthoneous material of the dismembered formation. The dismembered formations in the Chañaral vicinity and also in the Sierra del Tigre Fm. (BREITKREUZ 1986) could have been formed in an intracontinental upthrust zone. NE-dipping upthrusts also occur in the El Toco Fm. (HARRINGTON 1961, BREITKREUZ & BAHLBURG 1985).

No Late Paleozoic HP/LT-metamorphic rocks have yet been reported from Northern Chile indicating, as is the case in Central and Southern Chile, the existence of a subduction zone.

c) Geochemically, the alkaline and tholeiitic basalts and ultrabasic volcanic rocks near Chanaral display definite WPB-affinity. REE characteristics point to an intracontinental setting for the extrusion of the submarine lavas (PICHOWIAK et al. 1987) as do the constraints given by regional geology (see DALMAYRAC et al. 1980, MILLER 1984).

d) The available geological and geochemical data is not sufficient to rule out the possibility of the Late Carboniferous-Triassic magmatic rocks having formed in a magmatic arc. Nevertheless, the percentage of andesitic volcanic rocks, typical component of volcanic arc sequences, is very low in the north Chilean series. The few intermediate and basic volcanic intercalations to be found display geochemical characteristics of continental rift affinity (PICHOWIAK et al. 1987). Continental rift regime is assumed to fit the formation of the Late Permian-Triassic Mitu Group volcanic rocks (and associated calcalkaline plutonic bodies) of the Peruvian and Bolivian Eastern Cordillera (NOBLE et al. 1978, KONTAK et al. 1985).

Radiometric dating of the plutons around the latitude of Chañaral confirms simultaneous intrusive activity in the Coastal Cordillera and High Cordillera during the Permian (PANKHURST & BROOK 1987). If the High Cordilleran magmatism really did develop in the postulated magmatic arc, the intrusives of the Coastal Cordillera, characterized by relatively high crustal contamination (BERG & BAUMANN 1985, PANKHURST & BROOK 1987), must be seen as having formed in the forearc area. The intrusion of large acidic magmas of S type affinity in the forearc area of an active continental margin is impossible as the nescessary crustal thickness is not given. This is also the case assuming a Devonian-Early Carboniferous flat-angle subduction without magmatic arc activity to have existed in Northern Chile.

Supporters of the subduction model fail to offer any explanations for the subsequent shift of the 2000 km long 'magmatic arc' from the High Cordillera at least 100 km towards the west to the Coastal Cordillera at the beginning of the Jurassic.

Furthermore, the subduction model ignores the following geometrical problem: In the Late Paleozoic, Gondwana extended towards the west beyond the present continental margin (DALMAYRAC et al. 1980, MILLER 1984, KATO 1985). This extension must have persisted even during the Jurassic, when a magmatic arc in the Coastal Cordillera was active (BUCHELT & TELLEZ and BAEZA & PICHOWIAK, this vol.), as its forearc must have had an extension of between 100 and 200 km (DICKINSON & SEELY 1979). Under these circumstances, the Late Carboniferous-Triassic magmatism of the High Cordillera would have been located at a distance of 200-300 km from the trench. Subduction-related formation is thus highly improbable.

In the previous paragraphs we emphasized the many inconsistencies of a simple subduction model for the north Chilean Late Paleozoic. We are conscious of the hypothetical character of our ideas concerning the geotectonic development in Northern Chile during the Devonian-Permian (Fig. 3 and 4): During the Silurian the western part of the Central Andes probably constituted an elevated area, produced by the oclóyic orogeny. The sea did not transgress as far as the western Puna until the Early Devonian (ACENOLAZA et al. 1972, NIEMEYER et al. 1985). Subsidence during the Devonian-Early Carboniferous must have been continuous as more than 2700 m of mainly intertidal to shallow subtidal sediments were deposited in the Sierra de Almeida area during this time. A deep basin formed contemporaneously in the Coastal Cordillera in which thick turbiditic series were deposited. We presume that the western border of the flysch trough was formed by a structural high (BAHLBURG 1987a, BAHLBURG et al. 1987), a possible southern extension of the Peruvian Arequipa Massif (see DALMAYRAC et al. 1980, GODOY 1983, DALZIEL & FORSYTHE 1985). Recent crustal seismic and petrological investigations in the Coastal Cordillera might offer an explanation for the disappearance of this structural high: South of Antofagasta, lower crustal rocks (e.g. granulites) outcrop at the surface (oral. comm. R. RÖSSLING, Berlin) and below 20 km depth a velocity inversion was detected (oral comm. P. WIGGER, Berlin). One possible explanation of these phenomena given by RÖSSLING and WIGGER is subduction of a continental fragment (including light high level crust) under the Coastal Cordilleran block during Andean convergence. This led to uplift and erosion of parts of the suprastructure giving way to the exposure of lower crustal material.

The Middle Devonian collision of the Chilenia terrane with the Argentinian Proto-Precordillera (RAMOS et al. 1986) could possibly have induced a response in the continental area to the north of the collision zone in form of a dextral strike-slip cycle (sensu MITCHELL & READING 1986) and analogous to EISBACHER's (1985) reverse indenter model. The first, transtensional stage of the strike-slip cycle led to intracontinental rifting and to the initiation of flysch sedimentation in the Coastal Cordillera during the Late Devonian at the latest. Sedimentation previous to the flysch deposition must have taken place but ist not documented. The Early

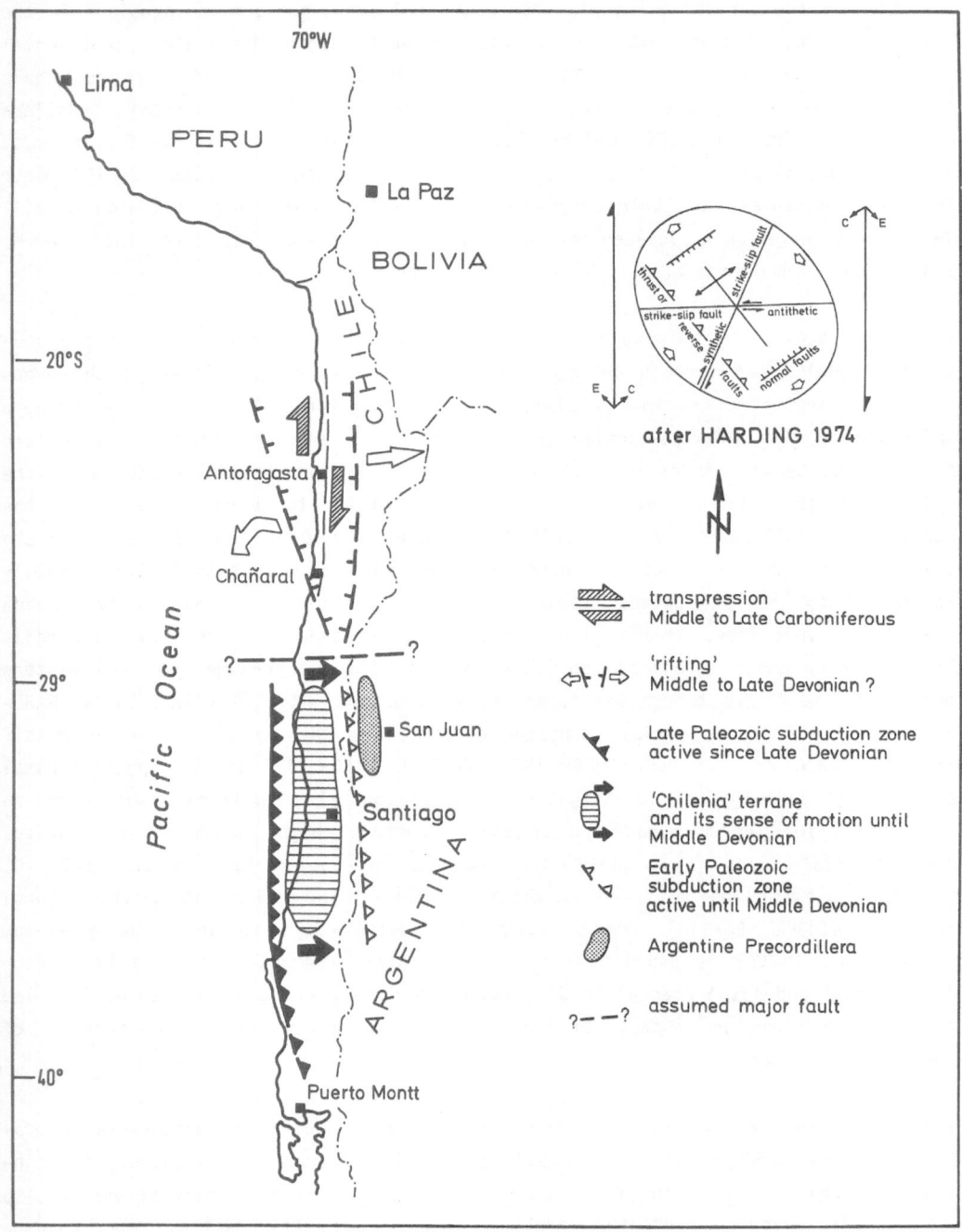

Fig. 3: Sketch of the possible geotectonic processes in Central and Northern Chile during the Devonian-Carboniferous (South of 29°S according RAMOS et al. 1986).

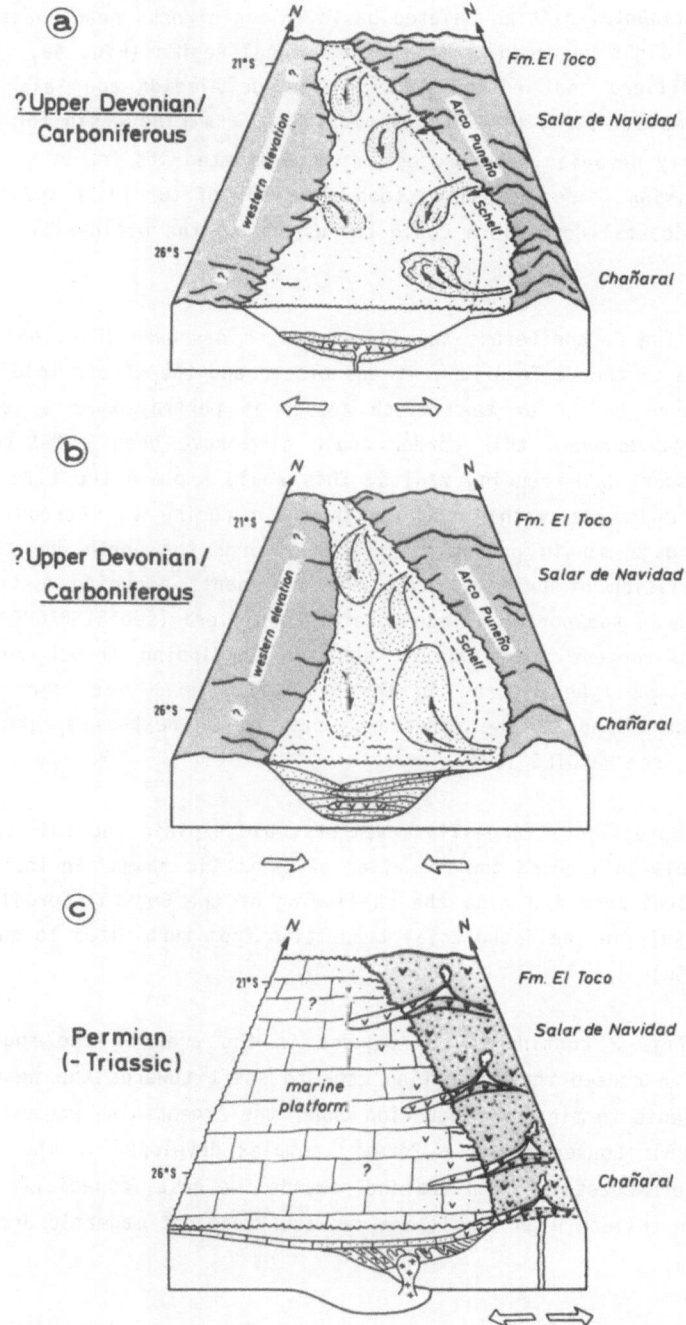

Fig. 4: Sedimentation, magmatism and tectonic processes in Northern Chile from the Upper Devonian to the Permian.

Carboniferous continental-rifting related basic volcanic rocks near Chañaral (PICHO-WIAK et al. 1987) might be products of this tensional regime (Fig. 4a). In the area of the High Cordillera and western Puna (Sierra de Almeida and Salar del Rincon areas, ACENOLAZA et al. 1972, NIEMEYER et al. 1985), a marine basin formed as early as during the Early Devonian. The process which initiated its formation previous to the Chilenia collision is not yet understood. The bulk of turbidite sedimentation in the area of the Coastal Cordillera could correspond to the basin-fill stage of the strike-slip cycle.

In the course of the Carboniferous the flysch trough narrowed (Fig. 4b) and in the North, in the area of the El Toco Fm., it was closed and the strata folded first. In Fig. 3, the western border of the flysch trough is portrayed as a straight NNW-trending structure. However, this border could also have been a N-S-trending en-echelon line of short NNW-trending faults. This would explain the time differences in compressional movements within the flysch basin during the Carboniferous. The western flysch trough margin probably migrated towards the north due to the east-ward-directed collision of Chilenia with the continent, inducing a transpressive strike-slip regime in the north Chilean Coastal Cordillera (see EISBACHER 1985). The NW-SE-trending, SW-vergent fold pattern and the NE-dipping thrust zones of the dismembered formation type exposed in the El Toco, Sierra del Tigre and in the Chanaral areas correspond to the structural pattern of dextral strike-slip zones (Fig. 3 top right, see HARDING 1974).

This assumed transpressive strike-slip movement would explain the thickening of the crust that favorably influenced the formation of anatectic magmas in the lower crust of the Coastal Cordillera and also the shallowing of the Coastal Cordilleran basin that manifests itself in the lithofacial transition from turbidites to shallow-water limestones (Fig. 4c).

As depicted in Fig. 3 continuous convergence in the area of the southern Andes during the Devonian caused the subduction zone to shift towards the west after the accretion of Chilenia terminated subduction under the Argentinian Proto-Precordilera (RAMOS et al. 1986). Consequently, a forearc complex developed in the area of the Central and southern Coastal Cordillera and towards the east, especially in the area of the Argentinian/Chilean High Cordillera, a Late Paleozoic magmatic arc was formed (HERVE et al. 1984).

While the subduction regime apears to have continued to exist during the Permian in the Southern Andes as documented e.g. by the accretion of Permian oceanic terranes (MPODOZIS & FORSYTHE 1983), in the north Chilean section of the Pre- and High Cor-dillera an elevated zone developed in the Late Carboniferous. It must have run parallel to the Pacific margin of Gondwana and was dominated by extensive acidic

volcanism and high level intrusions probably induced by continental rifting (see
also ZEIL 1981). The Peruvian/Bolivian Eastern Cordillera did not come under the
influence of this rifting regime until the Late Permian. The fact that both a convergent plate boundary in the south and an intracontinental rift system in the
north existed synchroniuously can perhaps best be explaind by a counter-clockwise
rotation of the oceanic plate(s) with respect to Gondwana; its rotation axis would
have been situated on the Chilean coast at about 29°S (Fig. 3).

Acknowledgement

This project is funded by the Deutsche Forschungsgemeinschaft (ref. no. Ze 6/37-2,
Gi 31/51-1,3). The translation was carried out by F. Lyons, Berlin.

References

ACENOLAZA,F.G., BENEDETTO,J.L. & SALFITY,J.A. (1972): El Neopaleozoico de la Puna
    Argentina: Su fauna y relación con areas vecinas.- An. Acad. brasil.
    Ciénc.(Supl.), 44, 5-20, Rio de Janeiro.
-, MILLER,H. & TOSELLI,A.J. (this vol.): The Puncoviscana Fm. (Late Precambrian-
    Early Cambrian) - Sedimentology, tectonomorphologic history and age of the oldest
    rocks in NW-Argentina.
BAEZA,L. (1984): Petrography and tectonics of the plutonic and metamorphic complexes
    of Limon Verde and Peninsula Mejillones, Northern Chile.- (unpubl.) Diss. Univ.
    Tübingen, 205pp..
- & PICHOWIAK,S. (this vol.): Ancient crystalline basement provinces in the North-
    Chilean Central Andes - relicts of continental crust development since the Mid-
    Proterozoic.
BAHLBURG,H. (1987a): Sedimentology, petrology and geotectonic significance of the
    Paleozoic flysch in the Coastal Cordillera of Northern Chile.- N. Jb. Geol.
    Paläont., Mh., H. 9, 1987, 527-559, Stuttgart.
- (1987b): Geochemical features of Devonian/Carboniferous flysch greywackes of the
    North Chilean Coastal Cordillera.- Zbl. Geol. Paläont., Teil I, 1987, H. 7/8,
    893-904, Stuttgart.
-, BREITKREUZ,C. & ZEIL,W. (1987): Paleozoic basin development in Northern Chile
    (21° -27°S).- Geol. Rdsch., 76,2, 633-646, Stuttgart.
-, - & - (this vol.): Geology of the Coquena Formation (Arenigian-Llanvirnian) in
    the NW Argentine Puna: Constraints on geodynamic interpretation.
BARTH,W. (1972): Das Permokarbon bei Zudánez (Bolivien) und eine Übersicht des Jung-
    paläozoikums im zentralen Teil der Anden.- Geol. Rdsch., 61, 249-270, Stuttgart.
BELL,C.M. (1982): The Lower Paleozoic metasedimentary basement of the Coastal Ranges
    of Chile between 25°30' and 27°S.- Rev. geol. Chile, 17, 21-29, Santiago.
- (1984): Deformation produced by the subduction of a Paleozoic turbidite sequence
    in Northern Chile.- J. geol. Soc. London, 141, 339-347.
- (1985): The Chinches Formation: An Early Carboniferous lacustrine succession in
    the Andes of Northern Chile.- Rev. geol. Chile, 24, 29-48, Santiago.
- (1987): The origin of the Upper Paleozoic Chañaral mélange of N Chile.- J. geol.
    Soc. London, 144, 599-610.
BERG,K. & BAUMANN,A. (1985): Plutonic and metasedimentary rocks from the Coastal
    Range of northern Chile: Rb-Sr and U-Pb isotopic systematics.- Earth planet. Sci.
    Lett., 75, 101-115, Amsterdam.
- , BREITKREUZ,C., DAMM,K.-W., PICHOWIAK,S. & ZEIL,W. (1983): The North-Chilean
    Coast Range - An example for the development of an active continental margin.-
    Geol. Rdsch., 72,2, 715-731, Stuttgart.

BREITKREUZ,C. (1985): Presentation of a marine volcano-sedimentary sequence of presumably Pre-Devonian age in the Sierra de Argomedo (24°45'S-69°22'W), Northern Chile.-IV. Congr. geol. Chile, Actas, 1, 1/76-88, Antofagasta.
- (1986): Das Paläozoikum in den Kordilleren Nordchiles (21°-25°S).- Geotek. Forsch., 70, 88 pp., Stuttgart.
- & BAHLBURG,H. (1985): Palaeozoic flysch series in the Coastal Cordillera of nor- thern Chile.- Geol. Rdsch., 74,3, 565-572, Stuttgart.
BUCHELT,M. & TELLEZ,C. (this vol.): The Jurassic La Negra Formation in the area of Antofagasta, Northern Chile (Lithology, petrology, geochemistry).
COIRA,B., DAVIDSON,J., MPODOZIS,C. & RAMOS,V. (1982): Tectonic and magmatic evolu- tion of the Andes of northern Argentina and Chile.- Earth Sci. Rev., 18, 303-332, Amsterdam.
DALMAYRAC,B., LAUBACHER,G., MAROCCO,R., MARTINEZ,C. & TOMASI,P. (1980): La chaine hercynienne d'amerique du sud structure et evolution d'un orogene intracrato- nique.-Geol. Rdsch., 69,1, 1-21, Stuttgart.
DALZIEL,I.W.D. & FORSYTHE,R.D. (1985): Andean evolution and the terrane concept.-in: HOWELL, D.G. (ed.): Tectonostratigraphic terranes of the Circum-Pacific region.- Circum-Pacific Council Energ. miner. resourc., Earth Sci. Ser., 1, 565-581, Houston.
DAMM,K.-W. & PICHOWIAK,S. (1981): Geodynamik und Magmengenese in der Küstenkordil- lere Nordchiles zwischen Taltal und Chañaral.- Geotek. Forsch., 61, 166pp., Stuttgart.
-, PICHOWIAK,S. & TODT,W. (1986): Geochemie, Petrologie und Geochronologie der Plutonite und des metamorphen Grundgebirges in Nordchile.- Berliner geowiss. Abh., (A) 66, I, 73-146, Berlin.
DAVIDSON,J., MPODOZIS,C. & RIVANO,S. (1981a): El Paleozoico de Sierra de Almeida, al oeste de Monturaqui, Alta Cordillera de Antogasta, Chile.- Rev. geol. Chile, 12, 3-23, Santiago.
-, - & - (1981b): Evidencias de tectonogenesis del Devónico Superior-Carbonífero Inferior al Oeste de Augusta Victoria, Antofagasta, Chile.- Rev. geol. Chile, 12, 79-86, Santiago.
-, RAMIREZ,C.F., GARDEWEG,M., HERVE,M., BROOK,M. & PANKHURST,R. (1985): Calderas del Paleozoico Superior - Triásico Superior y mineralización asociada en la Cordil- lera de Domeyko, Norte de Chile.- Communic., Univ. Chile, 35, 53-57, Santiago.
DICKINSON,W.R. & SEELY,R.D. (1979): Structure and stratigraphy of forearc regions.- Bull. Amer. Assoc. Petrol. Geol., 63, 2-31, Tulsa.
EISBACHER,G.H. (1985): Pericollisional strike-slip faults and synorogenic basins, Canadian Cordillera.- Soc. Econ. Paleont. Mineral. Spec. Publ., 37, 265-282, Tulsa.
FORSYTHE,R. (1982): The Late Paleozoic to Early Mesozoic evolution of southern South America: a plate tectonic interpretation.- J. Geol. Soc. London, 139, 671-682.
GARCIA,A.F. (1967): Geología del Norte Grande de Chile.- Soc. Geol. Chile, Simp. geosinclinal andino, 3, 138pp., Santiago.
-, PEREZ D'ANGELO,E. & CEBALLOS,E. (1962): El Ordovícico de Aguada de la Perdiz, Puna de Atacama, Provincia de Antofagasta.- Rev. Miner., Inst. Invest. Geol. Chi- le, 77, 52-61, Santiago.
GODOY,E. (1983): Avances en el conocimiento de las rocas devónicas aflorantes en Chile.- Rev. Técn. Yacim. Petrolíf. Fiscal. Boliv., 9, No.1-4, 111-114, La Paz.
HARDING,T.P. (1974): Petroleum traps associated with wrench faults.- Bull. Amer. Assoc. Petrol. Geol., 58, 1290-1304, Tulsa.
HARRINGTON,H.J. (1961): Geology of parts of Antofagasta and Atacama Provinces, Northern Chile.- Bull. Am. Ass. Petrol. Geol., 45, 169-197, Tulsa.
HELWIG,J. (1972a): Late Paleozoic stratigraphy and tectonics of the Central Andes.- An. Acad. brasil. Ciénc., 44, 161-171, Suplemento 1, San Francisco.
- (1972b): Stratigraphy, sedimentation, paleogeography, and paleoclimates of Car- boniferous ('Gondwana') and Permian of Bolivia.- Bull. Am. Assoc. Petrol. Geol., 56, 1008-1033, Tulsa.
HERVE,F., MUNIZAGA,F., GODOY,E. & AGUIRRE,L. (1974): Late Paleozoic K/Ar-ages of blueschists from Pichilemu, Central Chile.- Earth planet. Sci. Lett., 23, 261- 264, Amsterdam.
-, DAVIDSON,J., GODOY,E., MPODOZIS,C. & COVACEVICH,V. (1981): The late Paleozoic in Chile: Stratigraphy, structure and possible tectonic framework.- An. Acad. brasil. Ciénc., 53,2, 361-373, Rio de Janeiro.

-, KAWASHITA,K., MUNIZAGA,F.& BASSEI,M. (1982): Edades Rb-Sr de los cinturones metamórficos paredos de Chile Central.- III. Congr. geol. Chile, Actas, **2**, D116-135, Concepción.

-, KAWASHITA,K., MUNIZAGA,F. & BASSEI,M. (1984): Rb-Sr isotopic ages from the Late Paleozoic metamorphic rocks of Central Chile.- J. geol. Soc. London, **141**, 877-884.

-, MUNIZAGA,F., MARINOVIC,N., HERVE,M., KAWASHITA,K., BROOK,M. & SNELLING,N. (1985): Geocronología Rb-Sr y K-Ar del basamento cristalino de Sierra Limon Verde.- IV. Congr. geol. Chilen., Actas **3**, 4/235-253, Antofagasta.

ISAACSON,P., FISHER,L. & DAVIDSON,J. (1985): Devonian and Carboniferous stratigraphy of Sierra de Almeida Northern Chile, preliminary results.- Rev. geol. Chile, **25-26**, 113-121, Santiago.

KATO,T.T. (1985): Pre-Andean orogenesis in the Coast Ranges of Central Chile.- Bull. geol. Soc. Amer., **96**, 918-924, Boulder.

KONTAK,D.J., CLARK,A.H., FARRAR,E. & STRONG,D.F. (1985): The rift-associated Permo-triassic magmatism of the Eastern Cordillera: a precursor to the Andean orogeny.- in: PITCHER,W.S., ATHERTON,M.P., COBBING,E.J. & BECKINSALE,R.D.(eds.): Magmatism at a plate edge - The Peruvian Andes.- John Wiley & Sons, New York, Glasgow, London, 36-44.

KUMPA,M. & SANCHEZ,C. (this vol.): Geology and sedimentology of the Cambrian Grupo Mesón (NW-Argentina).

LIMARINO,C.O. & SPALLETTI,L.A. (1986): Eolian Permian deposits in West and Northwest Argentina.- Sedim. Geol., **49**, 109-127, Amsterdam.

MEGARD,F. (1978): Etude géologique des Andes du Pérou Central: contribution á l'étude des Andes.- 1. Mém., ORSTOM, **36**, 310 pp.

MILLER,H. (1984): Orogenic development of the Argentinian/Chilean Andes during the Paleozoic.- J. geol. Soc. London, **141**, 885-892.

MITCHELL,A.H.G. & READING,H.G. (1986): Sedimentation and tectonics.- In: READING, H.G. (ed.): Sedimentary environments and facies.- 2nd edition, Blackwell Sci. Publ., 471-519, Oxford.

MPODOZIS,C. & FOSYTHE,R. (1983): Stratigraphy and geochemistry of accreted fragments of the ancestral Pacific floor in southern South America.- Palaeogeogr. Palaeoclim. Palaeoecol., **41**, 101-124, Amsterdam.

-, HERVE,F., DAVIDSON,J. & RIVANO,S. (1983): Los granitoides de Cerros de Lila, manifestaciones de un episodio intrusivo y termal del Paleozoico Inferior en los Andes del Norte de Chile.- Rev. geol. Chile, **18**, 3-14, Santiago.

NARANJO,J.A. & PUIG,A. (1985): Hojas Taltal y Chanaral; Carta geológica de Chile; 1 : 250 000.- Serv. Nac. Geol. Mineria Chile, **62+63**, Santiago.

NIEMEYER,H. (1984): La megafalla Tucúcaro en el extremo sur del Salar de Atacama: Una antigua zona de cizalle reactivada en el Cenozoico.- Comunic., Univ. Chile, **34**, 37-45, Santiago.

-, URZUA,F., ACENOLAZA,F.G. & GONZALEZ,C. (1985): Progresos recientes en el conocimiento del Paleozoico de la Región de Antofagasta.- IV. Congr. geol. Chile, Actas, **1**, 1/410-438, Antofagasta.

NOBLE,D.C., SILBERMAN,M.L., MEGARD,F. & BOWMAN,H.R. (1978): Comendite (peralkaline rhyolite) and basalt in the Mitu Group, Peru: Evidence for Permian-Triassic lithospheric extension in the Central Andes.- J. Res. U. S. geol. Surv., **6,4**, 453 - 457, Washington.

OSORIO,R. & RIVANO,S. (1985): Parachiticae (Ostracoda) del Paleozoico Superior en la Formación Pular (Harrington, 1961), Quebrada de Pajonales, Vertiente Occidental de la Sierra de Almeida, Antofagasta.- IV. Congr. geol. Chile, Actas, **1**, 1/439-457, Antofagasta.

PACCI,D., HERVE,F., MUNIZAGA,F., KAWASHITA,K. & CORDANI,U. (1980): Acerca de la edad Rb-Sr precámbrica de rocas de la Formación Esquistos de Belén, Departamento de Parinacota, Chile.- Rev. geol. Chile, **11**, 43-50, Santiago.

PANKHURST,R.J. & BROOK,M. (1987): The isotope geochemistry of plutonic magmas in Northern Chile related to changing subduction tectonics.- IV. Meet. Europ. Union Geosci. Program. Suppl., p. 12, Strassbourg.

PICHOWIAK,S., BAHLBURG,H. & BREITKREUZ,C. (1987): Paleozoic volcanic and geotectonic evolution in northern Chile.- X. Congr. geol. Argent., **4**, 302-304, Tucuman.

RAMIREZ,R. & GARDEWEG,M. (1982): Hoja Toconao, Región de Antofagasta; Carta geológica de Chile; 1 : 250 000.- Serv. Nac. Geol. Mineria Chile, **54**, Santiago.

RAMOS,V., JORDAN,T.E., ALLMENDINGER,R.W., MPODOZIS,C., KAY,S.M., CORTES,J.M. & PALMA, M.A. (1986): Paleozoic terranes of the Central Argentine-Chilean Andes.- Tectonics, **5**, 855-880, Washington D.C..

RAYMOND,L.A. (1984): Classification of melanges.- Geol. Soc. Amer. Spec. Pap, **198**, 7 -20, Boulder.

SALFITY,J.A., OMARINI,R., BALDIS,B. & GUTIERREZ,W. (1975): Consideraciones sobre la evolución geológica del Precámbrico y Paleozoico del norte argentino.- II. Congr. Iberoamer. Geol. económ., **4**, 341-361, Buenos Aires.

SKARMETA,J. & MARINOVIC,N. (1981): Hoja Quillagua, Región de Antofagasta; Carta geológica de Chile; 1 : 250 000.- Inst. Invest. geol. Chile, **51**, Santiago.

ZEIL,W. (1981): Vulkanismus und Geodynamik an der Wende Paläozoikum/Mesozoikum in den zentralen und südlichen Anden (Chile-Argentinien).- Zbl. Geol. Paläont., I, **1981**, H. 3/4, 298-318, Stuttgart.

- (1983): Das Präkambrische Basement der Anden - Ein Überblick.- Zbl. Geol. Paläont., I, **1983**, H. 3/4, 246-254, Stuttgart.

B: ANDEAN EVOLUTION

I: Mesozoic-Cenozoic Basins

MARINE MESOZOIC PALEOGEOGRAPHY IN NORTHERN CHILE BETWEEN 21°-26°S

M. Gröschke, A. v. Hillebrandt, P. Prinz, L.A. Quinzio* & H.-G. Wilke
Institut für Geologie und Paläontologie, Technische Universität Berlin
Ernst-Reuter-Platz 1, D-1000 Berlin 10
*Departamento de Geociencias, Universidad del Norte,
Casilla 1230, Antofagasta

Abstract

In a narrow and long marginal sea orientated parallel to what is now the north Chilean Cordillera range, marine sediments were deposited from Late Triassic until Kimmeridgian times. The last marine deposits are formed by Cretaceous sediments that occur in only few areas and therefore do not reveal any relation to the Jurassic sedimentary basin. During the Early Jurassic a former (Late Triassic) bay expanded in all directions, especially towards the north and the west. In the Coastal Cordillera, terrestrial Triassic deposits are overlain by marine Hettangian series. Furthermore, a gradual transgression to the east is to be observed in the Jurassic marginal sea. In the Callovian, shallow water deposits of the eastern margin could be found only in the extreme north of the area studied. As no sediments of this age are known in NW-Argentina, it is to be assumed that the eastern coastal line of the Upper Jurassic sea was situated in the region of the High Cordillera.

Almost throughout the whole Jurassic there was volcanic activity in the west causing a magmatic arc (BUCHELT & ZEIL 1986), which separated the Mesozoic marginal sea from the ocean.

In the Hettangian and Early Sinemurian, this island chain might have still been situated in the west of the Coastal Cordillera. The easternmost signs of volcanic influence occur in upper Middle Jurassic beds.

The sequences studied can be correlated locally and on a worldwide scale with the help of ammonites. Inconsistency between the observation of eustatic sea level changes and major geodynamic processes during this period of time has been noted.
The paleogeographic situation changed almost completely during the Cretaceous, although little is yet known of it.

Lecture Notes in Earth Sciences, Vol. 17
H. Bahlburg, Ch. Breitkreuz, P. Giese (Eds.),
The Southern Central Andes
© Springer-Verlag Berlin Heidelberg 1988

Introduction

Today, Mesozoic outcrops are found mainly in the Coastal- and Pre-Cordilleras. In the Coastal Cordillera, volcanic rocks of middle Sinemurian to Oxfordian age are by far predominant; they make up the so-called La Negra-Formation (BUCHELT & ZEIL 1986). In the Pre-Cordillera, Mesozoic rocks are sometimes developed in marine facies from the Upper Triassic to the Kimmeridgian. Scarcely any marine sediments have been found in the Longitudinal Graben-Structure between the Coastal-and Pre-Cordilleras or east of the Pre-Cordillera (fig. 3a).

In order to obtain a paleogeographic reconstruction, an exact record of facies arrangements first had to be made. The few marine outcrops of marine Mesozoic rocks in the Coastal Cordillera and the Longitudinal Graben are of great importance, as an interfingering of volcanites and volcanoclastic rocks with marine sediments is expected in these regions. We studied most of the outcrops of marine Mesozoic rocks shown in fig. 3a. In the paleogeographic maps (figs. 3b - 5b) only a few facial units have been differentiated as a first step to the reconstruction of the sedimentary basin. This is not due to a lack of information on the sequences, but in order to enable the reader to follow developments with more ease.

To demonstrate the changes of facies in time and space, 5 mesozoic stages should be noted: Norian, Hettangian, Toarcian, Bajocian and Oxfordian. Either a large amount of detailed information is available on these stages or they can be distinguished by different facial developments.

On the other hand, one must take into consideration that even in the course of one stage, significant facial changes may occur. For the reasons mentioned and due to graphical ones, the drawings have been greatly simplified.

Stratigraphic results and fossil content of representative sections and areas have been published for example by BAEZA (1979), BOGDANIC (1983), CHONG & HILLEBRANDT (1985), HILLEBRANDT et al. (1986), JURGAN (1974), and QUINZIO (1987).

A small selection of 17 typical ammonites from Upper Triassic and Jurassic stages is shown in fig. 1. The frequency of ammonites is not uniform, neither in synchronous sediments nor in the vertical sequence and it can therefore be difficult to illustrate the thickness of sediments deposited in one stage.

Fig. 1: Some important Upper Triassic and Jurassic ammonites from Northern Chile.

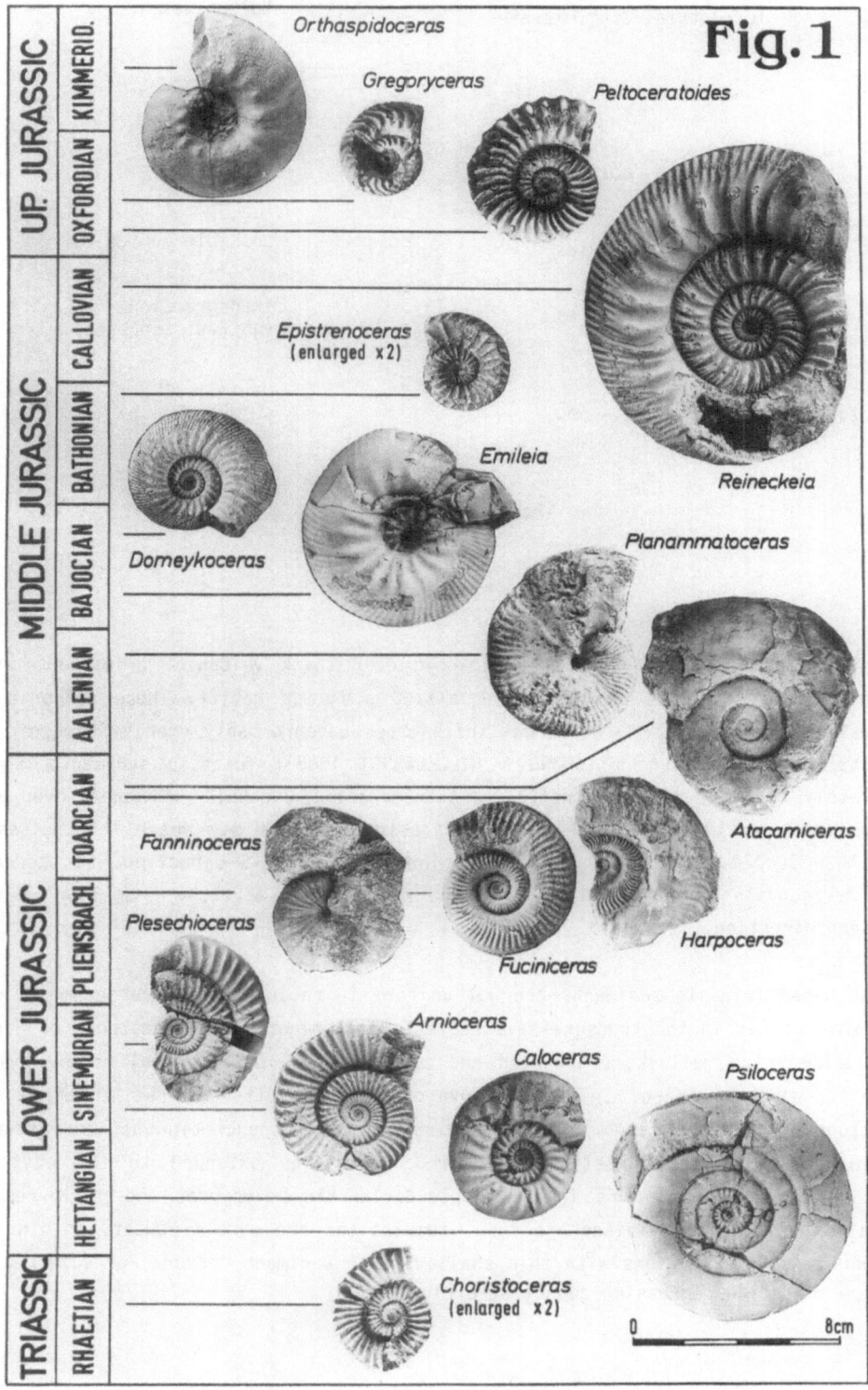

Fig. 1

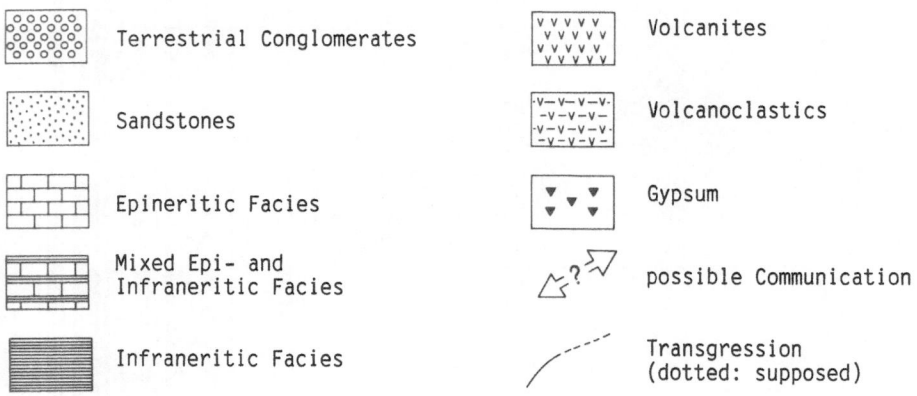

Fig. 2: Legend to figs. 3b - 5b.

Facies arrangements

Upper Triassic (fig. 3b)

During the Norian, a transgression occurred over thick volcanic series and terrestrial sediments, which had nearly equalized a former relief. These volcanic and
terrestrial series of very different thicknesses unconformably overly Paleozoic sediments and igneous rocks (CHONG & HILLEBRANDT 1985). Triassic sediments may be
completely absent, as demonstrated by transgressive Jurassic sediments over even
older rocks. In other sections, Permo-Triassic sediments may reach thicknesses of
more than 1000 m. Generally, thicknesses increase in a N-S direction. It therefore
seems reasonable to suppose that Triassic transgression advanced from S or SW in a
northern direction.

Marine Upper Triassic sediments are not uniform in facies. Nearshore sediments were
confirmed mainly in the transgressive facies. Conglomerates and sandstones of Norian
age yielded rich mollusk, brachiopod and coral faunas. In this shallow environment
ammonites are very rare. In the western outcrops an offshore sedimentation with
sandstones, siltstones and marls containing calcareous concretion horizons started
within the Norian. In Rhaetian time this facies was extended to the east and
ammonites are somewhat more frequent; pseudoplanctic pelecypods and plant remains
can also be found. Turbidites are rare, tempestites are more frequent. Within this
offshore area also regions with thin shallow water sediments occur. For some regions
an Upper Rhaetian regression cannot be excluded.

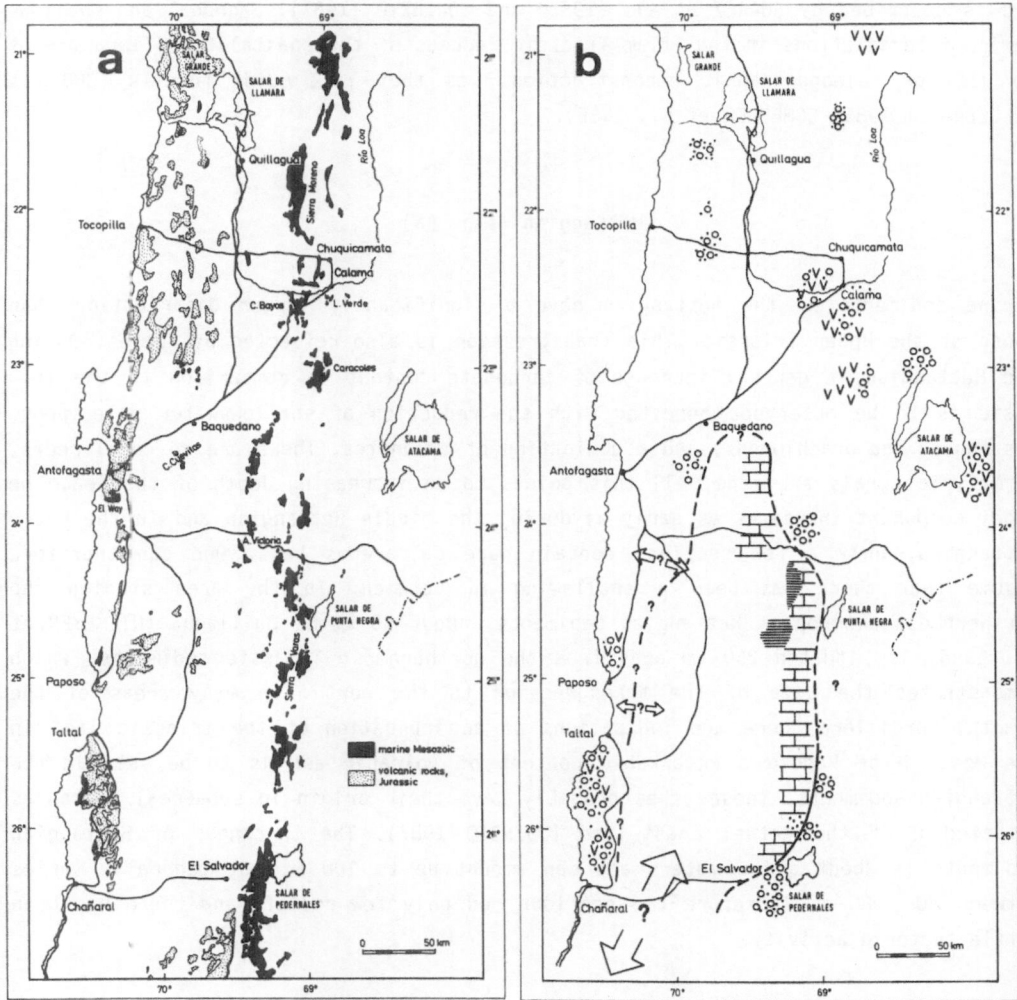

Fig. 3: a) Outcrops of marine Mesozoic sediments and Jurassic volcanites in the area studied.
b) Paleogeographic sketch map of the Permo-Triassic and marine Upper Triassic (Norian).

With these findings the existence of an Upper Triassic marginal sea can be assumed, of which the N-S extent must have been more than 300 km. The width in E-W direction exceeded the 35 km of the today's outcrops by far (HILLEBRANDT et al. 1986). The northern and eastern margin of this sea can be estimated, whereas nothing is known about the western coastline. In the region of the Coastal Cordillera, an emerged arc is described by SUAREZ et al. (1985) and QUINZIO (1987). Lagoonal and shallow-marine intercalations in the Permo-Triassic redbeds of the Coastal Cordillera are of no use for paleogeographic reconstructions, as they lack guide fossils (CHONG & HILLEBRANDT 1985, SCHEUBER et al. 1986).

Hettangian (fig. 4a)

Marine sediments of the Hettangian have a significantly larger distribution than those of the Upper Triassic. This transgression is also reflected by facies. During the Hettangian, a general increase of carbonate content in comparison to the Triassic is to be observed connected with the reduction of shallow-water pelecypods, gastropods and brachiopods, and a domination of ammonites. There are no coral reefs, hermatypic corals are rare. All this points to an increasing depth of the sea. The sandy component increases as early as during the Middle Hettangian and in the Upper Hettangian, only a few sections contain pure calcareous beds. One can therefore assume that there has been a shallowing in between. In the area studied the northernmost outcrop of Hettangian sediments, about 50 km NE Quillagua (NIEMEYER et al. 1985) is situated 250 km away from the northernmost Triassic sediments, which demonstrates the size of the transgression to the north. In many areas of the Coastal Cordillera there are indications of an inundation of the Triassic land in the west. N of Paposo a remarkable content of volcanic ash is to be seen in the Hettangian sediments. These ashes probably have their origin in subaereal volcanoes situated W of the actual coast-line (QUINZIO 1987). The thickness of Hettangian sediments is about a few meters and can amount up to 100 m, but generally varies between 20 - 80 m. Therefore the seafloor had only few relief, and there has been little tectonic activity.

Toarcian (fig. 4b)

Up to now, Toarcian sediments have been reported only from the Pre-Cordillera. Infraneritic and mixed epi- and infraneritic sediments occur repeatedly from north to south. These facial changes do not only reflect the distance from the coastline but also a morphological differentiation of the basin. Dark and bituminous carbonates in geode-bearing facies without benthic fossils occur e.g. W' Calama (BAEZA 1976, BIESE 1957) and SW' Salar de Punta Negra (QUINZIO 1987). On the other hand, shallow water carbonates, partly with corals, are less frequent and are found

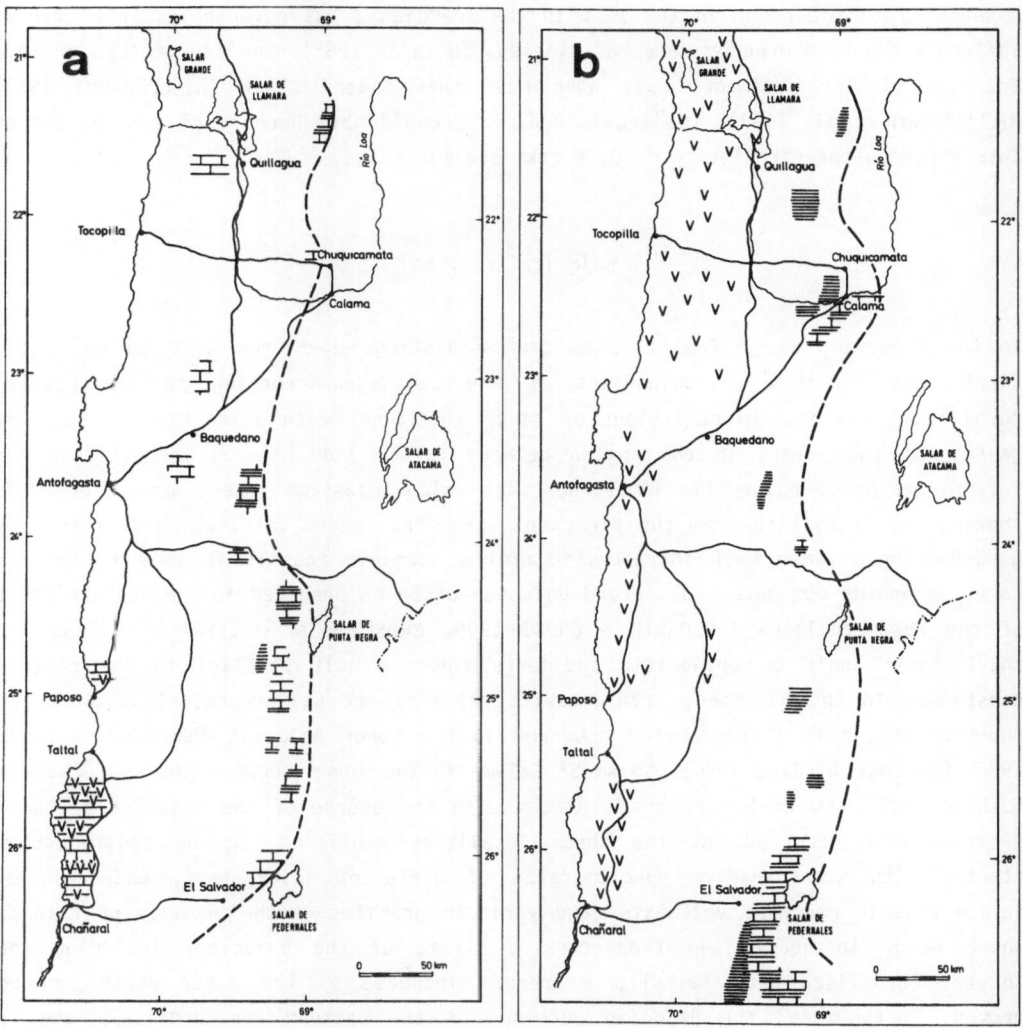

Fig. 4: Paleogeographic sketch map of the Hettangian (a) and Toarcian (b).

e.g. SE' Calama and south of 26°S (HILLEBRANDT 1971). The differentiation mentioned is obvious as sediment thicknesses in the Sinemurian vary on short distances from 10 to 150 m.

The La Negra Formation volcanism in the Coastal Cordillera started in the Upper Sinemurian and caused an island chain to form. This island chain obstructed the connection to the ocean in the west in the Toarcian , but nevertheless open marine influence can be proved by rare radiolarians (QUINZIO 1987). The source area of some Toarcian clastic sediments must have been this island chain (HILLEBRANDT 1973, HILLEBRANDT et al. 1986). The anoxic sediments could have been originated by either interrupted water circulation or by a fast sedimentation.

Bajocian (fig. 5a)

In the Bajocian sea, 3 facial units can be distinguished from west to east: the first unit consists of volcanites of the La Negra-Formation in the Coastal Cordillera, and has intercalations of sandy limestones with ammonites, pelecypods, gastropods and corals in the region between Taltal and Chañaral, constituting an interfingering of epineritic limestones with volcanoclastics. There are no outcrops showing the transition from the first unit into the second unit, which is formed by geod-bearing sediments of infraneritic origin. Benthic fossils in this facies are rare, ammonites dominate. The second unit can often be observed in the western parts of the Pre-Cordillera (BOGDANIC & CHONG 1985, GRÖSCHKE & HILLEBRANDT 1985). The third facial unit of epineritic sediments forms a belt parallel to the eastern coastline. In the Bajocian, transgressive deposits are well-represented. 50 km N' Chuquicamata, this transgression occurred in the Lower Bajocian (GRÖSCHKE & WILKE 1986, GRÖSCHKE & PRINZ 1986), 50 km S' Calama in the lower Upper Bajocian (JENSEN & QUINZIO 1979). 60 km SW' Salar de Atacama, in the course of the Bajocian a transgression also occurred. At the three localities mentioned, marine sedimentation started with transgressive conglomerates of different thicknesses, underlain by Permo-Triassic redbeds, volcanites and variscan granites. With the help of corals, which occur in nearly all transgressive strata of the Bajocian (including the Coastal Cordillera near Taltal), a gradual increase of the water depth can be proved. Furthermore, the Bajocian outcrops of the eastern Pre-Cordillera show a diachronic transition of the lithofacies from nearshore to offshore. In the upper parts of the Upper Bajocian pelagic sediments are common. The thicknesses of Bajocian sediments are generally uniform and amount to 100-150 m. It should however be mentioned that in the marginal parts, thicknesses are normally higher than in the center of the basin. Up to now, we had been undecided as to interpreting shallow water carbonates W' Calama that yield a corresponding fauna of big shelled pelecypods and hermatypic corals, as they had previously been dated as a regressive facies of Bajocian age (BIESE 1957, BAEZA 1976). In agreement with the general transgressive tendency during the Bajocian, a local uplift had to be supposed.

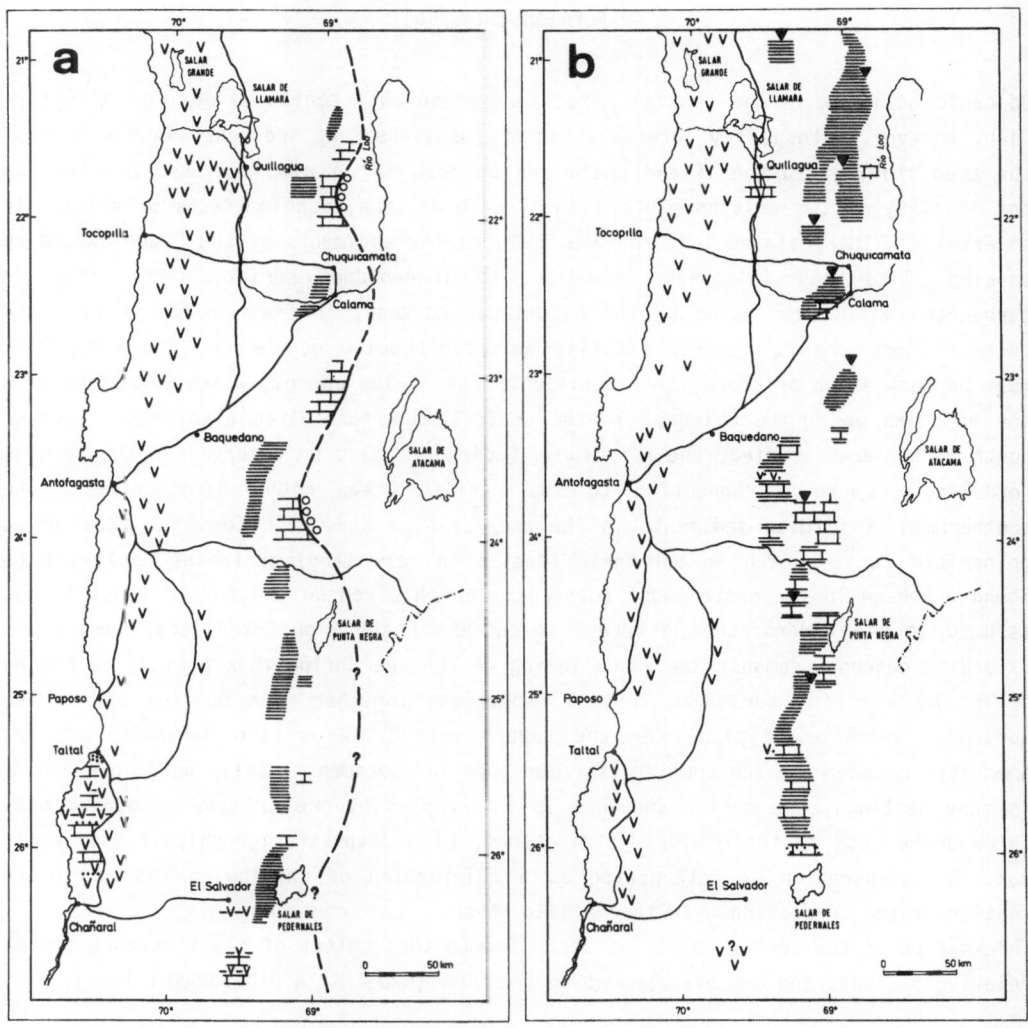

Fig. 5: Paleogeographic sketch map of the Bajocian (a) and Oxfordian (b).

Recent studies on the BIESE collection in Washington (Nat. Mus. of Nat. Hist.) by
one of the authors (v.H.) proved, that these strata have an Aalenian age.

Oxfordian (fig. 5b)

Volcanic activity in the Coastal Cordillera presumably continued during the Oxfordian, however, without any intercalation of fossil-bearing sediments in the west of the area studied. So the island chain can be seen as an almost closed barrier. At one locality of the western part of the Pre-Cordillera, conglomerates of volcanitic material are intercalated in carbonate beds. Marine sediments of the Lower Oxfordian consist, like most Callovian sediments, of thin-bedded series, partly rich in carbonate, which are caused by the bituminous content, in fresh outcrops of black colour. There are no benthic fossils, ammonite-bearing geodes are frequent. This must be seen as an offshore, infraneritic facies of low energy, which, compared with the Bajocian was quite uniform. In the central part turbidites also occur. Further north of the area studied, the bituminous facies can also be observed in the Coastal Cordillera, where a connection to the Pacific Ocean might have existed. The southernmost Oxfordian sediments in the area studied have been found at 26°S in an epineritic facies, rich in benthonic fossils. A real marginal facies could not be found anywhere in the easternmost outcrops. For this reason the former coastline is assumed to lie underneath the recent volcanoes of the High Cordillera. Almost all Oxfordian outcrops demonstrate a shallowing of the sea during this period, as can be proved by a rising carbonate content, sometimes together with oolites and oyster horizons, which are typical for the upper part of the marine Oxfordian strata. Ammonites became rare and coral-biostromes poor in specimens locally were observed.

In many sections, the marine sequence is interrupted by one or several gypsum beds which serve, due to their wide distributions, as a lithostratigraphical guide-horizon. The evaporation is interpreted as a culmination of shallowing and an interruption of the connection with the Pacific Ocean.

Thicknesses of the series vary, but according to the content of clastics they can be relatively great and may even amount up to 300 m (GRÖSCHKE & HILLEBRANDT 1985).

Kimmeridgian to Cretaceous

Only in a few areas of the southern Pre-Cordillera Kimmeridgian series with ammonites could be confirmed, although in most sections marine series overlie the Oxfordian gypsum beds.

These younger marine sediments are characterized by an extremely synchronous and heterochronous change of facies, so that no correlation can be made without enough guide-fossils. In the NE and E of the presented area, deltaic sandstones and redbeds are assumed to be Kimmeridgian deposits, whereas in the middle and southern part of the Pre-Cordillera, a transition from a calcareous-marlaceous cephalopod facies into an epineritic limestone facies with stromatolitic structures, oolites, pelecypods and gastropods is to be observed. This sequence is also overlain by redbeds. In the younger part of this sequence, volcanoclastic layers increase and the following true volcanites are at least of Cretaceous age. Marine Tithonian deposits could be found only 26°S, W Salar de Pedernales (PEREZ 1982).

Marine sediments with Cretaceous fossils are remarkably rare in northern Chile. In the southern part of the Pre-Cordillera pelecypod (Trigoniidae) yielding sandstones of the Lower Cretaceous locally occur (NARANJO & PUIG 1984). They are conformally overlain by redbeds, too. In this region an undisturbed section from Kimmeridgian to Cretaceous sediments was not observed.

Near El Way, 20 km S Antofagasta, a thick sequence of redbeds derived from a western height are overlain by marine series of silt- and sandstones, marls and carbonates (ALARCON & VERGARA 1964, JURGAN 1974, LEANZA & CASTELLARO 1955). Corals, pelecypods and echinids point to an epineritic environment, whereas the rare ammonites in the light-coloured carbonates would have required an open marine shelf. These ammonites date the marine sequence as Hauterivian to Albian. As no indication of a nearshore environment could be found in the thick and well-bedded white carbonates in the upper part of the section, it is to be assumed that the Lower Cretaceous sea could have been further extended.

Thin sandstone layers and shallow-marine carbonates, intercalated in redbeds in the High Cordillera have been mentioned by MARINOVIC & LAHSEN (1984) with a careful age estimation of Late Cretaceous. These sediments may have some relations to shallow marine series of the Upper Cretaceous in Bolivia that were connected with the open ocean via Peru.

Conclusions

As far as changes of the sea-level during the Upper Triassic and Jurassic in the north Chilean sedimentary basin are concerned, a comparison to eustatic curves (HALLAM 1978, 1981, VAIL et al. 1984) reveals a certain similarity.

There may be a small difference in a worldwide regression during the Bathonian that could not be proved by facial changes in northern Chile. 26°S, however, there is a hiatus during the Bathonian (HILLEBRANDT 1971).

In spite of all uncertainties connected with sea-level curves and our interpretation of facies, the similarity mentioned indicates low tectonic activity, at least during the Late Triassic and Jurassic. Transgression during the Early Cretaceous also complies with this observation. However, this phenomenon ceases in the Late Cretaceous as there seems to be no evidence of the world-wide transgression in the Cenomanian.

Consequently, a constant but not always uniform subsidence of the basin is presumable, in which sedimentation rates were compensated and the sea-level was generally not under an Andean-regional but a world-wide eustatic control. For the interpretation of geodynamic processes within the back arc-basin, magmatism seems to be of minor importance until the late Lower Cretaceous in contrast to the Coastal Cordillera, where volcanic activities had predominated since the middle Sinemurian. There are only very few outcrops of synsedimentary volcanites in the Pre-Cordillera. As late as at the beginning of the westward movement of South America during the Cretaceous, this part of the active continental margin enjoyed its own dynamic development.

References

ALARCON, B. & VERGARA, M. (1964): Nuevos antecedentes sobre la geología de la Quebrada El Way. - Ann. Fac. Cienc. Físicas y Matemáticas, 20/21, pub. 26: 101-128, 4 figs., 4 pls., 1 map, Santiago.
BAEZA, L. (1976): Geología de Cerritos Bayos y áreas adyacentes entre los 22°30' - 22°45' latitud Sur y los 68°55' - 69°25' longitud Oeste, II Región - Antofagasta, Chile. - Memoria de Tesis, Facultad de Ciencias, Departamento de Geociencias, Universidad del Norte, Antofagasta (unpublished).
BAEZA, L. (1979): Distribución de facies sedimentarias marinas en el Jurásico de Cerritos Bayos y zonas adyacentes, norte de Chile. - In: 2. Congr. Geol. Chileno, Actas, 3: H45-H61, 4 figs., Arica.
BIESE, W.A. (1957): Der Jura von Cerritos Bayos - Calama, Republica de Chile, Provinz Antofagasta. - Geol. Jb., 72: 439-494, 6 figs., 2 tabs., 6 pls., Hannover.
BOGDANIC, T. (1983): Antecedentes generales y bioestratigrafía del sistema Jurásico en la zona Preandina, entre los 24°30' y los 25°30' de latitud Sur y los 69°00' y 69°30' de longitud Oeste, II Región de Antofagasta, Chile. - Memoria de Tesis, Facultad de Ciencias, Departamento de Geociencias, Universidad del Norte, Antofagasta (unpublished).
BOGDANIC, T. & CHONG, G. (1985): Bioestratigrafía del Jurásico de la Zona Preandina Chilena entre los 24°30' - 25°30' de Lat. Sur. - In: 4. Congr. Geol. Chileno, Actas, 1: 1-38-1-57, 3 figs., Antofagasta.
BUCHELT, M. & ZEIL, W. (1986): Petrographische und geochemische Untersuchungen an jurassischen Vulkaniten der Porphyrit-Formation in der Küstenkordillere Nord-

chiles. - Berliner geowiss. Abh.(A)., 66: 191-204, 11 figs. 2 tabs., 1 pl.,
Berlin.
CHONG, G. & HILLEBRANDT, A.v. (1985): El Triásico Preandino de Chile entre los
23°30' y 26°00' de Lat. Sur. - In: 4. Congr. Geol. Chileno, Actas, 1: 1-162-1-
210, 4 figs., 1 tab., 4 pls., Antofagasta.
GRÖSCHKE, M. & HILLEBRANDT, A.v. (1985): Trias und Jura in der mittleren Cordillera
Domeyko von Chile (23°30'-24°30'S). - N. Jb. Geol. Paläont. Abh., 170, 2: 129-
166, 10 figs., Stuttgart.
GRÖSCHKE, M. & PRINZ, P. (1986): Geologische Untersuchungen in der nordchilenischen
Präkordillere bei 22°S. - N. Jb. Geol. Paläont. Mh., 1986, 7: 418-430, 4 figs., 1
tab., Stuttgart.
GRÖSCHKE, M. & WILKE, H.-G. (1986): Lithology and Stratigraphy of Jurassic sediments
in the North-Chilean Pre-Cordillera beween 21°30' and 22°S. - Zbl. Geol. Paläont.
Teil I, 1985, (9/10): 1317-1324, 2 figs., Stuttgart.
HALLAM, A. (1978): Eustatic Cycles in the Jurassic. - Palaeogeogr.,
Palaeoclimatol., Palaeoecol., 23 :1-32, 11 figs., 1 tab., Amsterdam.
HALLAM, A. (1981): A revised sea-level curve for the early Jurassic. - J. geol. Soc.
London, 138: 735-743, 3 figs., London.
HILLEBRANDT, A. v. (1971): Der Jura in der chilenisch-argentinischen Hochkordillere
(25° bis 32°30'S). - Münster. Forsch. Geol. Paläont., 20/21: 63-87, 5 figs.,
Münster.
HILLEBRANDT, A. v. (1973): Neue Ergebnisse über den Jura in Chile und Argentinien. -
Münster. Forsch. Geol. Paläont., 31/32, 167-199, 4 figs., 1 tab., Münster.
HILLEBRANDT, A.v., GRÖSCHKE, M., PRINZ, P. & WILKE, H.-G. (1986): Marines Mesozoikum
in Nordchile zwischen 21° und 26°S. - Berliner geowiss. Abh.(A), 66: 169-190, 4
figs., Berlin.
JENSEN, A. & QUINZIO, L.A. (1979): Geología del área de Pampa Elvira y contribución
al conocimiento del Jurásico marino entre los 23°03' y 23°30' latitud Sur y los
68°45' y 69°03' longitud Oeste. II Región de Antofagasta, Chile. - Memoria de
Tesis, Facultad de Ciencias, Departamento de Geociencias, Universidad del Norte,
Antofagasta (unpublished).
JURGAN, H. (1974): Die marine Kalkfolge der Unterkreide in der Quebrada El Way -
Antofagasta, Chile. - Geol. Rdsch., 63: 490-515, 16 figs., Stuttgart.
LEANZA, A.F. & CASTELLARO, H.A. (1955): Algunos fósiles cretácicos de Chile. -Rev.
Asoc. Geol. Argentina, 10, 3: 179-213, 4 pls., Buenos Aires.
MARINOVIC, N. & LAHSEN, A. (1984): Hoja Calama, Región de Antofagasta, 1: 250.000. -
Carta geol. de Chile, No. 58, Serv. Nac. Geol. Min., Santiago.
NARANJO, J. & PUIG, A. (1984): Hojas Taltal y Chañaral, Regiones de Antofagasta y
Atacama, 1:250.000, - Carta geol. de Chile, No. 62-63, Serv. Nac. Geol. Min.,
Santiago.
NIEMEYER, H., VENEGAS, R., BAEZA, L. & SOTO, H. (1985): Reconocimiento geológico del
sector sur-occidental del cuadrángulo Cerro Yocas, ubicado en la zona de falla
Quebrada Blanca - Chuquicamata, Región de Antofagasta. - 4. Congr. Geol. Chileno,
Actas, 4: 1-629-1-653, 5 figs., Antofagasta.
PEREZ, E. (1982). Bioestratigrafía del Jurásico de Quebrada Asientos, Norte de
Potrerillos, Región de Atacama. - Serv. Nac. Geol. Min., Chile, Bol., 37: 1-149,
17 figs., 20 pls., Santiago.
QUINZIO, L.A. (1987): Stratigraphische Untersuchungen im Unterjura des Südteils der
Provinz Antofagasta in Nord-Chile. - Berliner geowiss. Abh.(A), 87: 100 p., 32
figs., 5 pls., Berlin.
SCHEUBER, E., RÖSSLING, R. & REUTTER, K.-J. (1986): Strukturen der chilenischen
Küstenkordillere zwischen Paposo und Antofagasta. - Berliner geowiss. Abh.(A),
66: 209-224, 4 figs., 1 tab., 2 pls., Berlin.
SUAREZ, M., NARANJO, J.A. & PUIG, A. (1985): Estratigrafía de la Cordillera de la
Costa, al sur de Taltal, Chile: etapas iniciales de la evolución andina. -Rev.
Geol. Chile, 24: 19-28, 7 figs., Santiago.
VAIL, P.R., HARDENBOL, J. & TODD, R.G. (1984): Jurassic Unconformities, Chrono-
stratigraphy and sea-level changes from seismic Stratigraphy and Biostratigraphy.
- Mem. Amer. Assoc. Petrol. Geol., 36: 129-144, Tulsa.

TECTONIC FRAMEWORK AND CORRELATIONS OF THE CRETACEOUS-EOCENE SALTA GROUP; ARGENTINA

Rosa A. Marquillas & José A. Salfity
Universidad Nacional de Salta and CONICET
Buenos Aires 177, 4400 Salta, Argentina

Abstract

A regional analysis of the basin of the Salta Group (Northern Argentina) is given. In this context the problem of the tectonic framework is studied from the accumulation of the Pirgua Sub-Group (Early Cretaceous) up to the complete filling of the basin. A description of the paleogeographic development of central-western South America during Cretaceous times has been attempted. Correlations between different episodes of sedimentation and volcanism of this area are indicated from central Peru through the Bolivian and Chilean basins up to the Neuquén Basin (central-western Argentina, on the Chilean border). Four main branches that originate around a structural high (the Salta-Jujuy high) are located in tectonic depressions between the Condor, Michicola, Quirquincho-Pampear, Transpampean, Domeyko Cordillera (Chile) archs and the San Pablo high.

Introduction

The aim of the present work is to give a synthetical analysis of the Salta Group basin (Cretaceous-Eocene) in northern Argentina. However, it has been necessary to include a number of considerations on a much broader regional scale within the setting of the Central Andes and that of the southern cone of South America.

The paleogeographic development of the Salta Group basin has been analyzed within the context of the Cretaceous basins of South America, local interpretations being necessarily linked to a wide range of regional factors. Consequently, paleogeographic and stratigraphic considerations are based partly on personally obtained laboratory and field data, and partly on third-party information and interpretations.

The characteristics of the Salta Group (including paleogeographic evolution and correlations) have been studied by TURNER (1959), LEANZA (1969), MORENO (1970), REYES & SALFITY (1973), LENCINAS & SALFITY (1973), CASTAÑOS et al. (1975), REYES et al. (1976), CAZAU et al. (1976), SALFITY (1979, 1980, 1982), SALFITY & MARQUILLAS (1981), MARQUILLAS (1984, 1985, 1986) and SALFITY & ZAMBRANO (in press). Other works considering the correlation of the Cretaceous of central and southern South America

Lecture Notes in Earth Sciences, Vol. 17
H. Bahlburg, Ch. Breitkreuz, P. Giese (Eds.),
The Southern Central Andes
©Springer-Verlag Berlin Heidelberg 1988

are those of RUSSO & RODRIGO (1965), REYES (1972), CHERRONI (1977), SALFITY et al. (1985), RODRIGO & BRANISA (in press) and SEMPERE et al. (in press). A Spanish version of this paper was presented at the First Symposium of IGCP Project 242 "Cretaceous of Latin America" during the 8th Bolivian Geological Congress (La Paz, Bolivia, 1986).

Salta-Group breakdown used in this report

Units are listed according to their stratigraphic location, from top to bottom. The equivalents of some formation names in different parts of the basin are printed on the right.

SALTA GROUP (BRACKEBUSCH 1891, TURNER 1959)

  SANTA BARBARA SUB-GROUP (VILELA 1952, MORENO 1970)
   LUMBRERA FORMATION (MORENO 1970)
   MAIZ GORDO FORMATION (MORENO 1970)
   MEALLA FORMATION (MORENO 1970)

  BALBUENA SUB-GROUP (MORENO 1970)
   OLMEDO FORMATION (MORENO 1970) - TUNAL FORMATION (AMENGUAL in TURNER et al. 1979)
   YACORAITE FORMATION (TURNER 1959)
   LECHO FORMATION (TURNER 1959) - QUITILIPI FORMATION (SALFITY 1980)
                                  - PALA PALA FORMATION (SALFITY & MARQUILLAS 1981)

  PIRGUA SUB-GROUP (VILELA 1951, REYES & SALFITY 1973)
   LOS BLANQUITOS FORMATION (REYES & SALFITY 1973)
   LAS CURTIEMBRES FORMATION (REYES & SALFITY 1973)
   LA YESERA FORMATION (REYES & SALFITY 1973)

The Salta Group Basin and its tectonic setting

When the Salta Group is related to the Central Andes and the neighbouring eastern region (the Guaporé craton and the Atlantic coast, Fig. 2), it appears as quite evident that the basin must have been located in a highly unstable tectonic area in which conspicuous regional structures have disappeared. The northern end of the Pampean nesocraton (or Central Argentine Cratogene) lies buried under the southern part   of   the   basin,   southwest   of   the   El   Toro   lineament

(SALFITY 1985) (Figs. 4 and 5). The pre-Cretaceous basement north of the El Toro lineament is of Paleozoic age, thus coinciding with the Eastern Cordillera and the Sub-Andean belt.

There were positive topographic highs in the Bolivian-Argentine Eastern Cordillera ranges during the Cretaceous. On the other hand, the Sub-Andean System was a negative element, forming the Sub-Andean basin of Bolivia and Peru (Figs. 3 and 4). Likewise, the Argentine extension of the Sub-Andean System has been lodged into the Cretaceous basin of The Salta Group, i. e. south of the Michicola Arch (Fig. 4).

Figures 3 and 6 show the paleogeographic development of the Cretaceous basins of central-west South America, after SALFITY & ZAMBRANO (in press). A comparison between the development of the Salta Group basin (Fig. 3) and the major structural

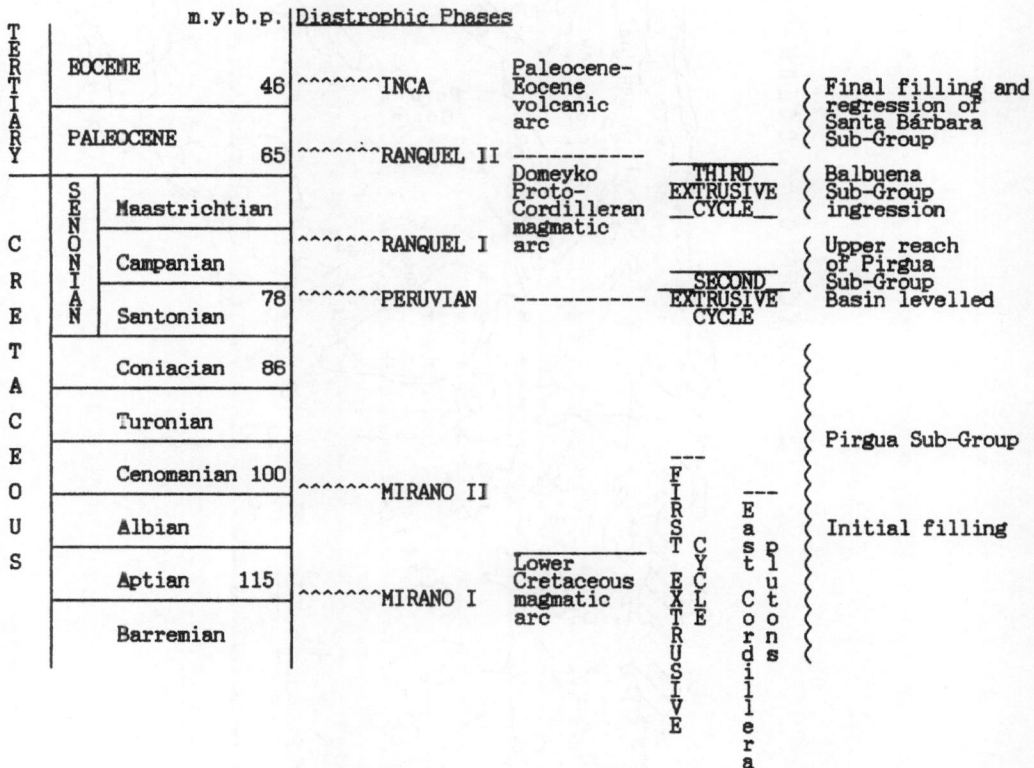

Fig. 1: The above chart shows the main diastrophic and magmatic phases that took place during Cretaceous-Eocene times in northern Argentina.

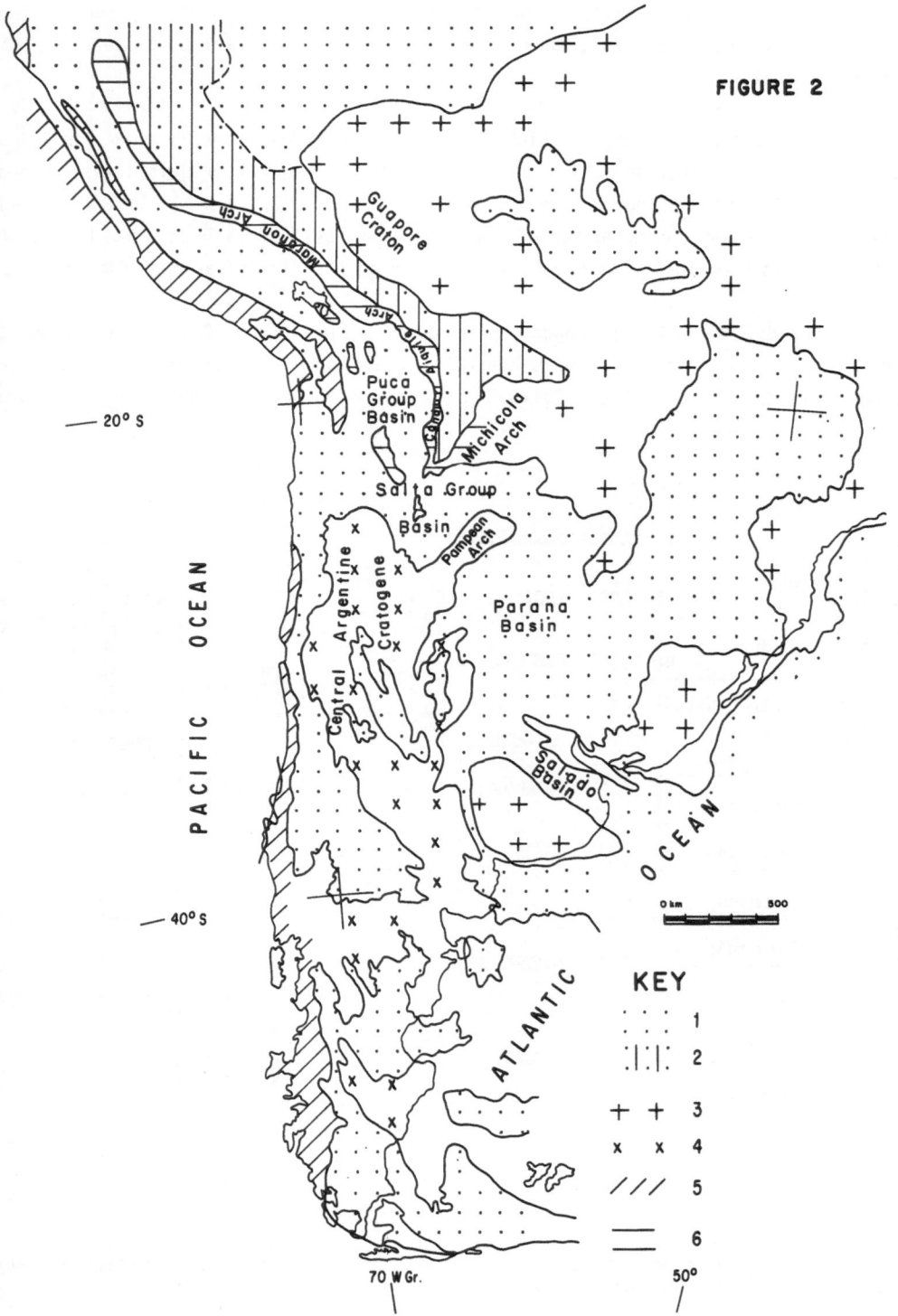

FIGURE 2

features shown in Figure 2 clearly shows how the northern end of the Pampean nesocraton remained upraised, thus conditioning the history of the whole southern portion of the basin. Two branches, the Pampean arch in the east and the Traspampean
arch in the west (PADULA & MINGRAMM 1968) (see Figs. 2 and 4), were formed by the
Pampean nesocraton during the Cretaceous.

The east flank of the Eastern Cordilleras in Bolivia and Argentina remained emerged,
forming part of the Aiquile/Condor Arch (see Figs. 2, 3 and 4), which is known as
the Marañon Arch in Peruvian territory. In Argentina the west flank of the Salta-
Jujuy high (Dorsal Salto-Jujeña) is part of the Eastern Cordillera, and the east
flank is part of the Sub-Andean belt, between the Ocloyan fracturation front and the
El Toro lineament. To the east of the Sub-Andean ridge the development of the Salta
Group basin took place along an extensive and elongated tectonic trench: The Olmedo
Sub-Basin, limited towards the north by the Michicola Arch, and to the south by the
Quirquincho Arch (Figs. 4 and 5). On the Argentine side of this sub-basin the
structure was governed by lineaments with a northeast-southwest direction (SALFITY
1985).

A similar situation might have taken place west of the basin, in the Argentine Puna
or Altiplano. One should note that the Sey Sub-Basin (SCHWAB 1984) partly follows
the El Toro lineament trend. North of this sub-basin the San Pablo high has remained
emerged (REYES 1972), as has the Traspampean Arch to the south (Figs. 4 and 5). The
Sey Sub-Basin was related to another, farther west: The Antofagasta or Purilactis
Sub-Basin, through the Huaytiquina swell (see Fig. 4) (SALFITY et al. 1985). Likewise, the west flank of the Antofagata or Purilactis Sub-Basin was governed by a
structural high that was part of an extensive Cretaceous-Eocene magmatic arc.

The Salta-Jujuy high has already been mentioned as forming part of the Eastern Cordillera. This ridge was encroached upon on all sides by the Balbuena Sub-Group
transgression (mainly the Yacoraite Formation) during the Maastrichtian, and was
thereby reduced to a small island compared to the size it had been before this
transgression took place (compare Figs. 4 and 5).

Yacoraite Formation limestones lie on several pre-Cretaceous units (Precambrian,
Ordovician, Devonian) along the flanks of the Salta-Jujuy high. The transgression
came across a quite heterogenous pre-Cretaceous basement, the youngest units being
exposed towards the east. On the other hand, there was no transgressive action by
the Yacoraite Formation on the basin rims formed by the Pampean and Traspampean

Fig. 2: Structural elements of the southern cone of South America during the Cretaceous system. 1) Areas with Cretaceous sedimentation, 2)Sub-Andean belt, 3) Precambrian shields, 4) Nesocratons, 5) Western and Coastal Cordilleras (Precambrian rocks
included), 6) Lower Paleozoic structural highs of the Eastern Cordillera.

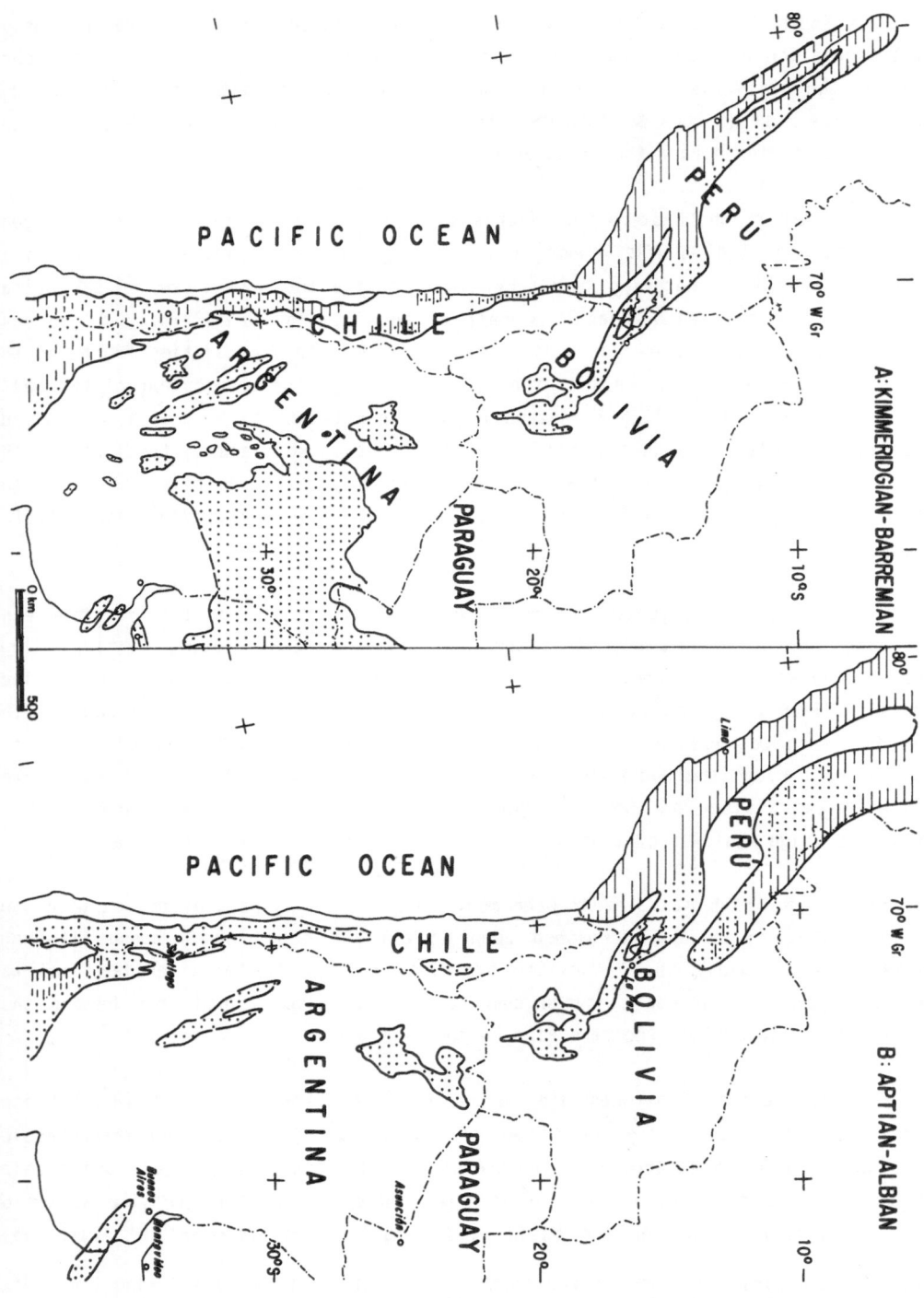

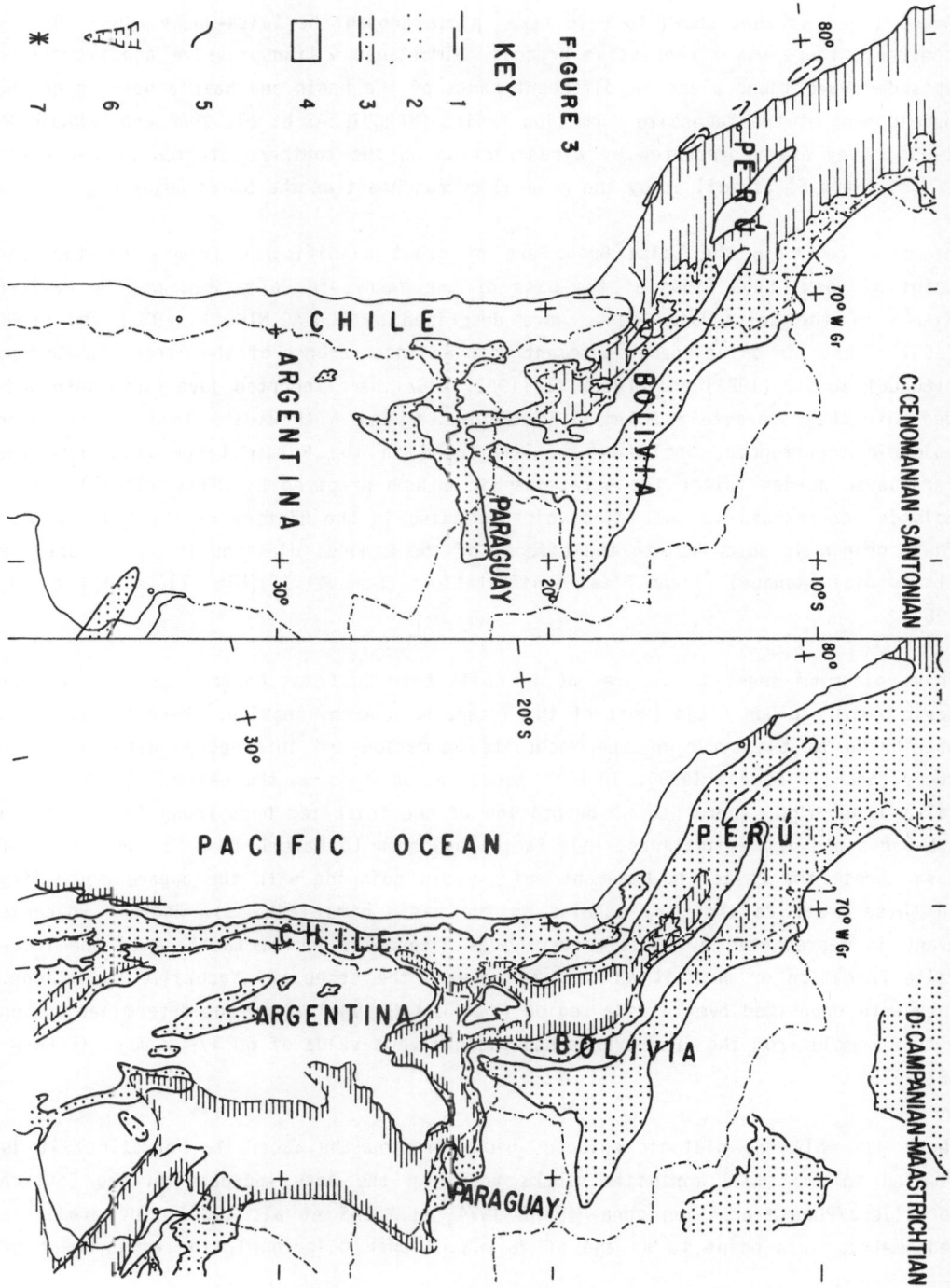

FIGURE 3

KEY

1
2
3
4
5
6
7

C: CENOMANIAN-SANTONIAN

D: CAMPANIAN-MAASTRICHTIAN

Archs (such as that shown to have taken place around the Salta-Jujuy high). On the contrary, there was a regressive process. Simultaneous transgressive and regressive episodes have taken place in different parts of the basin and have a bearing on the development of the Yacoraite Formation facies (MARQUILLAS et al. 1984 and MARQUILLAS 1985), they are represented by a restriction on the southern stretch of the basin (Figs. 4 and 5) as well as by the general encroachment on the Salta-Jujuy high.

Magmatic rocks in the Salta Group are of great significance from a stratigrafic point of view. Lava flows at the base of the Yacoraite Formation, on the eastern flanks of the Salta-Jujuy high, were described by SCHLAGINTWEIT (1937) and LYONS (1951); they lie on Paleozoic basement with no interference of the Pirgua Sub-Group. BIANUCCI et al. (1981) and BERCOWSKI (1982, 1987) have reported lava flows interbedded with the Yacoraite and Olmedo Formations. Figure 5 shows the location of these volcanic occurrences, including the one found in the Palmar Largo well near the Paraguayan border (after the stratigraphic column prepared by MÄDEL (1984)). These episodes correspond to the third volcanic pulse in the history of the Salta Group. Their origin is adscribed to the effects of the Ranquel diastrophic phase, both in its initial (Ranquel I) and final manifestations (Ranquel II)(Fig. 1)(SALFITY et al. 1984).

This volcanism seems to be present in Chile both in Lomas Negras (west of the San pablo high) and Don Alejo (west of the Traspampean Arch) section, where the deposits attributed to the cycle of the Yacoraite Formation are interbedded with extrusive rocks (SALFITY et al. 1985). In both cases, as well as in the eastern flank of the Salta-Jujuy high, there was no deposition of the thick red beds found in the Pirgua Sub-Group and its equivalents. This fact would seem to have helped to the uprise of lava across the Paleozoic basement which would coincide with the apparent mobility in these areas as they are located on the basin rims (Fig. 5). Another volcanic event is represented by tuffs occurring as intercalations at the base of the Yacoraite Formation or near it as well as between the Lecho and Yacoraite Formations. They were deposited over a wide region (MARQUILLAS 1985). K/Ar age determinations on a tuff sample from the Tres Cruzes sub-basin gave a value of 60 +/- 2 m.y. (FERNANDEZ 1975).

There was only one plutonic episode which affected the Yacoraite Formation. It is limited to two small monzonite stocks found on the Acay snowpeak in the Eastern Cordillera/Puna transition zone (MIRRE 1974, LLAMBIAS et al. 1986). The available radiometric data point to an age of 26 m.y. (Upper Oligocene). The emplacement of

Fig. 3: Paleogeographic evolution of the Cretaceous basins of central-western South America. 1) Marine and/or lacustre, 2) Continental, 3) Marine and continental, 4) Emerged areas, 5) Edges of structural basins, 6) Postulated extension of Cenomanian and Maastrichtian ingressions, 7) Cretaceous volcanism has been excluded.

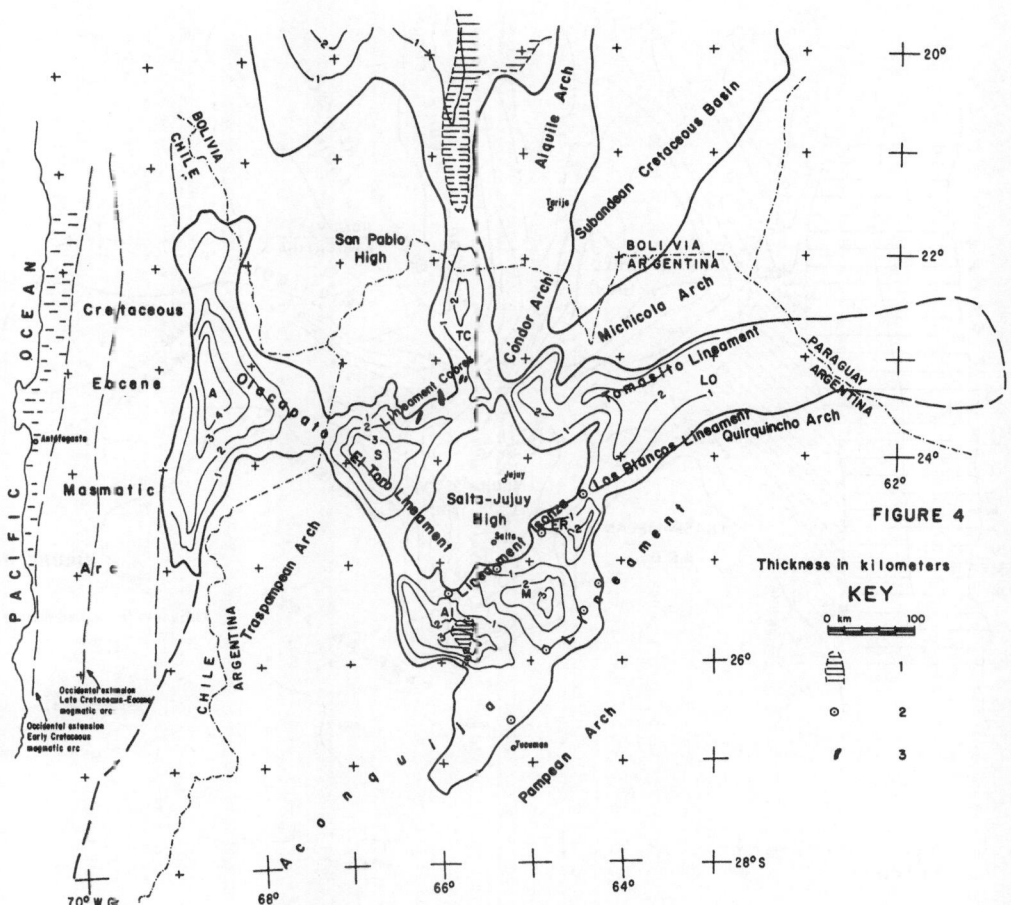

Fig. 4: Map of northern argentina and neighbouring regions at the close of the Campanian age; end of rift stage. 1) Dispersion of Upper Cretaceous volcanism (second cycle), 2) Lower Cretaceous volcanism (first cycle), 3) Cretaceous intrusives. Subbasins: TC: Tres Cruzes, LO: Lomas de Olmedo, ER: El Rey, M: Metán, Al: Alemanía, S: Sey, A: Artofagasta.

this Tertiary pluton coincides both with a rim of the Cretaceous basin and with an area where the thickness of the Pirgua Sub-Group was minimal, i.e. a location where magma emplacement must have been conditioned directly by the basement (Puncoviscana Formation, Precambrian). Furthermore, this plutonic occurrence coincides with the El Toro lineament. Thick beds of amphibole andesite intrude into units of the Salta Group in the Alemanía sub-basin (FRENGUELLI 1936, RUIZ HUIDOBRO 1949).

The two early effusive pulses of the Salta Group are included in the Pirgua Sub-Group (Fig. 7), very well represented towards the south of the basin (Fig. 4). The first one intercalates with the conglomerates of the La Yesera Formation: Isonza

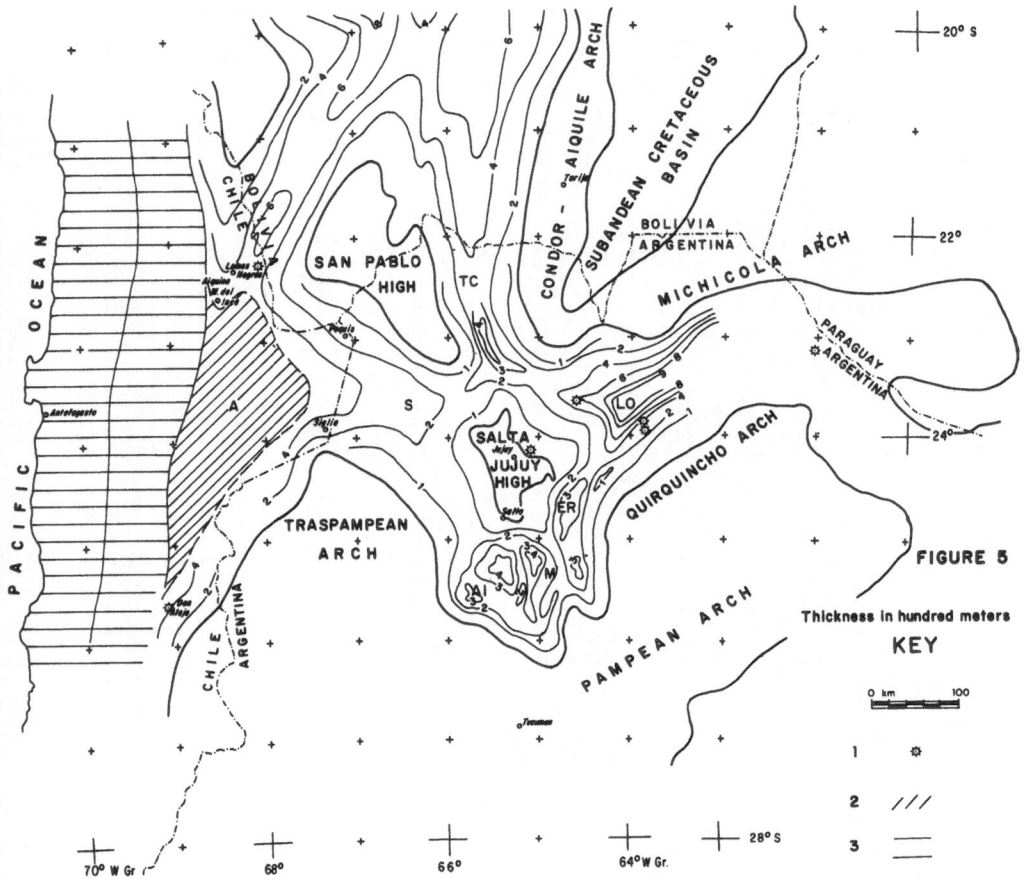

Fig. 5: Map of northern Argentina and neighbouring regions at the close of the Maastrichtian age (Paleocene partially included). 1) Upper Senonian-Paleocene volcanism (third cycle), 2) Known western limit of the Maastrichtian ingression, 3) Emerged areas (Jurassic-Lower Cretaceous magmatic arc). Sub-basins (see Fig. 4).

basalt, the radiometric ages range from 120-90 m.y. (REYES et al. 1976, VALENCIO et al. 1976). The volcanic foci that define the Isonza lineament along the southern border of the Salta-Jujuy high as well as those identified along the Aconquija lineament are very conspicuous (Fig. 4). The second effusive event is the one located in the upper part of the Las Curtiembres Formation: Las Conchas Basalt. These basalts crop out extensively in the Alemania Sub-Basin, their age is 78-76 m.y. (REYES et al. 1976, VALENCIO et al. 1976). This volcanic episode is clearly related to the Peruvian phase (Fig. 1) being the only one of this age in northwestern Argentina (MOYA & SALFITY 1982). The Cretaceous plutons of the region are located in the Ordovician basement at the southern end of San Pablo high. They define the Cobres lineament (Fig. 4) and their radiometric ages are 147-110 m.y. (HALPERN & LATORRE 1973). This reveals a simultaneous origin to that of the first volcanic cycle.

Correlation of the Salta Group with other units of central-western South America

The evolution of the Cretaceous basins of the central-western region of South America is rich in sedimentary, volcanic and tectonic episodes. Consequently, it seemed necessary to make a thorough analysis of the marine, continental and mixed Cretaceous units of this region (Figs. 3, 5, and 7), based on the work of SALFITY & ZAMBRANO (in press).

The development of marine sedimentation during the Lower Cretaceous took place in the Peruvian basins and in the Argentine-Chilean basin (Fig. 6), but with the difference that in this last basin sedimentation was interrupted before the beginning of the Cenomanian (compare Figs. 3A, 3B and 3C), due to the Mirano diastrophic phase. However, the Mirano phase had no effect on the evolution of the Cretaceous seas of the Eastern and Western Cordilleras of Peru, where sediments were deposited until the Santonian, when the seas started to retreat under the influence of the Peruvian phase.

The Cretaceous seas in Peru were directly linked to the Northern Andes, especially from Aptian times onward. On the other hand, no relationship existed with the southern Peruvian basins, i. e. those of the Puca Group in Bolivia and the Salta Group in Argentina, other than a finger of the sea which entered the region during the Cenomanian and invaded the Titicaca trench and the Bolivian Andean basin (Miraflores Formation) without reaching Argentine territory. In these last regions sediments contemporary with the Peruvian marine sedimentation were almost entirely continental: The Lower Puca Group and the Pirgua Sub-Group (Fig. 6).

Figure 6 shows that the Peruvian basins, during the time of the Peruvian phase (Late Santonian - Early Campanian), were almost exclusively filled with continental red beds and volcanic rocks of Upper Senomian and Eo-Tertiary age. Farther south, in Bolivia and Argentina, there are similar red beds (Upper Puca Group, part of Los Blanquitos Formation and Santa Barbara Sub-Group), but the interbedded limestones, shales and sandstones of El Molino and Yacoraite Formations are well developed. It is therefore evident that in the Eastern and Western Cordilleras of Peru, except for the Vilquechico Formation on the Peruvian Altiplano, no equivalent accumulation of the above mentioned calcareous facies has been deposited. On the other hand, these calcareous facies can be considered homologous and correlated with the units contained in the Parana and Salado Basins (Figs. 2 and 7). The need to establish a relationship between the Parana Basin and the Salta Group Basin through the northeast end of the Pampean Arch was suggested by RUSSO et al. (1979). The connection would seem to have come about at the beginning of the Balbuena Sub-Group deposition (SALFITY 1982, MARQUILLAS 1985). This being so, it is possible to complete the interpretation shown in Figure 7, where parts of the Las Chilcas and Mariano Moreno Formations are considered to be equivalents to the Balbuena Sub-Group. Naturally, this hypothesis remains subject to confirmation by underground studies and, eventually, microfaunal and sedimentary analysis.

As regards a connection between the Salta Group Basin and the Purilactis Formation basin in northern Chile, a great deal of stratigraphic information on the subject is available today, especially concerning northern Chile (MAKSAEV 1978, RAMIREZ & HUETE 1981, GARDEWEG & RAMIREZ 1985). There is also some literature on Argentina (SCHWAB 1984, MARQUILLAS et al. 1986, DONATO & VERGANI 1987). This information has made it possible to go back to BRÜGGEN's (1942) original interpretations, which were backed by GROEBER (1953), and ratify that relationship. Simultaneously, the shape of the Antofagasta or Purilactis Sub-Basin has been outlined (see Figs. 4 and 5). These illustrations show the state of the Salta Group Basin before the start of the initial Ranquel phase (Fig. 4), embodying the Balbuena Sub-Group and equivalents at the end of the transgression (Fig. 5).

The Antofagasta Sub-Basin (Fig. 4) received only red bed accumulations similar to the Pirgua Sub-Group, but with a different granular development: the Pirgua Sub-Group shows fining-upward sequences whereas the Purilactis Formation shows coarsening-upward sequences. Furthermore, the Purilactis Formation - the top of which shows an angular unconformity with Oligocene Miocene units (Tambores Formation) or Eocene units (Icanche Formation) - contains no interbedded limestones. The known sections are all made up of red beds, but it is not known whether part of these correspond to the Yacoraite Formation facies or not.

However, over the Huaytiquina swell, which is the link-up between the San Pablo high and the Traspampean Arch (Fig. 5), there are successions which can be correlated with the Yacoraite Formation, most of them lying on pre-Cretaceous basement. The Poquis section, southeast of the triple border intersection (between Chile, Argentina and Bolivia) is worthy of mention, where the Quebrada Blanca Formation, lying on Ordovician basement contains grey oolithic limestones with poorly preserved Cretaceous foraminifera (GARDEWEG & RAMIREZ 1985). Other sections on the Chilean side of the border which are linked to the Yacoraite Formation transgression are

Fig. 6: Correlation chart of Cretaceous continental and marine units in central-western South America. 1) Marine Clastics, 2) Marine limestones, 3) Continental clastics, 4) Continental volcanics, 5) Evaporites.

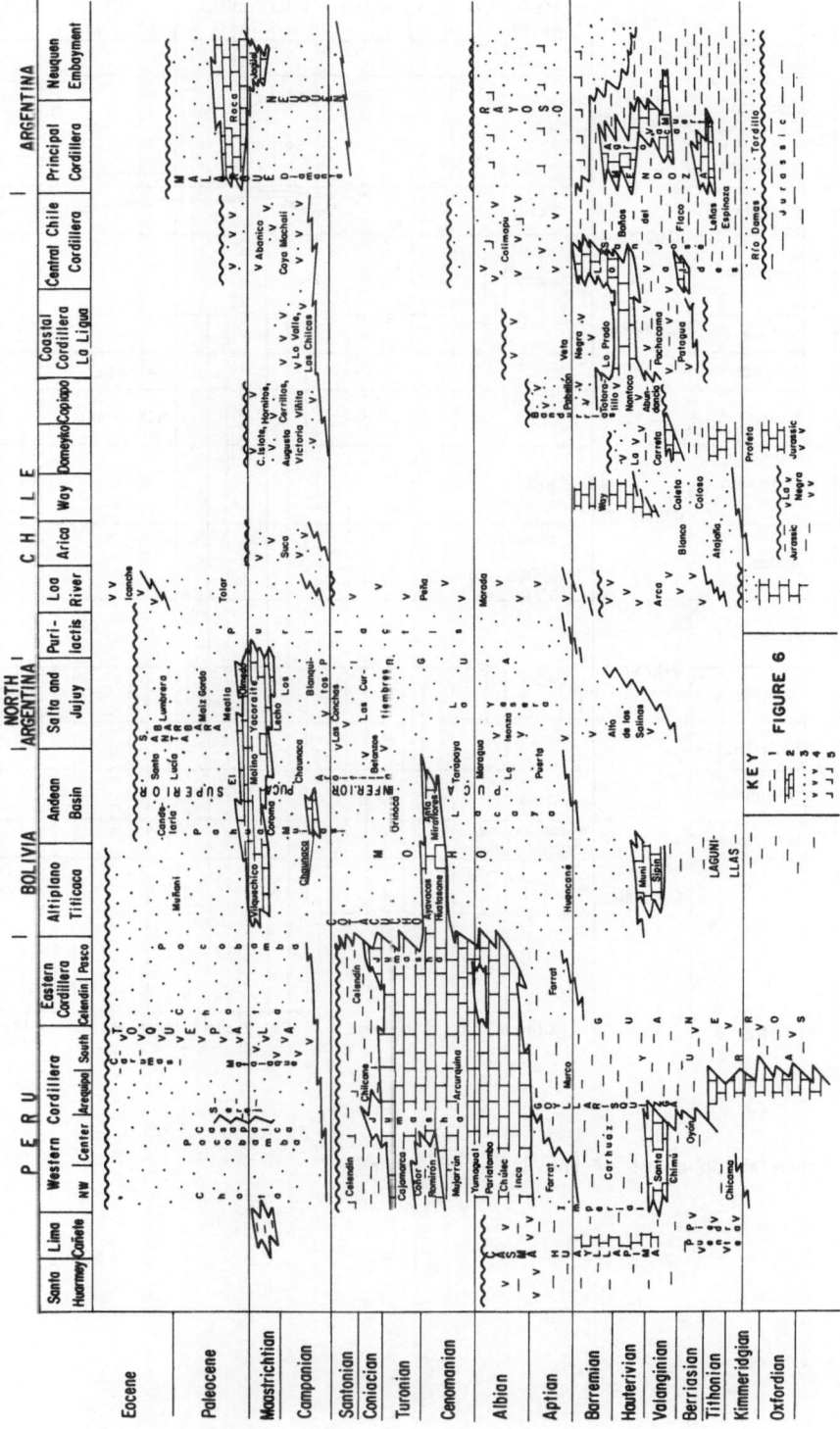

FIGURE 6

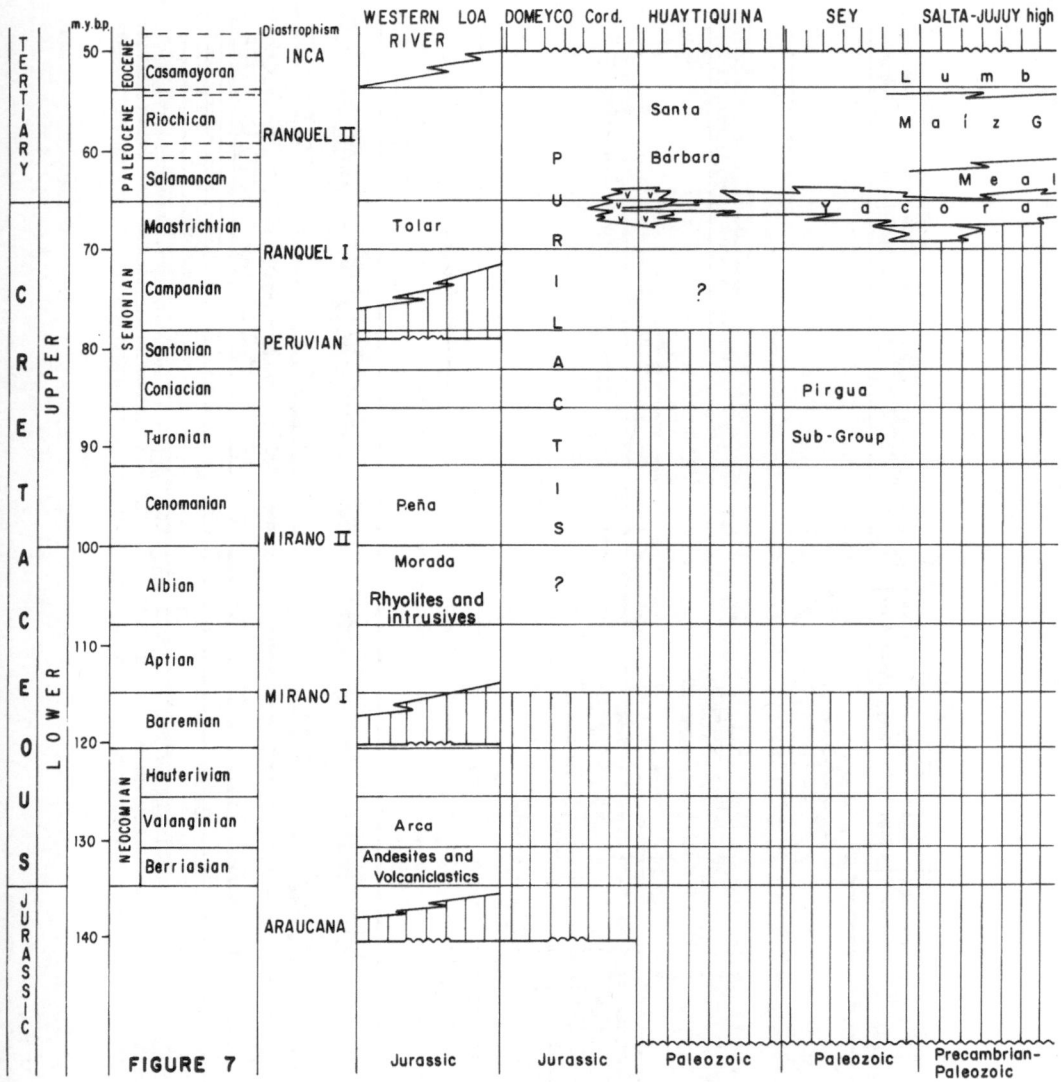

Fig. 7: Correlation chart of the Salta Group units.

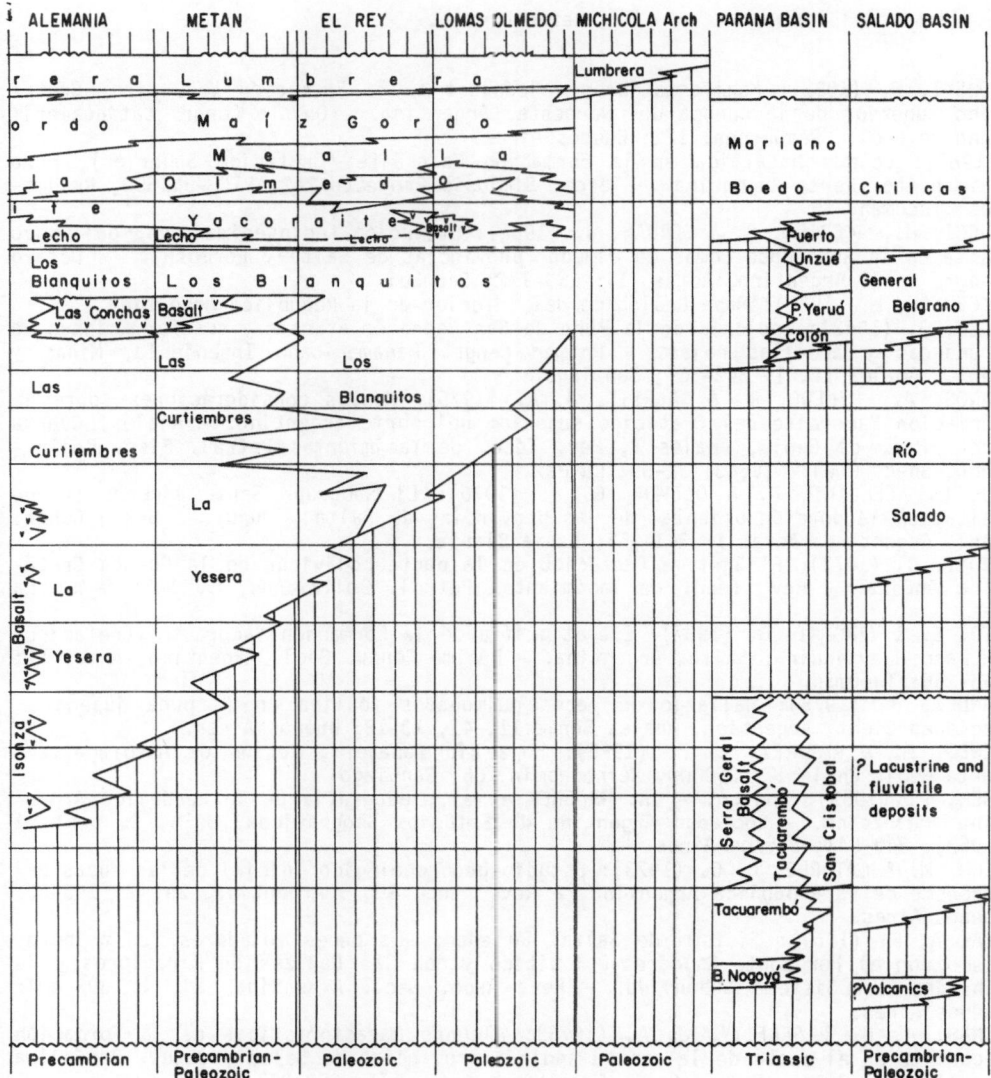

those of Lomas Negras, Aiquina, Morro del Inca, Siglia and Don Alejo (Fig. 5) (MARI-
NOVIC & LAHSEN 1984, SALFITY et al. 1985, GARDEWEG & RAMIREZ 1985). In general,
these all show a similar top and base stratigraphic relationship (Paleozoic lower
contact and a Tertiary top contact).

References

BERCOWSKI, F. (1982): Manifestaciones volcánicas en la Formación Yacoraite, Cretácico Superior de la Cuenca del Noroeste, Argentina. - Quinto Congr. Latinoameri-
cano de Geol., Resúmenes: 122; Buenos Aires.
-- (1987): Colada basáltica en la Formación Yacoraite (Cretácico Superior), pozo
Caimancito, Norte Argentina. - Tercer Simposio Proyecto 242, PICG-UNESCO, Resúme-
nes; Tucumán.
BIANUCCI, H., ACEVEDO, O. & CERDAN, J. (1981): Evolución tectosedimentaria del Grupo
Salta en la Subcuenca Lomas de Olmedo (provincias de Salta y Formosa). - Octavo
Congr. Geol. Argentino, Actas III: 159-172; San Luis.
BRACKEBUSCH, L. (1981): Mapa Geológico del interior de la República Argentina.
BRÜGGEN, J. (1942): Geología de la Puna de San Pedro de Atacama y sus formaciones de
areniscas y arcillas rojas. - Primer Congr. Panamericano Ingeniería, Minas y
Geol., Anales II, 1: 342-367; Santiago.
CASTAÑOS, A., PINEDO, R. & SALFITY, J. A. (1975): Nuevas consideraciones sobre la
Formación Yacoraite del Cretácico superior del norte argentino. - Anales, Cuarta
Conv. Nac. de Geol., Anales I, Rev. Técn. de Yacimientos Petrol. Fisc. Bolivi-
anos, Spec. Publ., IV, 3: 31-59; La Paz.
CAZAU, L., CELLINI, N. & OLIVER, G. J. (1976): El Subgrupo Santa Bárbara (Grupo
Salta) en la porción oriental de las provincias de Salta y Jujuy. - Sexto Congr.
Geol. Argentino, Actas I: 341-355; Bahia Blanca.
CHERRONI, C. (1977): El Sistema Cretácico en la parte boliviana de la Cuenca Cretá-
cica Andina. - Rev. Técn. de Yacimientos Petrol. Bolivianos, 6, 1-2: 5-46; La
Paz.
DONATO, E. & VERGANI, G. (1987): Estratigrafía de la Formación Yacoraite (Cretacico)
en Paso Huaytiquina, Salta, Argentina. - Décimo Congr. Geol. Argentino, Actas II:
263-266; Tucuman.
FERNANDEZ, J. (1975): Hallazgo de peces pulmonados fósiles en la puna jujeña. -
Sociedad Cient. Argentina, Anales Serie II, 41, 13-18; Buenos Aires.
GARDEWEG, M. & RAMIREZ, C. F. (1985): Hoja Río Zapaleri, Región de Antofagasta.-
Carta Geol. Chile 1 : 250000, Sernageomin, 66; Santiago.
GROEBER, P. (1953): Andico. - in: GROEBER et al.: Geografía de la República Argen-
tina. Mesozoico. - Sociedad Argentina de Estudios Geográficos, GAEA, 2, part. 1
(1952): 349-541; Buenos Aires.
HALPERN, M. & LATORRE, C. O. (1973): Estudio geocronológico inicial de las rocas del
noroeste de la República Argentina. - Rev. Asoc. Geol. Argentina, 28, 2: 195-20;
Buenos Aires.
LEANZA, A. F. (1969): Sistema de Salta. Su edad, sus peces voladores, su asincro-
nismo con el Horizonte Calcáreo-Dolomítico y con las Calizas de Miraflores y la
hibridez del Sistema Subandino. - Rev. Asoc. Geol. Argentina, 24, 4: 393-407;
Buenos Aires.
LENCINAS, A. N. & SALFITY, J. A. (1973): Algunas características de la Formación
Yacoraite en el oeste de la cuenca andina, provincias de Salta y Jujuy, República
Argentina. - Quinto Congr. Geol. Argentino, Actas III: 253-267; Carlos Paz.
LYONS, W. A. (1951): El filón-capa basáltico de la Formación Petrolífera de Jujuy.-
Rev. Asoc. Geol. Argentina, 6, 2: 114-132; Buenos Aires.
LLAMBIAS, E. J., SATO, A. M. & TOMSIC, S. (1986): Geología y caracteristicas químicas
del stock terciario del Nevado de Acay y vulcanitas asociadas, provincia de
Salta. - Rev. Asoc. Geol. Argentina, 40, 3-4: 158-175; Buenos Aires.
MÄDEL, F. (1984): Estratigrafía del tramo inferior del pozo descubridor Palmar Largo
x1. - Bol. Inform. Petrol., 1, 2: 109; Buenos Aires.
MAKSAEV, J. V. (1978): Cuadrángulo Chitigua y sector occidental del cuadrángulo cer-
ro Palpana, Región de Antofagasta. - Inst. Invest. Geol., 31, Santiago.
MARINOVIC, S. N. & LAHSEN, A. (1984): Hoja Calama, Región de Antofagasta.- Carta
Geol. Chile 1 : 250000, Sernageomin, 58; Santiago.
MARQUILLAS, R. A. (1984): La Formación Yacoraite (Cretácico Superior) en el río
Juramento, Salta: Estratigrafía y ciclicidad. - Noveno Congr. Geol. Argentino,
Actas V: 186-196; Bariloche.
-- (1985): Estratigrafía, sedimentología y paleoambientes de la Formación Yacoraite
(Cretacico Superior) en el tramo austral de la cuenca, norte argentino. -Doctoral
thesis, Univ. Nac. de Salta, unpublished.

135

-- (1986): Ambiente de depósito de la Formación Yacoraite (Grupo Salta, Cretácico-Eocénico), Norte argentino. - Primer Simposio Proyecto 242 PICG-UNESCO, 157-173, La Paz.
MARQUILLAS, R. A., ALONSO, R., GORUSTCVICH, S. & SALFITY, J. A. (1986): El Grupo Salta (Cretácico-Eogénico) en la Puna Argentina. - Resúmenes, Segundo Simposio Proyecto 242, PICG-UNESCO, Mendoza.
MARQUILLAS, R. A., BOSO, M. A. & SALFITY, J. A. (1984): La Formación Yacoraite (Cretácico Superior) en el norte argentino, al sur del paralelo 24°. - Noveno Congr. Geol. Argentino, Actas, II: 300-310; Bariloche.
MIRRE, J. C. (1974): El granito de Acay, intrusivo de edad terciaria en el ambiente de Puna. - Rev. Asoc. Geol. Argentina, 29, 2: 205-212; Buenos Aires.
MORENO, J. A. (1970): Estratigrafía y paleogeografía del Cretácico superior en la cuenca del noroeste argentino, con especial mención de los Subgrupos Balbuena y Santa Bárbara. - Rev. Asoc. Geol. Argentina, 24, 1: 9-44; Buenos Aires.
PADULA, A. & MINGRAMM, A. (1968): Estratigrafía, distribución y cuadro geotectónico-sedimentario del "Triásico" en el subsuelo de la llanura Chaco-Paranense. - Terceras Jornadas Geol. Argentinas, Actas I: 291-331; Comodoro Rivadavia.
RAMIREZ, C. & HUETE, C. (1981): Hoja Ollagüe, Región de Antofagasta. - Carta Geol. Chile 1 : 250000, Sernageomin, 40; Santiago.
REYES, F. C. (1972): Correlaciones en el Cretácico de la Cuenca Andina de Bolivia, Peru y Chile. - Rev. Tecn. de Yacimientos Petrol. Fiscales Bolivianos, 1, 2-3: 101-144; La Paz.
REYES, F. C. & SALFITY, J. A. (1973): Consideraciones sobre la estratigrafía del Cretácico (Subgrupo Pirgua) del noroeste argentino. - Quinto Congr. Geol. Argentino, Actas III: 355-385; Carlos Paz.
REYES; F. C., SALFITY, J. A., VIRAMONTE, J. G. & GUTIERREZ, W. (1976): Consideraciones sobre el vulcanismo del Subgrupo Pirgua (Cretácico) en el norte argentino. - Sexto Congr. Geol. Argentino, Actas I: 205-223, Bahia Blanca.
RUIZ HUIDOBRO, O. J. (1949): Estudio geológico de la región de los cerros Quitilipi y Pirgua (departamento de Guachipas, provincia de Salta). - Rev. Asoc. Geol. Argentina, IV, 1: 10-75; Buenos Aires.
RUSSO, A. & RODRIGO, L. A. (1965): Estratigrafía y paleogeografía del Grupo Puca en Bolivia. - Bol. Inst. Boliviano del Petróleo, 5, 3-4: 5-53; La Paz.
SALFITY, J. A. (1979): Paleogeología de la cuenca del Grupo Salta (Cretácico-Eocénico) del norte de Argentina. - Séptimo Congr. Geol. Argentino, Actas I: 505-515; Neuquén.
-- (1980): Estratigrafía de la Formación Lecho (Cretácico) en la Cuenca Andina del Norte Argentino. - Doctoral thesis, Univ. Nacional de Salta, Spec. Publ.
-- (1982): Evolución paleogeográfica del Grupo Salta (Cretácico-Eogénico), Argentina. - Quinto Congr. Latinoamericano de Geol., Actas I: 11-26; Buenos Aires.
-- (1985): Lineamientos transversales al rumbo andino en el noroeste argentino. - Cuarto Congr. Geol. Chileno, Actas II: 2/119-237; Antofagasta.
SALFITY, J. A., GORUSTOVICH, S. & MOYA, M. C. (in press): Las fases diastróficas en los Andes del Norte argentino. - Simposio Internacional de Tectónica Centro-Andina y Relación Recursos Naturales, La Paz.
SALFITY, J. A. & MARQUILLAS, R. A. (1981): Las unidades estratigráficas cretácicas del Norte de la Argentina. - in: VOLKHEIMER, W. & MUSACCHIO, E. A. (eds.): Cuencas Sedimentarias del Jurásico y Cretácico de América del Sur. Comité Sudamericano del Jurásico y Cretácico, 1: 303-317; Buenos Aires.
SALFITY, J. A. MARQUILLAS, R. A., GARDEWEG, M., RAMIREZ, C. & DAVIDSON, J. (1985): Correlaciones en el Cretácico Superior del norte de Argentina y Chile. - Cuarto Congr. Geol. Chileno, Actas IV: 1/654-667; Antofagasta.
SALFITY, J. A. & ZAMBRANO, J. (in press): Cretácico. - in: BONAPARTE, J. F. & TOSELLI, J. A. (coord.): Geología de América del Sur, Univ. Nac. de Tucumán, Tucumán.
SCHLAGINTWEIT, O. (1937): Observaciones estratigráficas en el norte argentino. -Bol. de Informaciones Petroleras, 14: 1-49; Buenos Aires.
SCHWAB, K. (1984): Contribución al conocimiento del sector occidental de la cuenca sedimentaria del Grupo Salta (Cretácico-Eogénico) en el noroeste argentino. - Noveno Congr. Geol. Argentino, Actas I: 586-604; Bariloche.
SEMPERE, T., OLLER, J., CHERRONI, C., ARANIBAR, O., BARRIOS, L., BRANISA, L., CIRBIAN, M. & PEREZ, M. (1987): Un ejemplo de cuenca carbonatada en un contexto distensivo de retroarco: paleodinámica del Cretácico terminal en la República de Bolivia (Formación El Molino y equivalentes). - Tercer Simposio Proyecto 242 "Cretácico de América Latina" PICG-UNESCO, Resúmenes; Tucumán.

TURNER, J. C. M. (1959): Estratigrafía del cordón de Escaya y de la sierra de Rinco-
    nada (Jujuy). - Asoc. Geol. Argentina, Rev., 13 (1958), 1-2: 15-39; Buenos Aires.
TURNER, J. C. M., MENDEZ, V., LURGO, C., AMENGUAL, R. & VIERA, O. (1979): Geología
    de la región noroeste, provincia de Salta y Jujuy, República Argentina. -Séptimo
    Congr. Geol. Argentino, Actas I: 367-387; Neuquén.
VALENCIO, D. A., GIUDICE, A., MENDIA, J. E. & OLIVER G., J. (1976): Paleomagnetismo
    y edades K/Ar del Subgrupo Pirgua, provincia de Salta, República Argentina. -
    Sexto Congr. Geol. Argentino, Actas I: 527-542; Bahia Blanca.
VILELA; C. R. (1951): Acerca del hallazgo del horizonte calcáreo Dolomítico en la
    Puna Salto-Juneña y su significado geológico. - Rev. Asoc. Geol. Argentina, 6, 2:
    100-107; Buenos Aires.
-- (1952): Acerca de la presencia de sedimentos lacustres en el valle Calchaquí. -
    Rev. Asoc. Geol. Argentina, 7, 4: 219-227; Buenos Aires.

THE CENOZOIC SALINE DEPOSITS OF THE CHILEAN ANDES BETWEEN 18°00' and 27°00' SOUTH LATITUDE

Guillermo Chong Díaz
Departamento de Geociencias, Universidad del Norte,
Casilla 1280, Antofagasta, Chile

Abstract

During Cenozoic times, various continental evaporite deposits formed in Northern Chile. The salars have to be considered as dynamo-sedimentary entities, the formation of which was controlled by a combination of geological, morphological, hydrological, and climatical factors. Volcanic activity and leaching of volcanic products are the most important contributors of chemical components of the Cenozoic evaporits in Northern Chile.

Introduction to saline deposits in the Central Andes

The Andes reach their wider dimension, some 700 km, between approximately 14° and 27° south latitude. From a geomorphological point of view, this region is characterized by a large number of endorreic basins many of them with a marked tectonic control. These basin systems are important local base levels for the drainages of the Pacific watershed. Outstanding examples are those of the Altiplano Boliviano and the Depresión Central of Chile which are situated mainly in the Pampa del Tamarugal area.

In Tertiary and Quaternary times, this region has witnessed intense volcanic and sedimentary activity. It included wide lacustrine systems that later on evolved to evaporitic basins. This process is still at work today and lakes, saline lakes and evaporitic basins can be profusely observed. In Argentina, Bolivia, Chile and Peru, the evaporitic basins including the saline bodies are called "salares", a term with a wide synonymy all over the world (i.e.: salinas, salt-flats, alkali-flats, salt-pans, playas, salt ponds, marsh-pan, takir, kavir, sebkhas, vloer, among many others) (STOERTZ & ERICKSEN 1974, SURIANO et al. 1980, CHONG 1984).

A salar s. str. is an evaporitic-detritic body sedimented in the lowest part of closed basins of desert or semi-desert environments. These basins, or basin systems, can be of tectonic origin or can form through volcanic activity (lava-flows acting as dumps, cauldrons or even craters). Its size can be a few square kilometres up to thousands of square kilometres. In simple terms, a salar forms by surface and underground inflows and subsequent water discharge through evaporation and evapo-transpiration. They are very "dynamic" sedimentary units due to climatic factors, tectonics and through a real "saline tectonic" caused by water movements. Its surface consists of saline or saline detritic crusts and efflorescences which show many different structures reflecting both these saline dynamics. Salars have ponds and underground water levels. Water in the basin and inflows form complex hydrologic systems. Water quality ranges from brackish to real brines and a hydraulical gradient concentrates the highest salinity in the centre. However, some salars can be absolutely dry. Especially in the older deposits, a zonation can frequently be observed: clastic material is graded according to grain size at the periphery and salts according to solubility in the basin itself (Fig. 1).

SALAR SCHEME

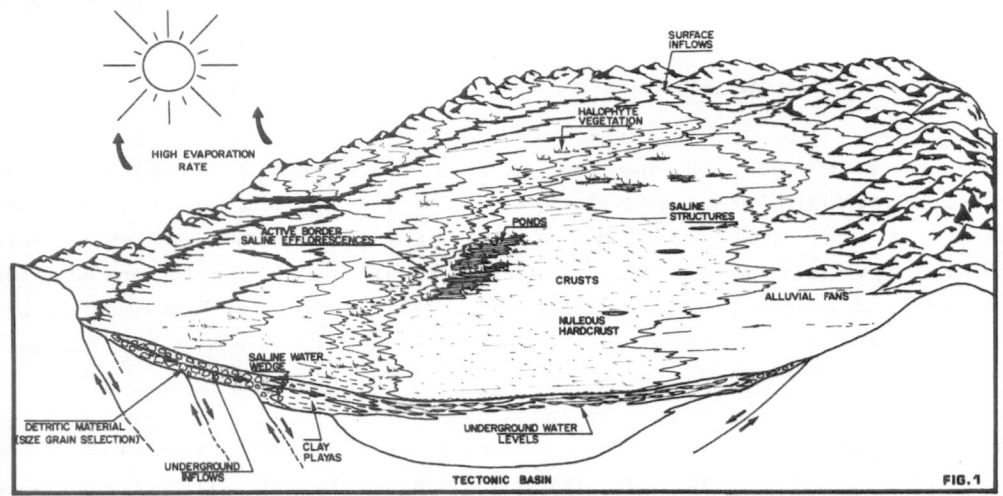

Fig. 1: Schematic model of a salar, displaying tectonic, sedimentary, hydrological, climatical and biological factors of its formation.

The "saline domain" of Northern Chile

Only a narrow belt of the described region known as Norte Grande de Chile, lies in Chile between 17° and 27° south lat. Its surface is roughly 250.000 km², extending from sea level to volcanic peaks at more than 6000 m above sea level; two areas can be distinguished in this region: the western band, known as Desierto de Atacama in s. str., with a reputation as one of the most arid areas of the world, and the eastern one with semi-arid characteristics which is called the Altiplano (+ Puna + High Andes) (Fig. 2).

Throughout this region there is a wide "Saline Domain" including about 10.000 km² of evaporitic basins of the salar and saline lake types, with an area of hydrological basins some three times larger in size. We have to add tenths of thousands of square kilometres of saline crusts and soils to these basins that act like a blanket on the Coastal Range, the Depresión Central and the pre-Andean slopes, the band of nitrate deposits and associated salt, the evaporites interbedded with sedimentary sequences (i. e. Cordillera de la Sal) and evaporitic horizons sedimented in marine "sebkha" environments (i. e. Península de Mejillones).

Many of these deposits (mainly salars) have been exploration targets in the last two decades due to their economic content of water and industrial minerals like lithium, potassium, iodine, sodium sulphate and boron among others (DINGMAN 1967, ERICKSEN et al. 1976, 1977). Salars are difficult to study because of their remote and inaccessible location and due to the fact that expertise from many different disciplines is needed to understand them (geology, chemistry, mineralogy, geomorphology, hydrology, biology and climatology - to name the main ones). In Chile this situation is even more complicated because the deposits like the Salar de Atacama, the Salar Grande, the Nitrate Deposits and the Lago Chungara, are unique in the world (VILA 1976) (Fig. 3).

Overview of the Cenozoic saline deposits in Northern Chile

In "Norte Grande de Chile" the conditions for the formation of saline deposits are optimal, the main ones being climate, geomorphology, geology and Cenozoic volcanism.

The climate is desert to semi-desert with some local areas of extreme aridity. However, this situation must be considered with caution since the meteorological stations are scarce, and reliable information is difficult to come by and does not allow one to make comparisons or to arrive at definite conclusions. We have to consider, therefore, that although aridity is quite evident and impressive thoughout the region, this aridity has very conspicuous features because the desert receives water through two exceptional meteorological phenomena. One of these is the dripping fog called "Camanchaca", considered to be a local coastal fog but that reaches more than 100 km inland. The second one is the "Invierno Boliviano" with plenty of rainfall, snow and electric storms between December and March in the Altiplano zone. The effects of this "winter" can reach very far to the west (i. e. up to the eastern part of the Coastal Range) as direct rainfall or as floods and mud flows. The regional slope to the west concentrates the surface and underground drainages in the local base levels of the arreic bassins, especially in those of the Depresión Central (Fig. 2). The stratigraphic record shows that this region has been arid, except for sporadic changes, since at least Upper Jurassic times. It is also evident that

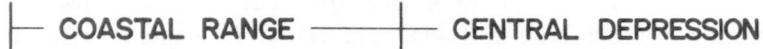

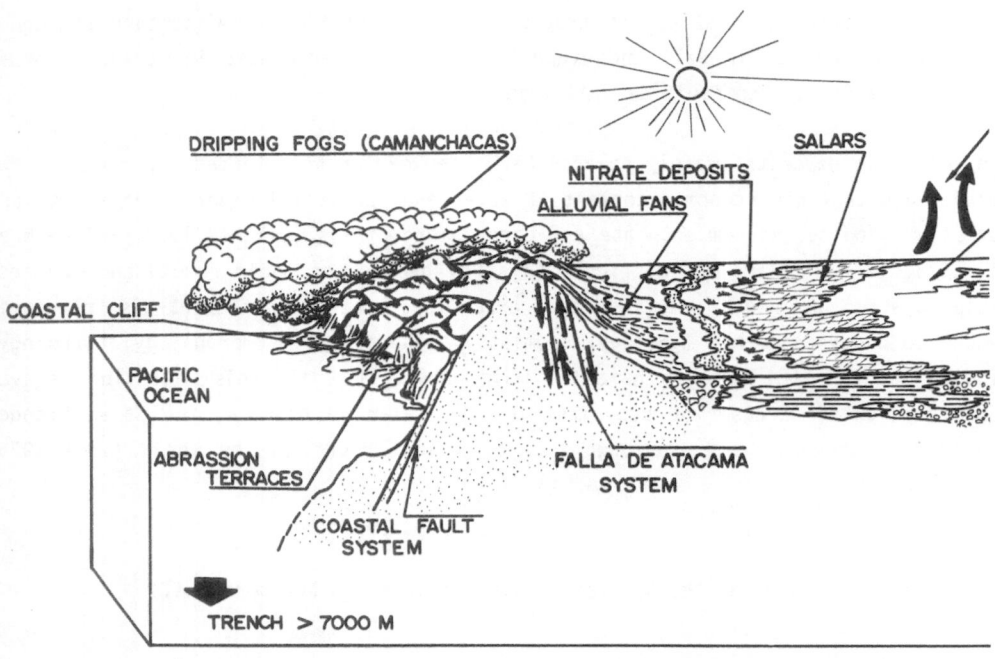

Fig. 2: Morphostructural cross section through the North Chilean Andes: recent climatic variation and salinal formation.

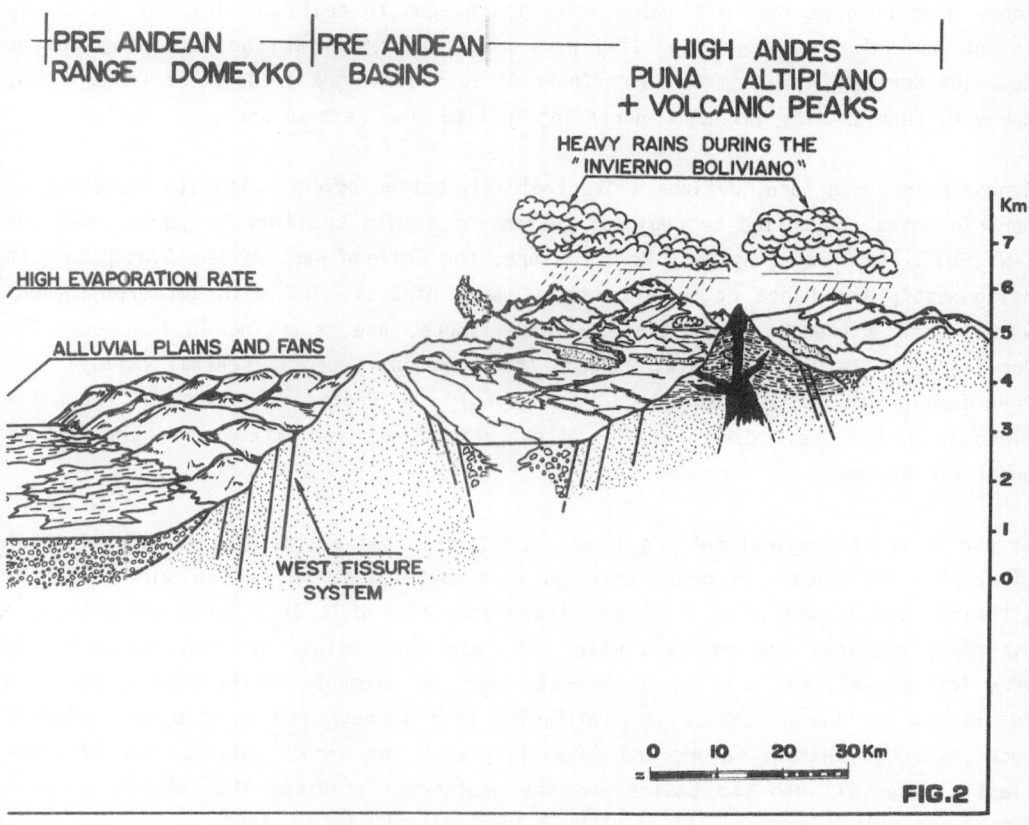

FIG.2

there have been periods with more water (i. e. due to deglaciation). In relatively
recent geological times, coinciding with other phenomena as the appearing of the
Humboldt Current, the climate in northern Chile became more arid. It is to be shown,
however, that aridity has been increasing in time from west to east..

Geomorphology, in turn, defines a "tectonically active desert" with its largest geo-
morphic units controlled by regional systems of faults trending in general N-S and
E-W (Falla de Atacama System; West Fissure, the Chilean part of the Olacapato-Toro
"alignment"; the Punta Negra alignment, among others). The main geomorphological
units, with a trench deeper than 7000 m offshore, are, starting in the west: the
Cordillera de la Costa (Coastal Range); the Depresión Central (Central Valley); the
Preandean Ranges (i. e. Cordillera de Domeyko); the Preandean Basins and the Altiplano (Fig. 2). Their distribution and characteristics facilitate the formation of
endorreic basins.

As far as the geological setting is concerned, it shows a wide distribution of stratigraphic and igneous records, both in time and space. We have a wide range of
lithology and abundance of outcrops, therefore, that will feed large quantities of
different chemical and detritic materials into the basins. Cenozoic volcanism is
very influencial, not only in the direct supply of elements to the basins, but also
in setting the basins themselves. Volcanism acts through the leaching of volcanic
rocks, through thermal waters and fumaroles, with the direct contribution of pyroclastic material into the basins and the geothermal gradient that should allow a
better and major leaching (ZEIL & PICHLER 1967, PICHLER & ZEIL 1969).

A great variety of clastic and chemical material is discharged into the basins
depending on the geological framework. Most important contributions are those of
Cenozoic volcanic activity. In short, the chemical components of saline deposits are
provided mainly by leaching, erosion and weathering of rocks, marine spray, dripping
fogs ("Camanchacas"), photochemical reactions and biological (microbiological) activity. Water and to a lesser degree wind are the main agents of transport and of
"saline tectonics".

It is difficult to classify these deposits because of the many and different parameters that can be used. Life is not made easier by the uniqueness of some of the
deposits in Chile. A main problem is the lack of general geological, geochemical and
hydrological data. In addition, one cannot find correlations with similar deposits
in Argentina or Bolivia in the literature.

The main guidelines that can be used to attempt a classification are the geological
setting, geographical location, geological age, mineralogy; nature of brines and/or
crusts, or economic aspects. It seems very clear to us that the classification of

saline deposits, including salars, will be controversial in the future and more or less definite decisions will be made only after many years of specific investigation. Meanwhile we use a preliminary scheme, modified from CHONG (1984)(s. a. Fig. 3):

1. Continental environments
   1.1 In basins of the High Andes:
      - Andean Saline Lakes and Andean Salars
   1.2 In Preandean basins:
      - Preandean Salars
   1.3 In the Depresión Central:
      1.3.1 Nitrates and Associated Salts:
          1.3.1.1 Alluvial
          1.3.1.2 In rocks
          1.3.1.3 In salars
      1.3.2 Sodium Sulphate Deposits
      1.3.3 Ulexite Deposits
      1.3.4 Salars s. str.
      1.3.5 Salars of the Pampitas Area
      1.3.6 "Domes"
   1.4 In basins of the Coastal Range:
      - The Salar Grande
   1.5 Miscellaneous deposits with different geographical setting:
      1.5.1 Saline soils or Saline Regoliths
      1.5.2 Gypsum Horizons ("Panquecues")
      1.5.3 Evaporites interbedded with continental sediments
   1.6 Playas
2. Transitional environments
   2.1 "Sebkhas"

Characteristics of the main deposits

Andean saline lakes and salars

We speak of an Andean Salar when the surface of its saline basin is covered to 50 % or more by solid facies (crusts, efflorescences, fine grain detritic material); in turn, a lake is a body of water in which the surface of "free water" is wider than 50 % of the basin. At times this difference is purely academic, since the basins are completely flooded during some seasons. Both types of deposits are quite similar and show the different stages in the evolution of a lake starting with fresh or brackish

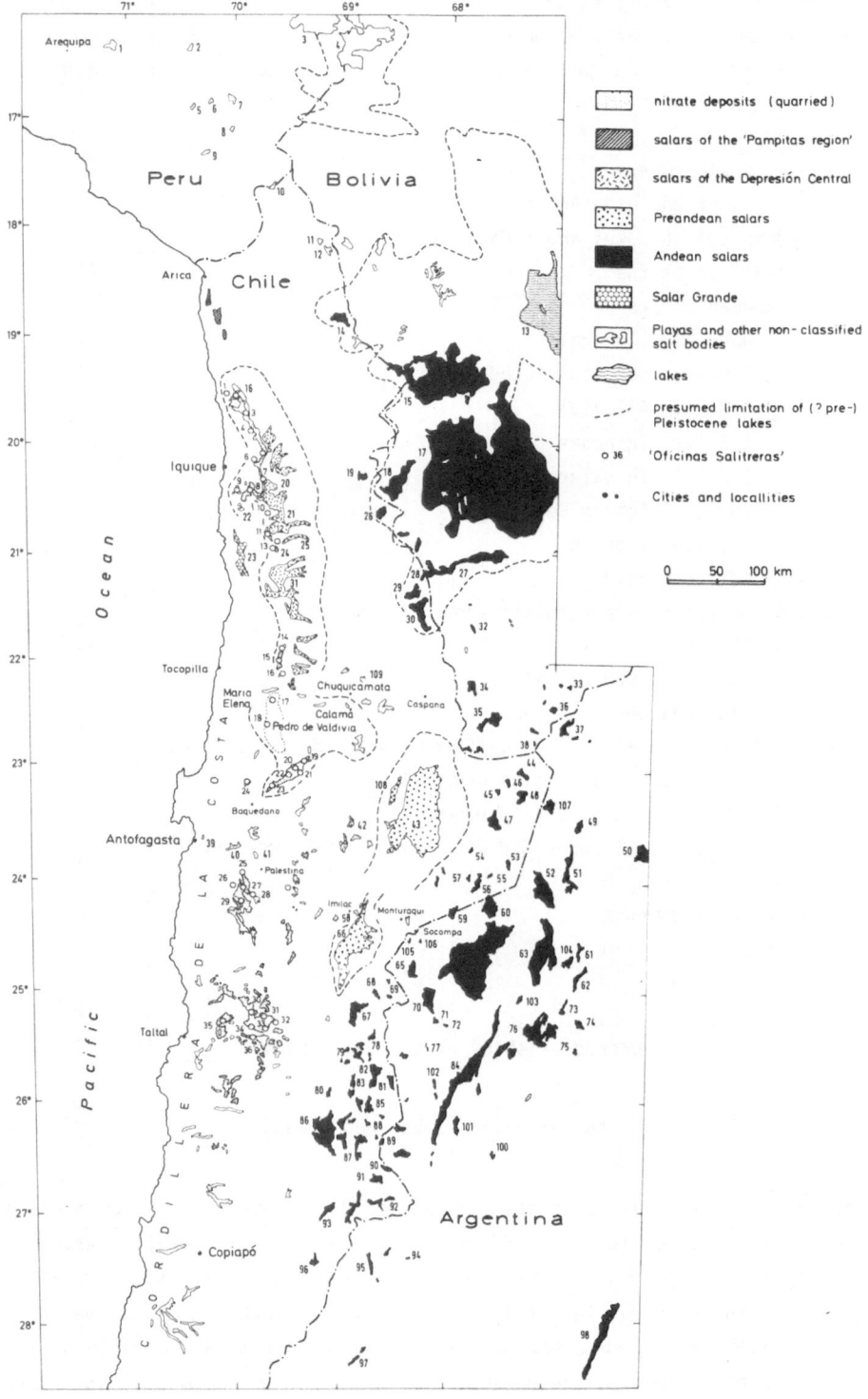

71°   70°   69°   68°

Arequipa

Peru   Bolivia

Arica

Chile

17°
18°
19°
20°
21°
22°
23°
24°
25°
26°
27°
28°

Iquique

Tocopilla

Maria Elena
Chuquicamata
Calama
Pedro de Valdivia
Caspana

Antofagasta

Baquedano

Palestina
Imilac
Monturaqui
Socompa

Taltal

Copiapó

Argentina

Ocean

Pacific

CORDILLERA DE LA COSTA

	nitrate deposits (quarried)
	salars of the 'Pampitas region'
	salars of the Depresión Central
	Preandean salars
	Andean salars
	Salar Grande
	Playas and other non-classified salt bodies
	lakes
	presumed limitation of (? pre-) Pleistocene lakes
o 36	'Oficinas Salitreras'
• •	Cities and locallities

0   50   100 km

salars and salt-lakes

(1 - 109)

1.- LAGUNA SALINA
2.- LAGUNA UCUMARINI
3.- LAGO TITICACA
4.- LAGO UINAMARCA
5.- LAGUNA SUCRE
6.- LAGUNA VIZCACHA
7.- LAGUNA DE LORISCOTA
8.- LAGUNA VILACOTA
9.- LAGUNA ARICOLA
10.- LAGUNA BLANCO
11.- LAGO COTACOTAN
12.- LAGO CHUNGARA
13.- LAGO POOPO
14.- SALAR DE SURIRE
15.- SALAR DE COIPASA
16.- SALAR DE OBISPO
17.- SALAR DE UYUNI
18.- SALAR DE EMPEXA
19.- SALAR DE HUASCO
20.- SALAR DE PINTADOS
21.- SALAR DE BELLAVISTA
22.- SALAR DE SORONAL
23.- SALAR GRANDE
24.- SALAR DE LAGUNAS
25.- SALAR DE SUR VIEJO
26.- SALAR DE COPOSA
27.- SALAR DE CHIGUANA
28.- SALAR DE OLLAGUE
29.- SALAR DE SAN MARTIN
30.- SALAR DE ASCOTAN
31.- SALAR DE LLAMARA
32.- LAGUNA PASTOS GRANDES
33.- LAGUNA ARENAL
34.- LAGUNA COLORADA
35.- SALLAR CHALLVIRI
36.- LAGUNA DE CURUTU
37.- LAGUNA DE VILAMA
38.- LAGUNA ZAPALERI
39.- SALAR DEL CARMEN
40.- SALAR DE NAVIDAD
41.- SALAR MAR MUERTO
42.- SALAR ELVIRA
43.- SALAR DE ATACAMA
44.- SALAR DE TARA
45.- SALAR DE PUJSA
46.47.- SALAR DE AGUAS CALIENTES
48.- SALAR DE QUISQUIRO
49.- SALINA OLAROZ
50.- SALINAS GRANDES
51.- SALAR DE CAUCHARI
52.- SALINA DEL RINCON
53.- SALAR EL LACO
54.- LAGUNA MISCANTI
55.- LAGUNA TUYAJTO
56.- SALAR DE AGUAS CALIEN-
TES Y PURISUNCHI
57.- SALAR DE TALAR
58.- SALAR DE IMILAC
59.- SALAR DE PULAR
60.- SALINA DE INCAHUASI
61.- SALAR DE PASTOS GRANDES
62.- SALAR CENTENARIO
63.- SALAR POCITOS
64.- SALAR DE ARIZARO
65.- SALINA DE LLULLAILLACO
66.- SALAR DE PUNTA NEGRA
67.- SALAR DE PAJONALES
68.- SALAR DE AGUAS CALIENTES
69.- LAGUNA DE LA AZUFRERA
70.- SALAR DE RIO GRANDE
71.- LAGUNA ARCHIBARCA
72.- LAGUNA VERDE
73.- SALAR RATONES
74.- SALAR DIABLILLOS

75.- LAGUNA DEL HOMBRE MUERTO
76.- SALAR DEL HOMBRE MUERTO
77.- LAGUNA DE LOS PATOS
78.- SALAR CE GORBEA
79.- SALAR DE AGUA AMARGA
80.- SALAR DE INFIELES
81.- SALAR DE LAS PARINAS
82.- SALAR DE LA ISLA
83.- SALAR DE LAS AGUILAS
84.- SALAR DE ANTOFALLA
85.- SALAR GRANDE
86.- SALAR DE PEDERNALES
87.- SALAR DE PIEDRA PARADA
88.- LAGUNAS BRAVAS
89.- LAGUNA WHEELRIGHT
90.- LAGUNA ESCONDIDA
91.- SALAR WHEELRIGHT
92.- LAGUNA VERDE
93.- SALAR DE MARICUNGA
94.- LAGUNA FRIAS
95.- SALINA DE LA LAG VERDE
96.- LAGUNA DEL NEGRO FCO
97.- LAGUNA BRAVA
98.- SALAR DE PIPANACO
99.- LAGUNA SANTA ROSA
100.- SALAR CARACHIPAMPA
101.- SALAR DE LA MINA
102.- SALAR EL FRAILE
103.- SALAR TALILLAR
104.- SALAR POZUELOS
105.- LAGUNA TECAR
106.- LAGUNA SOCOMPA
107.- SALAR DE JAMA
108.- SALAR DE LLANO DE
LA PACIENCIA
109.- SALAR CE TALABRE

oficinas salitreras

(1 - 37)

1.- TRINIDAD
2.- JASPAMPA, PACCHA
3.- SAN ANTONIO, COMPAÑIA, OREGON ENQUETA, SACRAMENTO, SAN PA-TRICIO, HERVATSKA, CAROLINA, CALIFORNIA, SANTA RITA, PORVE-NIR, SAN FRANCISCO, CAMIÑA, CONSTANCIA, SANTA CAROLINA, ANGELA, RECUERDO, LA PALMA, AGUADA, CONCEPCION, VICTORIA, HUASCAR, LA PATRIA, AGUADA, REDUCTO.
4.- MERCEDES, SALVADORA, DEMOCRA-CIA, AGUA SANTA, UNION, JOSEFI-NA, BARCELONA, ABRA, PROGRESO, PUNTUNCHARA, ROSARIO DE NE-GREIROS, PRIMITIVA, TRES MARIAS VALPARAISO, SAN JORGE, ROSA-RIO DE HUARA, PUNTILLA DE HUARA, SANTA ROSA DE HUARA, AMELIA Y AURORA, JOSEFINA, MAROUSSIA
5.- RAMIREZ, SANTIAGO, MAPOCHO, SAN DONATO, PEÑA GRANDE, SAN JOSE, KERIMA
6.- PEÑA CHICA.
7.- CALA CALA, BUEN RETIRO, CAR-MEN BAJO, LA SERENA, SAN MA-NUEL, SARA, TARAPACA.
8.- SEBASTOPOL, CHOLITA, PAPOSO, SAN ENRIQUE, TALUNA, SANTA CLARA, LA PERLA, ESMERALDA, RESTAURACION, IQUIQUE, SAN LO-RENZO, SANTA LUCIA, ARGENTINA SAN PABLO, VIRGINIA, SANTA CLA-RA, DIANA, VIS, ADRIATICO
9.- CONDOR, PIRINEOS, PROVIDENCIA, LA GLORIA
10.- DURRERA.   11.- ALIANZA, SCA-NONIA, BUENA VENTURA, PAN DE AZUCAR   12.- LA GRANJA
13.- NORTH LAGUNAS, CENTRAL LA-GUNAS, SOUTH LAGUNAS.
14.- SANTA FE   15.- IBERIA, GRUTAS PROSPERIDAD, RICA AVENTURA, BUENA ESPERANZA EMPRESA
16.- PEREGRINA, SANTA ISABEL
17.- COYA, ASTORECA, MARIA ELENA
18.- PEDRO DE VALDIVIA
19.- ACONCAGUA.   20.- CARMEN.
21.- FILOMENA, CURICO, MARIA
22.- LUISIS, CANDELARIA, ANITA, CECI-LIA, LEONOR
23.- AGUSTIN EDWARDS, EUSONIA, JO-SE SANTOS OSSA, CARMELIA, ARELIA, LASTENIA, FRANCISCO, PUELMA, CELIA, FLORENCIA, RIVIERA
24.- LA RIOJA.   25.- CASTILLA
26.- PAMPA RICA.   27.- LA AMERICANA.
28.- AVANZADA
29.- EUGENIA, PETRONILA, MARIA TE-RESA, SAN GREGORIO, LA VALPA-RAISO
30.- SALINITAS, MORENO, CHILE, ATACAMA
31.- ALIANZA, SANTA CAROLINA, LAUTARO
32.- LILITA, BALLENA, DELAWARE
33.- ESPERANZA, GHYSELA
34.- BRITANIA, FLOR DE CHILE
35.- SANTA LUISA, PORTEZUELO, ALEMANIA
36.- MIRAFLORES, TRICOLOR
37.- PISSIS, COCHRANE

Fig. 3: Distribution of the most important salars, saline lakes and saline deposits in S-Peru, N-Chile, SW-Bolivia and NW-Argentina (after CHONG 1984).

water (i. e. Chungara) and ending with a real saline lake (i. e. Lejía, Miscanti). Eventually, this leads to the development of a salar s. str. (i. e. Surire, Ascotán). The original lakes were larger, and paleocoast lines are a typical feature at their edges. Andean Lakes and Salars are located in the High Andes at 4000 m a.s.l. or more. Without exception they are affected by their Cenozoic volcanic environment which is illustrated by the fact that thermal springs occur even inside the basins. The climatic conditions are very rugged; temperature can fall to -35° C, and rainfall can reach tenths of millimetres within hours. The water inflows are brackish (up to 3500 ppm salts) and in the basins they become brines (50.000 ppm salts or more). Major ions in these brines are Na^+, Ca^{++}, Mg^{++}, K^+ with SO_4^{--}, Cl^- and $B_2O_3^{--}$; Sr, Li, Cs, Rb, As are minor or trace elements. Some of the more common minerals are gypsum, halite, thenardite and ulexite. Mineralogy, however, recognizes in addition more than 40 minerals, some of rare occurrence.

Preandean salars

Located in tectonic basins, they are the largest of the country, are older than most and have a diverse geology. Almost all inflows come from the High Andes. Due to the age of these salars, however, the brines can become enriched with up to 350.000 ppm salts and ions. Mineral composition is the same as that found in the High Andean deposits. A typical feature is the zonation of detritic material (according to its grain size) at the periphery of the salars and of the salts (according to their solubility) in the basin itself. Except for the Salar de Atacama where the thickness of the halite central nucleus has been shown to reach 1000 m in places, the thickness of the saline bodies is generally unkown.

Nitrates and associated salts; thernardite and ulexite deposits

In the Atacama Desert between 17°30' and 26°00' Lat. and along the western and eastern borders of the Depresión Central and Coastal Range respectively, one can recognize a 700 km long band which is ten to tenths of km wide. This band is characterized by an exotic saline assemblage of sulphates, chlorides, nitrates, borates, carbonates, iodates and chromates with Na, Ca, Mg, K as major cations. They are called Nitrate Deposits since nitrates are exploited in this assemblage. Considering the abundance of other salts, however, one should speak of Nitrate Deposits "and associated salts". Salt ore appears as filling material in cavities and pores, or as the cement of detritic sediments (locally with exclusively volcanic detritic material) forming the so called Alluvial Deposits. When ore appears like veins or irregular bodies of very pure nitrate, they are called Deposits in Rocks. Finally,

nitrates can also reach exploitable grade in the crusts of some salars of the Depresión Central (Deposits on Salars). The genesis of some of the rare salts in these
deposits (i. e. nitrates, chromates, iodates, perchlorate) is still unclear. The
concentration mechanisms of ores and their age still require explanation in spite of
more than one hundred years of geological research in this area. Our evidence suggests a major volcanic origin of elements, a mechanism of concentration and deposi-
ting through water, fractional "vertical" crystallization and its subsequent preservation because of the arid environment and a possible "sealing" effect at the surface (ERICKSEN 1983).

In the upper parts of the alluvial nitrate deposits one finds centimetre-sized
lenses which consist of monomineral horizons of sulphates (thenardite), borates
(ulexite) or complex nitrates (humberstonite). In places these horizons can be
interpreted as salts that leached from the upper levels. Thenardite and ulexite are
exploited intensively, and it seems that they are concentrated in the northern and
southern parts of the nitrate belt (Aguas Blancas and El Toco, respectively).

Salars of the Depresion Central

These salars are located on the western border of the Depresión Central and form the
distal part of large alluvial fan systems. They border on the eastern part of the
Coastal Range. Their inflows originate from the High Andes and in their crusts and
brines we find mainly sulphates and chlorides of Na, Ca, Mg and K. Borates and other
salts are also present. Some typical features of these salars are their dry crusts,
the relatively large total thickness of their deposits, the occurrence of shallow
underground water and different types of basins (i. e. Salar de Pintados, Salar
Yungay).

Salars of the Pampitas

Between 18°00' south lat. (approximately Arica) and 19°30' south lat. (approximately
Pisagua) the Depresión Central is deeply dissected by tectonic valleys with E-W
trend. We can observe partial plains ("Pampitas"), therefore, which are remnants of
the former homogeneous Depresión Central and which are now separated by canyons with
depths of more than 700 metres. Before the dissection, saline deposits of the salar
type were sedimented in synchrony with the older stages of the salars of the
southern Depresión Central. These deposits reached only an "embryonic" stage, since
they were abruptly cut by the drainage of primary fluvial systems. Some of their
charateristic features are shallow crusts consisting of sodium chloride, thickness

in the range of metres, a typical cover of anhydrite nodules and their close relationship with volcanic ashes (i. e. Pampa Chiza; Pampa Camarones).

"Domes"

To the north of the Quillagua Valley (21°37' south/68°35' west) there are some irregular halite bodies in outcrops of more than 50 km² (i. e. Lomas de la Sal). Their
geographical distribution is irregular but they are a more or less local phenomenon.
We have evidence to show that they are in fact massive halite (thickness at least
50 m) with insignificant amounts of gypsum-anhydrite (? possibly with an irrelevant
gypsum-anhydrite cap?). The outcrops seem to resemble saline domes but further
research is needed to clarify this point. An alternative hypothesis would be that
these outcrops derive from salar halite nuclei similar to the ones found in the
Salar de Atacama of today.

The Salar Grande

This is the only saline body of this type in the world. It is a tectonically active
basin filled with almost pure halite (more than 95 % NaCl) covering an area of
4x35 km minimum and a measured depth of, at least, 100 m. The only other minor constituents (less than 1 %) are some thenardite and possibly gypsum. It includes an
isolated horizon of thenardite with interspersed euhedral crystals. This is a
"fossil" salar without brine levels or interstitial water between halite crystals
except for the occasional fluid-filled, millimetre-sized bubbles. Several theories
have been discussed to explain the genesis of the Salar Grande.

Saline soils or/and regoliths

Along the Coastal Range a pink to light brown saline blanket covers large areas with
peaks and slopes in addition to the basins. Its thickness can reach up to 40 cm. It
consists of gypsum, minor halite and possibly other saline constituents. Salts are
disseminated in a frail mixture of fine grain sediments (clays and silts). This
crust is homogeneous, but includes some open cavities, metres to decametres in diameter, with halite at the bottom. We think that this crust is formed by salt deposits through dripping fogs ("camanchacas") associated with a process of destruction
of the elastic material by saline attrition. There is a close relationship between
the areas with this saline blanket and the location and permanence of dripping fogs.

"Panqueques"

In the Coastal Range, the Depresión Central, and in some locations of the pre-Andean zone we found a hard, light brown, homogeneous gypsum horizon (similar to a B soil horizon). The thickness of this crust is 40-60 cm. We suggest that it was formed through leaching and differential crystallization starting with liquid precipitation (or possibly dripping fogs in recent geological times). This crust has a close relationship with nitrate deposits to the degree that miners created the name "panqueque" for it. We think that these "panqueques" aid the preservation of nitrate deposits by acting like a seal and thus reduce the effects of weathering and erosion.

Evaporites interbedded with continental sediments

In Lower Tertiary sequences we find large quantities of halite and gypsum and other minor saline constituents which are interbedded with mainly sandstones, shales and volcanic ashes. Part of these salts are disseminated in the sediments through leaching and weathering processes. Sedimentation took place in braided rivers and in alluvial plains under severe arid conditions. The best examples are the sequences of the San Pedro Formation in the Cordillera de la Sal - San Bartolo area and in some isolated outcrops north of the Salar de Punta Negra.

Playas

We define playas as those basins or alluvial plains where fine grained materials (clays, silts) have been deposited and still are sedimented. These sediments represent the finer fraction of the non-soluble material. Salt content is irrelevant but can outcrop with sporadic floods which cover the basins with centimetre thick efflorescences. They are located everywhere between the coast and the High Andes.

Sebkhas

In Miocene-Pliocene times a general regression took place along the coast of northern Chile. Terraces of marine fossiliferous sediments are testimony of this. In certain locations one can observe gypsum/anhydrite bodies interbedded with sandstones. Most likely, these evaporites were sedimented in transitional environments of the marine "sebkha" type.

Biological and paleontological aspects

Salars and their basins are regions with abundant biological activity, certainly due to the fact that these areas are true oases in large desert zones. Macro- and micro-organisms occur in direct relation to water salinity, and there are different forms of life adapted to survive under extreme conditions (some plants and many algae and bacteria). As in the case of Lago Chungara, we might find places with exceptional and unique forms of life.

The paleontological aspects of these deposits are quite unknown. Remnants of Megha-terium have been found on Salar de Bellavista, fossil logs in salars of the Pampitas area. The ostracodes of the saline sequences of Lower Tertiary age are presently studied, and recently we have found ostracodes associated with diatomites in upper Tertiary-Quaternary basins. We expect in the next future to investigate the sedimen-tary sequences of the Río Loa, María Elena and Quillagua basins with emphasis on fossils.

Economic aspects of saline deposits

Industrial and mineral rocks have been exploited in Chile for more than a century and this exploitation has played and continues to play an important role in the economic history of the country. Some time ago, Chile has held the first place in the world as boron, nitrate and iodine producer. Today the Chilean contribution is irrelevant compared to global production, but Chile is still exporting lithium from brines, boron, nitrates, iodine, sodium sulphate and diatomites, and potash (in the next future). There are extensive resources of sodium chloride and calcium sulphate. The future of a chemical industry using these ores, and others without present use (Mg), or with emerging potential (REE) is very promising. The use of salars in high-ly sophisticated technologies as there are runways for space ships, horizontal antennae, fluid hydrocarbon reservoirs, radioactive material reservoirs or energy producers has not been considered so far.

Acknowledgement

This article summarizes results of different research programs on saline deposits of the Andes, most of them in Chile, carried out by the author in the last ten years and supported by the Humboldt Stiftung (Federal Republic of Germany). It also reports on work in progress which is supported by the Fondo Nacional de Investiga-ción Científica y Tecnológica of the Chilean Government (FONDECYT) and the Universi-dad del Norte (Chile).

References

CHONG, G. (1984): Die Salare in Nordchile - Geologie, Struktur und Geochemie. - Geotekt. Forsch. **67**, 146 pp., Stuttgart.

DINGMAN, R. (1967): Geology and groundwater resources of the northern part of the Salar de Atacama, Antogafasta Province. - U.S. Geol. Survey, Bull. **1219**, 49 pp., Washington.

ERICKSEN, G. E. (1983): The Chilean Nitrate Deposits. - Amer. Sci. **71**, 366-374., New Haven.

ERICKSEN, G. E., CHONG, G. & VILA, T. (1976): Lithium resources of Salars in the Central Andes. - U.S. Geol. Survey, Prof. Paper **1005**, 66-74., Washington.

ERICKSEN, G. E., VINE, J. & BALLON, A. (1977): Lithium rich brines at Salar de Uyuni and nearby Salars in southwestern Bolivia. - U.S. Dept. of Interior Geol. Survey Project Report. Bolivian Invest. (IR), Bol. **7.**, Washington.

PICHLER, H. & ZEIL, W. (1969): Die quartäre "Andesit"-Formation in der Hochkordillere Nord-Chiles. - Geol. Rdsch. **58** (3), 866-903., Stuttgart.

STOERTZ, G. E. & ERICKSEN, G. E. (1974): Geology of Salars in northern Chile. - U.S. Geol. Survey, Prof. Paper **811**, 65 pp., Washington.

SURIANO, J. M., KIMSA, J. F. & BRODTKORB, A. (1980): Características geoquímicas generales de agua y salmueras de la Puna Argentina. - Acad. Nac. Ciencias, Córdoba, Miscelánea, n. **63**, 38 pp., Córdoba.

VILA, T. (1976): Modelo de distribución y origen de algunos elementos en salmueras de depósitos salinos andinos, Norte de Chile. - Actas Primer Congr. Geol. Chileno, Tomo II, E65-E82, Santiago.

ZEIL, W. & PICHLER, H. (1967): Die känozoische Rhyolith-Formation im mittleren Abschnitt der Anden. - Geol. Rdsch. **57**, 48-81., Stuttgart.

GEOMORPHOLOGICAL WEST-EAST-SECTION
THROUGH THE NORTH CHILEAN ANDES NEAR ANTOFAGASTA

Gerhard Abele
Geographisches Institut, Universität Mainz
Saarstraße 21, D-6500 Mainz

Abstract

The extremely arid western slope of the Andes near Antofagasta is an
area of young tectonic movements on the one hand, but of very low erosion
rates on the other. Therefore the relief is characterized by very old
erosional forms, but fresh and clearly visible tectonic forms. This
is quite evident from the following characteristics of a west-east-
section from the Coastal Cordillera to the High Cordillera: 1. The great
height and steepness of the western escarpment of the Coastal Cordillera, devoid of deep valleys. 2. The preservation of the miocene and
pliocene caliche deposits with highly soluble nitrates in the eastern
part of the Coastal Cordillera and of the landforms covered by them.
3. The preservation of the nearly continuous "dam" of the Coastal Cordillera, holding back the sediments from the Precordillera and the High
Cordillera in the Longitudinal Depression. 4. The existence of comparatively deep endorheic salar basins directly at the foot of the very
high mountains of the High Cordillera. 5. The preservation of thin but
vast miocene and pliocene ignimbrite and tuff sheets even on the western
slope of the High Cordillera.

Introduction

In the Antofagasta area of northern Chile (Fig. 1) one of the driest
deserts (perhaps the driest) and the highest slope on earth (nearly
15 000 m between the High Cordillera, Llullaillaco 6723 m, and the Atacama Trench with a depth of 8066 m) come together. Because of the extreme and long-term aridity erosion has not kept up with the young tectonic uplift of the Andes thus preserving the high average altitude of
this part of the Cordillera. The bottom of the sea in the Atacama Trench
in turn is so deep not only because of subduction, but also because
of the poor supply of continental sediments. Accordingly the sediment
cover in the trench is very thin, getting thicker to the north and south,
where the climate has been more humid and the dissection of the cordillera more active (HAYES 1974, p. 586, ZEIL 1979, p. 30 and ZIEGLER et
al. 1981, p. 256).

Lecture Notes in Earth Sciences, Vol. 17
H. Bahlburg, Ch. Breitkreuz, P. Giese (Eds.),
The Southern Central Andes
© Springer-Verlag Berlin Heidelberg 1988

The higher the Central Andes were lifted up the more effective became the barrier effect towards the more humid air east of them (WEISCHET 1966, p. 6 f), the more steady became the East Pacific Anticyclone, above all its eastern fringe (TREWARTHA 1961, p. 28 ff), and the more constant became the southerly wind along the North Chilean coast thus increasing the upwelling effect in the Pacific Ocean as well as the divergence effect (LYDOLPH 1957 and BRYSON and KUHN 1961) in the atmosphere. These effects brought forth an even greater aridity on the western flank of the cordillera, which in turn hampered erosion even more.

The combined influence of young tectonic movements and long-term aridity did not only affect the North Chilean Andes on the whole but also in its different parts as will be demonstrated in the following west-east-section from the Coastal Cordillera through the Longitudinal Depression, the Precordillera (Cordillera Domeyko), one of the intra-andine basins (Salar de Atacama) to the High Cordillera (Fig. 1)[1].

1. The Coastal Cordillera

In its central and eastern parts the Coastal Cordillera is characterized by conspicuously smooth landforms. On the eastern slope they are partly covered by the widespread caliche deposits containing the once economically important saltpeter. The deposition of these highly soluble nitrates starting in miocene (ERICKSEN 1983) and their preservation until today is ample proof for long-term aridity. Certainly there existed periods of a moister climate, but they were not humid and/or long enough in order for the saltpeter deposits to dissolve and for the landforms covered by them to be eroded (ERICKSEN 1983). The great extension of the caliche on the surface therefore gives evidence of the very old age of the relief. The smooth landforms end abruptly in the west at the very steep pacific slope of the Coastal Cordillera, formed by marine erosion. According to MORTIMER (1980, p. 7) this abrasion was decisively supported by a long-term subsidence in the Atacama Trench ("tectonic erosion" according to ZIEGLER et al. 1981, p. 253). The impressive escarpment would have been worn down and dissected to a greater extent under more humid conditions (MORTIMER and SARIC 1972), all the more as presently there is no active cliff-formation at the foot of the slope itself along great parts of

[1] Field studies in 1985 and 1987 were financially supported by the Deutsche Forschungsgemeinschaft. In 1985 a cross country vehicle of the "Forschergruppe Mobilität aktiver Kontinentalränder der FU und TU Berlin" was put at my disposal. In 1987 I had the opportunity to accompany the group led by Prof. Görler (Institut für Geologie, FU Berlin) for several days.

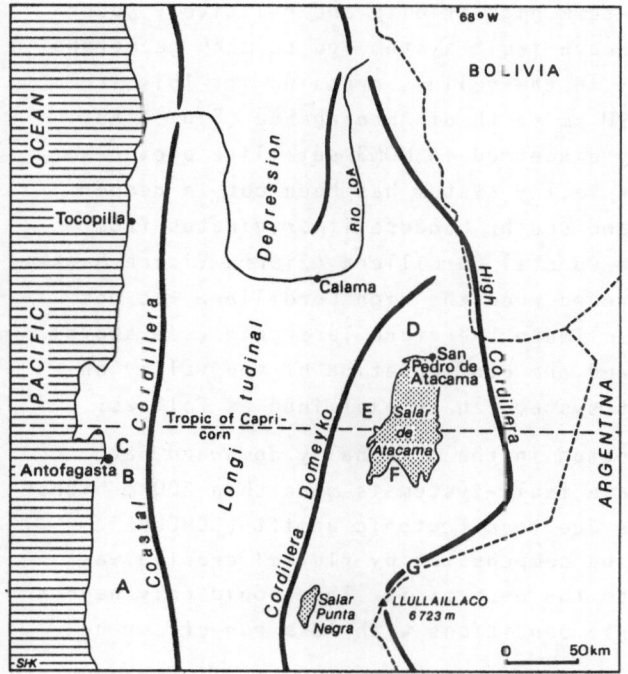

Fig. 1   Location map of the Andes
          near Antofagasta

the North Chilean coast.
There are even marine
terraces of the penulti-
mate or perhaps ante-
penultimate interglacial
(RADTKE 1985) and very
old alluvial cones
(PASKOFF 1978/79, p. 12)
at the foot of the escarpment. Retrocession of the
steep slope can therefore
not be very active at
these places. In the long
run it was active enough,
however, to produce the
conspicuously sharp edge
which separates the es-
carpment from the smooth
forms further east. The
great height of this slope
of up to more than 1000 m
is not only due to the tec-
tonic uplift but also to the lack of wearing down on the surface of the
Coastal Cordillera.

In spite of the height and steepness of the escarpment the short and narrow gorges cut into it are not deep, and north of the mouth of Río Loa
(Fig. 1), where rain is even more scarce and the Coastal Cordillera is
less high, there is nearly no autochthonous incision at all. The even
greater lack of rainfall in the Coastal Cordillera north of Río Loa is
due to the fact that the divergence effect increases to the north towards Arica, because the High Cordillera draws much nearer to the coast.

The extreme lack of subaerial erosion in the Coastal Cordillera led to
the fact that relief is very closely adapted to the young tectonic movements (MORTIMER and SARIC 1972). This is quite evident at the Atacama
fault (CHONG 1984, p. 30) stretching north-south in the hinterland of
Antofagasta and causing a young vertical displacement of pediments, bajadas and slopes of several meters. Even here the very clear shape of
the fault scarp is not only a consequence of the young tectonic movement, but also of the climatical conditions favouring preservation.

The combined influence of long-term high aridity and relatively young tectonic uplift west of the Atacama fault system led to many geomorphological consequences, above all in the valleys crossing it: This is the case with a valley system 90 km south of Antofagasta (Fig. 1 A), whose E-W-course can be clearly discerned in MOMS satellite pictures (SCHEUBER 1987, p. 140 f). This valley system has been cut in deeper than other ones further north and south, because it originates from a particularly high area in the Coastal Cordillera (Sierra Vicuña Mackenna, 3114 m). It was probably eroded when the High Cordillera was not as high as it is today and the climate therefore less arid (see above). The differences of the height and the configuration of the valley on both sides of the Atacama fault system can be explained as follows:

1. The fact that the valley bottom in the originally downward part of the valley west of the Atacama fault system is more than 200 m higher than in the part east of it, is due to a tectonic uplift (SCHEUBER 1987, p. 140), which could not be compensated by fluvial erosion west of the fault or sedimentation to the east of it. This could only happen by a climatic change to more arid conditions with less run-off or no river at all as nowadays.

2. West of the Atacama fault system the surface of the Coastal Cordillera is higher and the incision of the valley deeper than east of it. Thus the uplift of the western part of the Coastal Cordillera must have already been active during the formation of the valley, i.e. during the more humid period. Fluvial run-off was strong enough at that time so that the uplift could be matched by erosion.

3. The part of the valley west of the Atacama fault system directly joins that east of it. Therefore there cannot have been any horizontal shifting along this fault system (SCHEUBER 1987, p. 140) since the beginning of the formation of the valley.

About 9 km west of the Atacama fault and 1550 m above sea level the generally smooth landforms of the valley system end abruptly to be continued by Quebrada Remiendos with its steep gradient to the sea, due to the tectonic uplift of the Coastal Cordillera. The great inclination of this gorge allowed fluvial erosion even under arid conditions. Thus the young drainage system has been capable of "intruding" into the old one by headward erosion. Additionally and from a more general point of view the old valley may have lost its lower course by the recession of the western escarpment due to "tectonic erosion" (ZIEGLER et al. 1981) and marine abrasion (MORTIMER 1980, p. 20).

There are similar valleys crossing the Atacama fault and losing their
lower course to a younger drainage system further north. In Quebrada La
Negra south-east of Antofagasta (Fig. 1 B) the young headward erosion
has reached the area east of the Atacama fault. That is the reason why
there is no reversed gradient at the fault system itself. North-east of
Antofagasta the erosion in Quebrada Caracoles (Fig. 1 C) was not active
enough to do so. Therefore a playa was dammed up east of the Atacama
fault (MORTIMER 1980). In contrast to the old valley east of Quebrada
Remiendos this drainage system, coming from the Longitudinal Depression,
carried enough material to fill the depression east of the fault. Thus
a widespread accumulation plain was formed continuing to the south towards the drainage area of Quebrada La Negra.

2. The Longitudinal Depression

The Longitudinal Depression has been filled with sediments from the
Precordillera and High Cordillera up to more than 1000 m above sea level.
In its northern part (Pampa del Tamarugal) radiometric dates assign
this aggradation to the oligocene and miocene (PASKOFF and NARANJO
1983; NARANJO and PASKOFF 1985). The sediments could only have attained
such a high level because the Coastal Cordillera acted as a continuous
dam (MORTIMER and SARIC 1972, p. 167). The lack of deep incisions in
the Coastal Cordillera near Antofagasta is due to the aridity of the
High Cordillera so that apart from the Río Loa no river reaches the sea
(Fig. 1). The Río Loa manages to do so because it obtains enough water
in its long N-S-course between the High Cordillera and Precordillera
(WEISCHET 1970, p. 230), where evaporation is reduced by the permeable
ignimbrites. Its discharge has not been high enough, however, to cut
a deep valley into the Coastal Cordillera. Downward erosion has been
additionally hampered by the fact that the Río Loa has not found the
nearest access to the sea. After a S-N-detour of about 100 km in the
Longitudinal Depression it has lost much of its gradient.

South of the Río Loa, where the High Cordillera is at its driest, no
river at all crosses the Precordillera, the Longitudinal Depression
and the Coastal Cordillera.

3. The Precordillera (Cordillera Domeyko)

In contrast to the High Cordillera there is no young volcanism in Cordillera Domeyko. Although attaining great heights (3500-4250 m) it is

devoid of deep valleys. Relief is dominated by vast pediments and bajadas, which partly come down from about 3000 m above sea level without
being dissected. Rivers flow down only on rare occasions and then in
shallow channels, finding their way in the depressions where the bajadas
meet, and ending in wide playas. This was the case after the exceptional
rains in 1987, when many avenidas (debris flows, mudflows and rivers)
flowed down to the Llano de la Paciencia (Salar de Atacama basin).

From all the slopes of Cordillera Domeyko, its eastern flank, descending to the intraandine Salar de Atacama basin is the most prominent
one. Its stepped pediment and bajada levels can be used to reconstruct
different stages of geomorphological development (see fig. 2, which
is only the first attempt of a geomorphological model, designed by com-
bining profiles between D and E in fig. 1. The geological content of
the model is only meant to be a help to explain the geomorphological
sequence and therefore should not be taken too seriously). The oldest
stage can be traced by the conspicuously smooth surface, covering great
parts of the Cordillera Domeyko at about 3000 - 3300 m above sea level.
In the east this surface ends at a sharp edge, giving way to the western
slope of the salar de Atacama basin. At this edge there are subangular
gravel deposits up to 150 m thick, overlying the folded Purilactis
Formation in angular unconformity (Tambores Formation according to
RAMIREZ and GARDEWEG 1982, p. 28-30 and MARINOVIC and LAHSEN 1984, p.
40-42). Considering their great N-S-extension of at least 150 km with
only few and short interruptions, and their interfingering with fine-
grained and evaporitic San Pedro formation in the present Salar de Atacama basin in the east (MARINOVIC and LAHSEN 1984, p. 41), they can
only have been transported from the west. This can be confirmed by paleocurrent evidence (FLINT 1985, p. 536). As Tambores Formation was deposited together with the evaporites of San Pedro Formation and as its
bedded sediments are similar to the bajada sediments accumulated under
present arid conditions, it is highly probable that it was also deposited in bajadas. Accordingly the surface of unconformity underlying
the gravels must have once been a pediment inclined to the east.

Provided that these assumptions, as well as the datations and correlations of MARINOVIC and LAHSEN (1984, p. 41) and RAMIREZ and GARDEWEG
(1982, p. 29) are right, the following stages of the geomorphological
development can be distinguished:

1. Formation of a pediment in angular unconformity to the folded Purilactis Formation dipping towards the evaporites in the east (fig. 2, profile 1).

2.   Subsidence in the present Salar de Atacama basin and in the area
of the pediment west of it (fig. 2, profile 2). Accumulation of San
Pedro and Tambores Formation (according to MARINOVIC and LAHSEN 1984,
p. 41, during oligocene or lower miocene). Due to the large quantity
of coarse sediments a higher area must have been eroded over a long
period of time further west. There are especially big crystalline boul-
ders in Tambores formation near Quimal (4278 m) so that it is assumed
that a precursor of this mountain existed at that time.

3.   Uplift of the pediments and bajadas west of the present salar, increase of their inclination in the eastern parts and erosion of a
great part of Tambores Formation (fig. 2, profile 3). In contrast to
this, the inclination of the western part of the pediment and bajada
area diminished until it was horizontal. At some parts it even has a
slight dip to the west. Assuming that the old pediments and bajadas
had the same angle as those of today ($\pm 3°$) the dip must have changed
here about 3°.

4.   Formation of a new pediment and bajada system according to the higher
gradient of the eastern part of the slope (fig. 2, profile 4). Thereby
the pre-tambores pediment has been worn down to a somewhat lower level.
Thus west of the new pediment a scarp was left behind, mostly consisting
of Purilactis Formation in the lower and Tambores Formation in the
higher parts (El Bordo). At some places the pre-tambores pediment was
dissected by selective erosion bringing forth hogbacks. Thereby a wide
depression was formed in the Quebrada del Diablo area between the hogbacks of the Cerros de Purilactis and the Cordón Barros Arana. This
depression is not a simple valley, but it belongs to the new pediment
and bajada system instead, as can clearly be seen at its surface of
unconformity and its bajada sediments covered by the ignimbrites of
the following stage.

5.   Ignimbrites spread over vast areas of nothernmost Cordillera Domeyko
(BRÜGGEN 1950, p. 134) including the wide depression near Quebrada del
Diablo (DINGMAN 1963, p. 23). Thereby the bajadas of the preceding
stage (no. 4) were covered concordantly (fig. 2, profile 5; photo 1).
According to MARINOVIC and LAHSEN (1984, p. 54) this cover belongs to
Sifón ignimbrite with a radiometric age of $8,5 \pm 0,25$ m.y. B.P. (BAKER
1977, p. 458). Thus a miocene pediment and bajada area has been preserved until today. This excellent preservation was certainly favoured by
the bajada sediments covering the ignimbrites. The large ignimbritic
blocks in these upper bajada sediments were eroded in the higher parts of
Sifón ignimbrite.

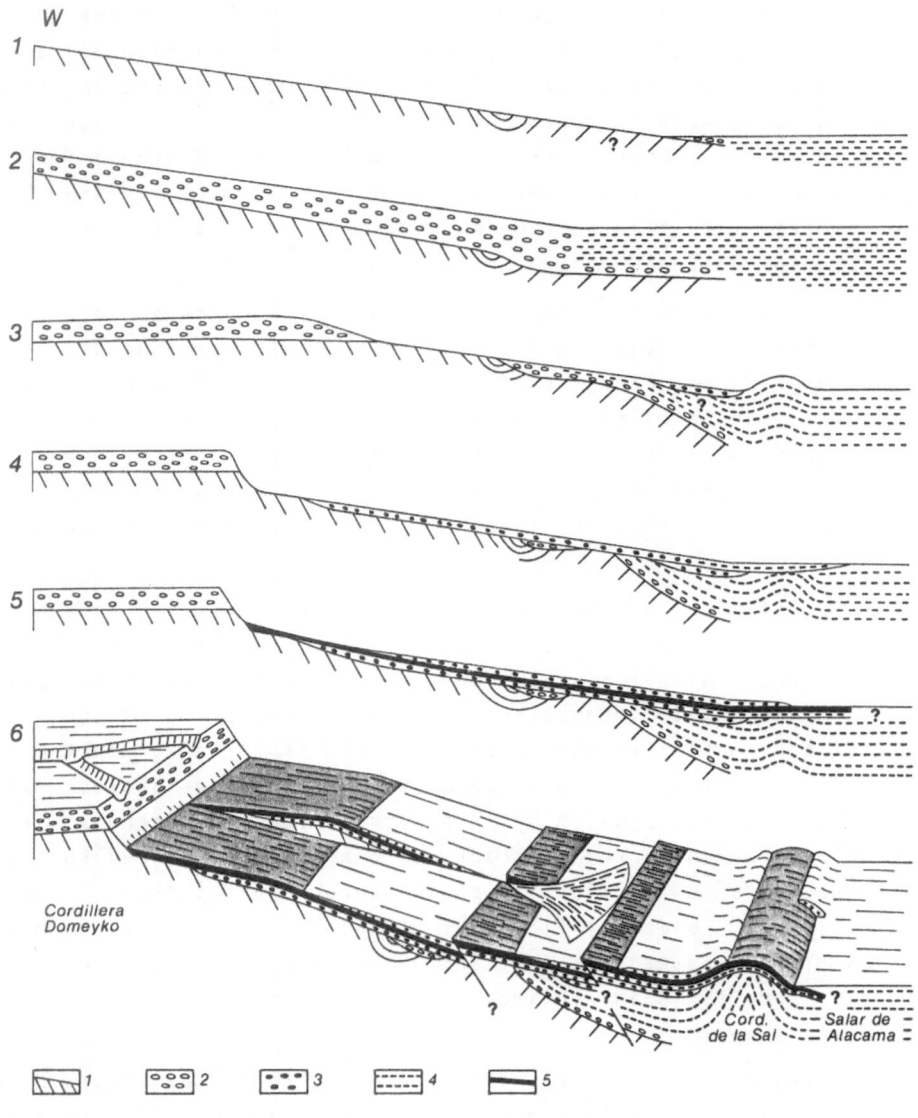

Legend: 1. Purilactis Formation
2. Tambores Formation
3. bajada sediments younger than Tambores Formation
4. San Pedro Formation and fine-grained material of younger age
5. Sifón ignimbrite

Fig. 2: Geomorphological development at the eastern slope
of northern Cordillera Domeyko, attempt of a model.

Photo 1   Sifón ignimbrite concordantly overlying bajada sediments,
          which in turn overlie the folded Purilactis Formation in
          angular unconformity near Quebrada del Diablo. The gradient
          of the bajada and the igrimbrite sheet increases towards the
          Salar de Atacama (in the background together with the Cor-
          dillera de la Sal).

6.   Tilting of the pediments and bajadas of stage 5 towards the salar.
This can be clearly seen near Quebrada del Diablo (fig. 2, profile 6).
Here the gradient has increased from about 3° (if we assume the average
inclination of the present pediments and bajadas) to 5° in the upper
parts and to 6° in the middle parts. Further down the bajadas and ignimbrites have been even dislocated by faults. The increase of the inclination of the slope has led to the dissection of the upper parts
of the pediments and bajadas (photo 1) and to a further accumulation
in the lower part of the bajadas. Thus a new pediment and bajada level
is presently formed. Further south the dissection of the old level as
well as the formation of the new pediments and bajadas have arrived
at a more advanced stage than at profile 6.

At other places Sifón ignimbrite was certainly eroded to a greater extent. It has partially vanished for instance northwest of Barros Arana
at about 3 200 to 3 300 m above sea level. There it was eroded together
with its basement rock by autochthonous streamlets. The valleys thus

formed are so close to each other that the original adaptation of the
ignimbrite to the preexisting landforms is much less conspicuous than
that of pliocene Tucúcaro ignimbrite (see below).

West of Cordón Barros Arana a pit without any outlet and a depth of
about 100 to 150 m has been formed between Tambores Formation in the
west and Purilactis Formation in the east. The horizontally concave
curvature of the slope of the pit on the Tambores side is formed like
a valley head. But instead of a continuation of the valley towards the
east there is the Cordón Barros Arana. Here an incision about 150 m
deep interrupts the cordón. So the eastern block with Purlactis Formation was probably lifted up, thus causing a reversal of the originally
eastward gradient in a valley coming from Tambores Formation. Before
this uplift of about 100 m, Tambores Formation must have formed the
highest part of the El Bordo ridge as it still does today further south.

On the whole the eastern flank of the northern Cordillera Domeyko has
been arched up since the oligocene or miocene. Thereby the older pediment
and bajada levels have been tilted and successively replaced by newer
ones. Thus in the upper parts at some places three or more levels of pediments can be dicerned, the younger ones wearing down the older ones.
Thereby the asymmetrical divide of El Bordo with its steep flank to the
east was gradually pushed to the west, thus beheading the westward
channels still clearly visible at the divide (fig. 2, profile 6).

4. The Salar de Atacama Basin

There is a great difference between the north and the south of the
Salar de Atacama basin. The cordillera south of the salar is even more
arid than that north of it; therefore no river and no alluvial cone
enters from the south. In the north, however, there is enough precipitation to bring forth Río San Pedro with its tributaries Río Grande
and Río Salado. Like in the area of upper Río Loa, run-off is favoured
by the long N-S-course of the river system, where evaporation is reduced
by the pervious ignimbrites. Thus Río San Pedro supplies the water for
San Pedro de Atacama, the most important oasis of the area. It even
continues to flow further south towards the center of the basin, accumulating a comparatively large alluvial plain.

Amazingly Río San Pedro has eroded its N-S-valley north of San Pedro
de Atacama near the axis of the Cordillera de la Sal anticline, avoid-
ing the depression of the Llano de la Paciencia in the west and the

Llano Vilama in the east. The Río Salado, its tributary, flows in a tectonic depression in its upper course and enters the Cordillera de la Sal anticline in a narrow gorge. The rivers had certainly been flowing south along their present course before the Cordillera de la Sal was lifted up. At that time they approximately had found their way between the bajadas coming down from the Cordillera Domeyko in the west and the alluvial fans of the rivers from the High Cordillera in the east. Afterwards they managed to keep their course by compensating the uplift of Cordillera de la Sal by erosion. The tilted gravel deposits on both sides of the anticline (fig. 2, profile 6) give evidence that the bajadas as well as the alluvial fans have been also lifted up.

Although the southern part of the Salar de Atacama basin is especially dry there is a well developed valley system in Cordón Chinquilchoro (northern part of Cerro Lila; Fig. 1 F), which protrudes from the south into the salar plain. This is the more surprising as the western and eastern rim of the Salar de Atacama basin are nearly devoid of deep valleys in spite of their greater height. Certainly the valleys have been eroded during a more humid climate, helped by the steepness of the flanks of Cordón Chinquilchoro.

Under the present climatical conditions there is little further development of the valley system of the Cordón Chinquilchoro. This can be confirmed by Tucúcaro ignimbrite which entered the valleys and is closely adapted to the present landforms. The distribution of this ignimbrite and its radiometric age of $3,2 \pm 0,3$ m.y. B.P. has been determined by RAMIREZ and GARDEWEG (1982, p. 56). It is really striking that the pliocene ignimbrite today still follows even very small fluvial incisions. Only in narrow and deep valleys with a comparatively high catchment area the ignimbrite has been eroded to a greater extent. But even there it still forms terraces up to only 20 m above the present valley bottom. Because of the great conformity of the pre-ignimbritic landforms of Cordón Chinquilchoro with those of today even the influence of pleistocene climate could not have been considerable. This also applies to the pediments and bajadas on the eastern flank of Cordón Chinquilchoro, which are covered by the Tucúcaro ignimbrite. In the upper parts of the pediments and bajadas these ignimbrites are overlain by sediments up to a few meters thick and in the lower parts a small number of channels eroded quebradas into the ignimbrites up to 15 m deep. The erosion of these quebradas was conditioned by the formation of a nearly vertical scarp at the lower end of the bajada. This geo-

morphological feature lies exactly where the ignimbrite and the salar
come together. It cannot be a fault scarp because its course is very
sinuous. The horizontal interfingering between the Tucúcaro ignimbrite
and the salar is even more conspicuous at La Penînsula, which forms
the northern end of Cordón Chinquilchoro. There the ignimbrite protrudes
into the salar, forming a flat plain which is delimited by a very sharp
edge (MORAGA et al. 1974, p. 15) about 15 m above the salar surface.
A formation of the scarp as a cliff alongside an old lake can also be
excluded, because at some places there is not only one scarp but a
steplike sequence of up to three scarps. The height of these steps is
so irregular that they could never have been formed by abrasion. Probably the scarps of the ignimbrites and bajadas surrounding Cordón Chinquilchoro have been formed by subrosion, i.e. the solution of the
evaporites unterlying the sediments of the bajadas and the permeable
ignimbrites which covered the south-eastern part of the salar. When-
ever tectonic subsidence occurred at a given place of the salar,
the overall salar-surface sank according to the new volumetric capacity
of the basin. This also applied to the parts of the salar, which were
covered by ignimbrites and bajadas. Being above the level of the brine
in the salar, the salt dissolved by rain-water percolating through the
ignimbrite and the bajada sediments. Therefore the ignimbritic and sedimentary cover gradually subsided. The scarps were left behind at the
fringe of the salar, where there was less or no more salt. Thus the
ignimbrites surrounding the eastern side of Cordón Chinquilchoro mark
more or less the level of their original deposition. They are like a
sheet of ice, still clinging to the river bank after the water has subsided. Such scarps have not been preserved at the foot of the bajadas
and alluvial plains of Cordillera Domeyko and the High Cordillera because sedimentation has since been much more active, the average altitude of these mountains being higher than that of Cordón Chinquilchoro.

5. The High Cordillera

The combination of young tectonic uplift, young volcanism and long-
term aridity led to a series of consequences characteristic of the High
Cordillera near the Salar de Atacama:

1. Taking into account that the High Cordillera rises up to more than
6000 m, it is really striking that immediately at its foot deep basins
with internal drainage (Salar de Atacama 2300 m and Salar Punta Negra
2950 m) have been preserved. This is not only due to young tectonic
subsidence of the basins but also to persisting dryness. Under more

humid conditions the basins would have been totally filled with sediments and an outlet towards the sea would have been cut into the continuous "dam" formed by Cordillera Domeyko (lowest gap about 600 m
above the Salar de Atacama).

2. The valleys are not cut deep into the western slope of the High
Cordillera, because there was neither enough water nor time for erosion.
Superficial run-off and fluvial erosion were additionally hampered by
the permeability of the ignimbrites and tuffs covering the western flank
of the High Cordillera.

3. The alluvial cones and plains at the foot of the High Cordillera
therefore only attain moderate size.

4. Thus the salar surface extends very close to the foot of the High
Cordillera.

5. The miocene, pliocene and pleistocene tuffs, ignimbrites (BAKER
1977, RAMIREZ and GARDEWEG 1982 and MARINOVIC and LAHSEN 1984) and
intercalated fluvial deposits (FLINT 1985) were accumulated concordantly over vast areas of the western flank of the High Cordillera. This
could have only happened under long-term arid conditions and on a slope
less high and less steep than today, otherwise the ignimbrites and fluvial sediments would have been canalized in preexisting valleys. Moreover they would have been eroded a great deal after their deposition.
It is therefore not only because of geological reasons but also because
of climatical reasons that the ignimbrites in the North Chilean, Peruvian and Bolivian Andes cover a very large area. According to ZEIL and
PICHLER (1967, p. 49) it is one of the largest ignimbrite areas in the
world. The fact that the tuffs and ignimbrites lie on a great plane,
inclined towards the Salar de Atacama basin, supports the view that
they were deposited on pediments and bajadas. It is open to question,
however, whether the highly fluid ignimbrites could have been deposited
on a plane with an inclination of 3°, which is the average angle of
the actual pediments and bajadas. (The ignimbrites overlying the bajada
sediments in the Quebrada del Diablo area, see above, support such an
assumption). In any case the present inclination of the ignimbrite and
tuff sheets, which by far exeeds 3°, has been attained by an uplift
of the High Cordillera. Thus the gradient of pleistocene Cajón ignimbrite
(radiometric age 1,0 ± 0,2 m.y. B.P., MARINOVIC and LAHSEN 1984, p.
64) east of San Pedro de Atacama increases from 4° in the lower parts
to 5° above 3400 m. According to RUTLAND and GUEST (1965) and FISCHER
(1985) the High Cordillera was lifted up during the whole period of

ignimbrite formation in miocene, pliocene and pleistocene.

6. The only deep valley on the western slope of the High Cordillera is that of Río Grande, because it obtains enough water due to the long N-S-course of its tributaries. Where erosion sank through the resistant ignimbrites and reached the subjacent and very unstable San Pedro Formation, many rockslides have occurred on both sides of the valley. These are due to the steep gradient of the slope, which in turn is a consequence of the lack of other slope forming processes under the predominant arid conditions.

7. The fact that the highest volcanos of the earth are situated in the North Chilean Cordillera (ZEIL 1986, p. 119) is not only due to geological reasons, but also to long-term aridity. Under more humid conditions the altiplano, which serves as a basis for the volcanos, would have been dissected before the formation of the cones, and the cones themselves would have been eroded to a certain extent by streams and glaciers. Moreover, eruptions of a strongly glaciated volcano would not have contributed so very much to the vertical growth of the cone, but to the formation of lahars, accumulating mudflow fans at the foot of it. The extremely large debris avalanche from Socompa volcano (Fig. 1 G; FRANCIS et al. 1985) is certainly due to the great size of the original cone.

8. Due to the dry climate, the extension of the pleistocene glaciation was comparatively small. At the foot of Cerro Toco (5604 m), east of San Pedro de Atacama, the lowest glacial striae can be found on the surface of an old lava stream at 4830 m. Further down at about 4500 m there are the lowest erratic blocks upon the ignimbrite surface.

References

BAKER, M.C. (1977): Geochronology of upper Tertiary volcanic activity in the Andes of North Chile. - Geologische Rundschau 66, p. 445-465.
BÖRGEL, R. (1983): Geomorfología. Geografía de Chile, tomo II. Santiago
BRÜGGEN, J. (1950): Fundamentos de la geología de Chile. Santiago
BRYSON, R.A. and KUHN, P.M. (1961): Stress-differential induced divergence with application to littoral precipitation. - Erdkunde 15, p. 287-294
CHONG, G. (1984): Die Salare in Nordchile - Geologie, Struktur und Geochemie. Geotektonische Forschungen 67. Stuttgart
DINGMAN, R.J. (1963): Cuadrángulo Tulor, Prov. de Antofagasta. Carta Geológica de Chile 1:50 000. Santiago
ERICKSEN, G.E. (1983): Meditaciones sobre el origen de los depósitos chilenos de nitrato. - Minerales 38, no. 163, p. 5-15
FISCHER, K. (1985): Grundzüge der jungtertiären Geomorphogenese der mittleren Anden. - Erdkunde 39, p. 248-259

FLINT, S. (1985): Alluvial fan and playa sedimentation in an Andean
    arid closed basin: the Paciencia Group, Antofagasta Province, Chile.
    Journal Geol. Soc. London, vol. 142, p. 533-546
FRANCIS, P.W., GARDEWEG, M., RAMIREZ, C.F. and ROTHERY, D.A. (1985):
    Avalancha catastrófica de detritos del volcán Socompa. - Comunica-
    ciones 35, Departamento de Geología, Univ. de Chile, Santiago, p.
    69-71
HAYES, D.E. (1974): Continental Margin of Western South America. -
    The Geology of Continental Margins, Ed.: C.A. Burk and C.L. Drake.
    New York, p. 581-590
HOLLINGWORTH, S.E. (1964): Dating the Uplift of the Andes of Northern
    Chile. - Nature 201, p. 17-20
LYDOLPH, P.E. (1957): A comparative analysis of the dry Western Lit-
    torals. - Ann. Assoc. Am. Geographers 47, p. 213-230
MARINOVIC, N. and LAHSEN, A. (1984): Carta Geológica de Chile, Escala
    1:250 000, Hoja Calama, Región de Antofagasta. Santiago
MORAGA, A., CHONG, G., FORTT, A. and HENRIQUEZ, H. (1974): Estudio
    geológico del Salar de Atacama, Provincia de Antofagasta. Instituto
    de investigaciones geológicas, Boletín 29, Santiago
MORTENSEN, H. (1927): Der Formenschatz der nordchilenischen Wüste.
    Abh. d. Ges. d. Wiss. zu Göttingen, Berlin
MORTIMER, C. (1980): Drainage evolution in the Atacama desert of
    northernmost Chile. - Revista Geológica de Chile 11, p. 2-28
MORTIMER, C. u. SARIC, N. (1972): Landform evolution in the coastal
    region of Tarapacá Province, Chile. - Revue de Géomorphologie dyna-
    mique 4, p. 162-170
NARANJO, J.A. and PASKOFF, R.P. (1981): Estratigrafía de los depósitos
    cenozóicos de la región de Chiuchiu Calama, desierto de Atacama.
    Revista Geológica de Chile 13-14, p. 79-85
NARANJO, J.A. and PASKOFF, R. (1985): Evolución cenozóica del piede-
    monte andino en la Pampa del Tamarugal, Norte de Chile (18°-21° S).
    IV congreso geológico chileno, 1985, Antofagasta, p. 5-149 - 5-164
PASKOFF, R. (1978/79): Sobre la evolución geomorfológica del gran acan-
    tilado costero del Norte Grande de Chile. - Norte Grande, Inst. de
    Geogr., Univ. Católica de Chile, 6, p. 7-22
PASKOFF, R. and NARANJO, J.A. (1983): Formation et évolution du piémont
    andin dans le désert du Nord du Chili (18-21° latitude Sud) pendant
    le Cénozoique supérieur. C. R. Acad. Sc. Paris, t. 297 (14 novembre
    1983), série II, p. 743-748
RADTKE, U. (1985): Chronostratigraphie und Neotektonik mariner Terras-
    sen in Nord- und Mittelchile - erste Ergebnisse. -IV Congreso Geo-
    lógico Chileno, Antofagasta, p. 4-436 - 4-457
RAMIREZ, C.F. and GARDEWEG, M.P. (1982): Carta Geológica de Chile,
    Escala 1:250 000, Hoja Toconao. Región de Antofagasta. Santiago
RUTLAND, R.W.R., GUEST, J.E. and GRASTY, R. (1965): Isotopic Ages and
    Andean Uplift. - Nature 208, p. 677 f.
SCHEUBER, E. (1987): Geologie der nordchilenischen Küstenkordillere
    zwischen 24° 30' und 24° S, unter besonderer Berücksichtigung duk-
    tiler Scherzonen im Bereich des Atacama-Störungssystems. - Diss.
    Fachbereich Geowissenschaften, FU Berlin
TREWARTHA, G.T. (1961): The Earth's Problem Climates. Madison
WEISCHET, W. (1966): Zur Klimatologie der nordchilenischen Wüste.
    Meteorologische Rundschau 19, p. 1-7
WEISCHET, W. (1970): Chile, seine länderkundliche Individualität und
    Struktur. Darmstadt
ZEIL, W. (1959): Junger Vulkanismus in der Hochkordillere der Provinz
    Antofagasta (Chile). - Geologische Rundschau 48, p. 218-232
ZEIL, W. and PICHLER, H. (1967): Die känozoische Rhyolith-Formation
    mittleren Abschnitt der Anden. - Geologische Rundschau 57, p. 48-81

ZEIL, W. (1979): The Andes, a geological review. Berlin, Stuttgart

ZEIL, W. (1986): Südamerika. Stuttgart

ZIEGLER, A.M., BARRETT, S.F. and SCOTESE, C.R. (1981): Paleoclimate, sedimentation and continental accretion. - Phil. Trans. R. Soc. Lond. A 301, p. 253-264

II: Mesozoic-Cenozoic Magmatism and Tectonics

THE JURASSIC LA NEGRA FORMATION IN THE AREA OF ANTOFAGASTA, NORTHERN CHILE (LITHOLOGY, PETROGRAPHY, GEOCHEMISTRY).

Michael Buchelt

Institut für Geologie und Paläontologie, Technische Universität Berlin
Ernst-Reuter Platz 1, D-1000 Berlin-10

Carlos Tellez Cancino

Facultad de Ciencias, Departamento de Geociencias, Universidad del Norte
Casilla 1280, Antofagasta, Chile

ABSTRACT

Based on their lithology and their geographical position, the rocks of the La Negra Formation in the investigated area could be subdivided into the following three
groups: First, the Cuevitas-group, which overlies marine sediments of the Hettangian
and consists of HK-calc-alkaline basaltic to andesitic volcanics (partly submarine),
pyroclastics and intercalations of marine and terrestrial sediments. Second, the La
Chimba-group, composed of calc-alkaline basalts to basaltic andesites (in the lower
parts submarine) intercalated by terrestrial sediments. Third, the La Negra-group,
which represents a widespread distribution of co-existing lava successions with thicknesses up to 1400m.

Magmagenetically, the volcanics have a polymagmatic origin. Although most of the
rocks show a basic character, true basalts are not present. The predominant series
of calc-alkaline to HK-calc-alkaline basalts, basaltic andesites and andesites are
enriched by the LIL-elements Rb, K, Ba and Th, probably originating from the dehydra-
ting subducted oceanic crust. The concentrations of the HFS-elements Ta, Nd, P, Hf
and Zr show enrichments which are developed in a back-arc volcanism with affinities
to within-plate basalts.

Since the Lower (?) Sinemurian the volcanic arc has migrated 40-50km trenchward
to the West. Geochemical parameters point to a change from a more alkaline to a calc-
alkaline trend, indicating a more evolved stage of the back-arc volcanism. The most
intense volcanic activity in the "retreating" back-arc at an active continental margin
was during the Middle- to Late Jurassic in the present coast area.

INTRODUCTION

The descent of the oceanic Nasca plate below the western part of South America
since the Upper Triassic (?) is marked by tectonic events and magmatic activity. The
volcanism in the Coast Range of Chile, which was associated with the interaction of
the converging oceanic and continental plates, started during this time and peaked in
the Early to Late Jurassic.

The volcanic rocks were described first by Charles DARWIN (1838) as a "Porphyrite
Formation". Later they were explored in detail by GARCIA (1967), who investigated

Lecture Notes in Earth Sciences, Vol. 17
H. Bahlburg, Ch. Breitkreuz, P. Giese (Eds.),
The Southern Central Andes
© Springer-Verlag Berlin Heidelberg 1988

the stratigraphic relations and the lithology. He described the complete successions of the lavas and pyroclastites in the following terms: "Formación La Negra"....."mantos de lavas porfiríticas, particialmente amigdaloideas, de color gris oscuro, en parte, verdosas". According to this author and subsequently to COIRA et.al (1982) and HILLEBRANDT et.al (1986), the base of the La Negra Formation overlies marine sediments of the Pliensbachian/Toarcian. The upper limit is still uncertain. Due to a rapid uplift of the Coast Range since the Mid-Cretaceous, an erosion unconformity exists, but there is some evidence that the volcanic activity continued into the Lower Cretaceous (ZEIL, 1979).

The rocks described lithologically, petrographically and geochemically in this paper were sampled along five traverses in the area of Antofagasta (Fig. 1):
1. Cerros de Cuevitas
2. Cerro Mantos Blancos
3. Quebrada La Chimba
4. Antofagasta- Estación Portezuelo
5. Quebrada La Negra
In addition, some samples were collected in the Quebrada El Buey, the roof section of the La Negra Formation.

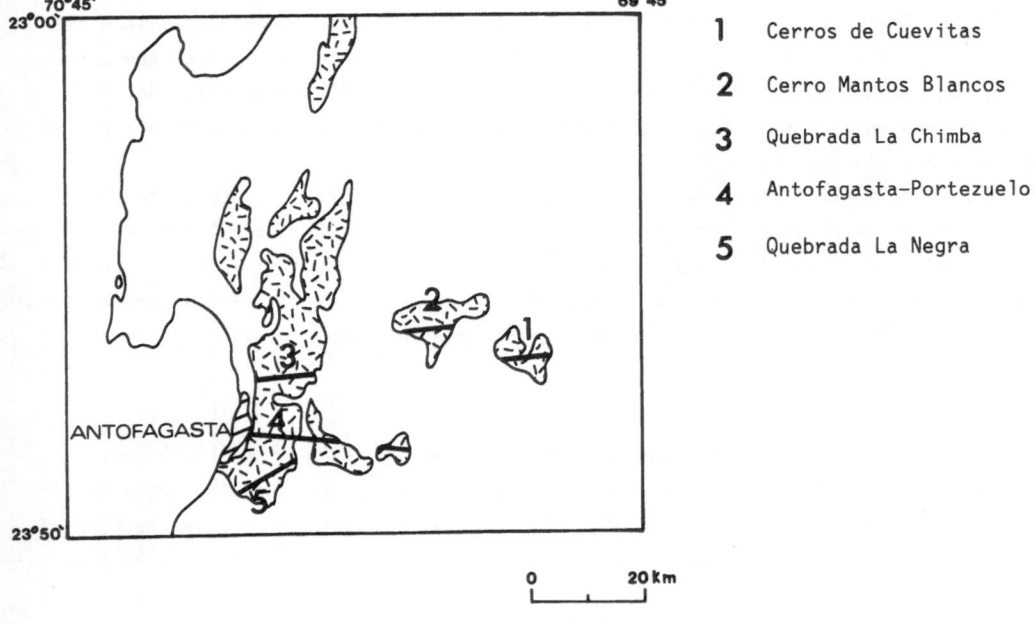

1   Cerros de Cuevitas

2   Cerro Mantos Blancos

3   Quebrada La Chimba

4   Antofagasta-Portezuelo

5   Quebrada La Negra

Figure 1: Geological sketch-map of the La Negra Formation in the investigated area with the localities of the sampled traverses.

GEOLOGICAL SETTING

   While the recent situation (Late Cretaceous to today) is that of an active conti-
nental-margin-system with an "intracratonic" magmatic arc, the early stages (Early
Jurassic to Early Cretaceous) were characterized by a magmatic back-arc basin. Early
geodynamic and geochemical interpretations of this geological setting reported a typical island-arc system, suggesting a zonation of the volcanics into tholeiitic-,
calc-alkaline- and alkaline rocks with shoshonitic affinities (LOSERT, 1974, PALACIOS,
1978, 1982). PICHOWIAK & BREITKREUZ (1984) observed dike-rocks related to the La Negra Formation and pointed out the existence of an ensialic setting as well as geochemical parameters of within-plate affinities.

   Based on their lithology and their geographical position, the rocks of the La Negra Formation in the investigated region can be subdivided into the following three
groups:
1. The lower part: The Cuevitas-group consists of up to 400m lavas with andesitic and
   basaltic-, tuffs with dacitic composition. The first volcanic rocks occur in the
   Lower to Middle Sinemurian. They overlie marine sediments of the Hettangian (Psilo-
   ceras sp.). Tuffs of the Middle Sinemurian are intercalated by marine sediments
   (Arnioceras sp.). Intercalations of sandstones and conglomerates occur in the upper
   part of this group, Fig. 2.
2. The middle part: The La Chimba-group includes the volcanites of the Quebrada La
   Chimba, the area of the Cerro Mantos Blancos, the traverse between Antofagasta
   and the Estación Portezuelo as well as the volcanites in the east part of the Quebrada La Negra. The lava successions represent a total thickness of more than 7500m.
   They consist predominantly of basaltic andesites and basalts. On the east side of
   the Antofagasta-Portezuelo traverse, pillow lavas with a basaltic andesitic composition can be observed, while towards the West intercalations of terrestrial deposits (epiclastical sandstones) gradually increase, Fig. 2. Also to the West tuff-
   layers occur with a cement of subaerial origin.
3. The upper part: The La Negra-group in the South of Antofagasta consists of mainly
   andesitic volcanites accompanied by minor occurences of volcanites with basaltic
   andesitic composition. Red sandstones and tuffs are increasingly intercalated towards the top of this group. These successions display a thickness of more than
   2000m, Fig. 2. The middle and the upper parts are distinguished by the occurence
   of dikes with doleritic fabric.

Although not completely exposed, the three groups are estimated to have a thickness
of more than 10 000m.

   No typical mineral assemblages signifying a zeolithe facies or even a greenschist
facies caused by a burial metamorphism could be observed in any of the studied samples.
The minerals which occur, epidote, clinozoisite and chlorite, are products of hydrothermal and metasomatic activity. In general, the subsequent hydrothermal alteration

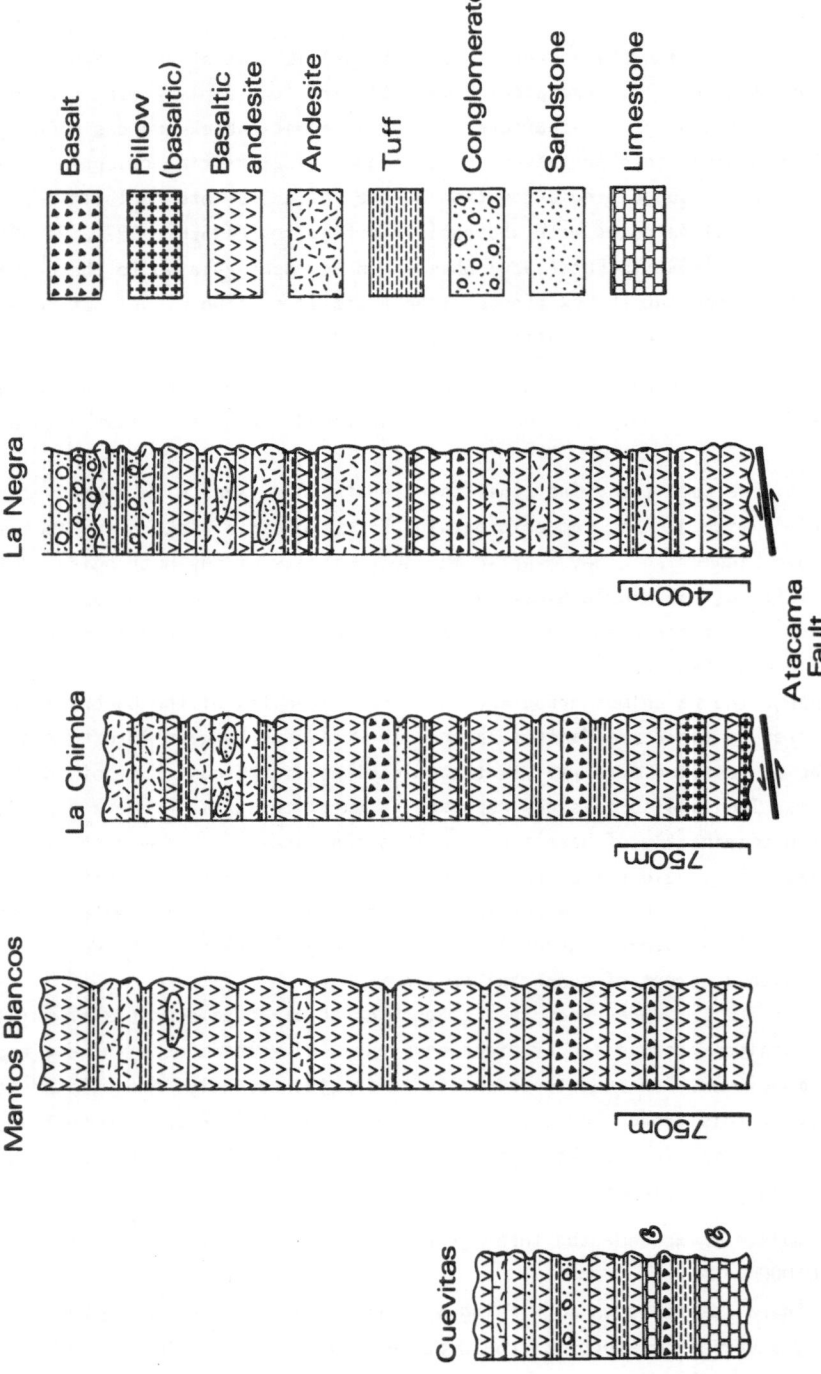

Figure 2: Schematic stratigraphic sequences in the sampled traverses.

(propylitization) in the volcanic rocks of the La Negra Formation is widespread and in some places very intense.

Although the Coast Range in the area of Antofagasta is controlled by intensive fracture tectonics, the thickness of more than 10 000m is unlikely to have been produced by tectonic events. The tectonic events resulted in NNW/SSE and NNE/SSW-striking faults (Fig. 3), which triggered tension faults striking E-W. The volcanic activity was probably connected with prolonged periods of extension. In the field, structures often appear as step folds while the dips of the lava flows show a steepening. In the middle part of the Antofagasta-Portezuelo traverse, only two cases of eastward-dipping upthrusts are evident.

Caused by uplift since the Cretaceous, the rocks of the La Negra Formation reach from today's sea-level up to an altitude of 1200m.

Although feeders of the volcanoes could not be found, field observations indicate that many of the lava successions co-exist. Particularly the area of the middle and upper parts (in the East and South of Antofagasta, Fig. 3) can be seen as a widespread distribution of co-existing lava successions partly showing considerable true thicknesses of up to 1400m.

PETROGRAPHY

The investigated rocks of the La Negra Formation consist about 85% of lava flows, 10% of tuff-layers (including eruptive breccias) and 5% of thin layers composed predominantly of sandstones and conglomerates. The single lava flow shows thicknesses of up to 80m, whereas the tuff-layers are only a few meters thick.

The binding material of the consolidated pyroclastic rocks found in the lower parts of the Cuevitas-group shows a submarine origin, as opposed to the subaerial origin of the two other groups.

A striking feature of many orogenetic volcanic rocks is their crystal-rich nature. Macroscopically, the volcanites show a grey to greenish-brownish matrix with laths of plagioclase- and pyroxene-phenocrysts; amphiboles (and minor biotite) are less abundant. The matrix assemblages are dominated by plagioclase and pyroxene. The fabric of the volcanites is porphyric or microporphyric.

Based on the chemical criteria, the classification of the volcanic rocks follows the scheme of PECCERILLO & TAYLOR (1976), Fig. 4). The lavas of the La Negra Formation are built up of mainly basaltic andesites with a silica mode of 52-56wt.%. Basalts and andesites are less abundant and shoshonites and absarokites are even less common.

MINERAL COMPOSITION

Plagioclase is the most frequent phenocryst, demonstrating mostly a labradorite -bytownite composition within the basalts and basaltic andesites. They possess oscillatory zoning with bytownite in the center. Labradorite is characteristic of the ande-

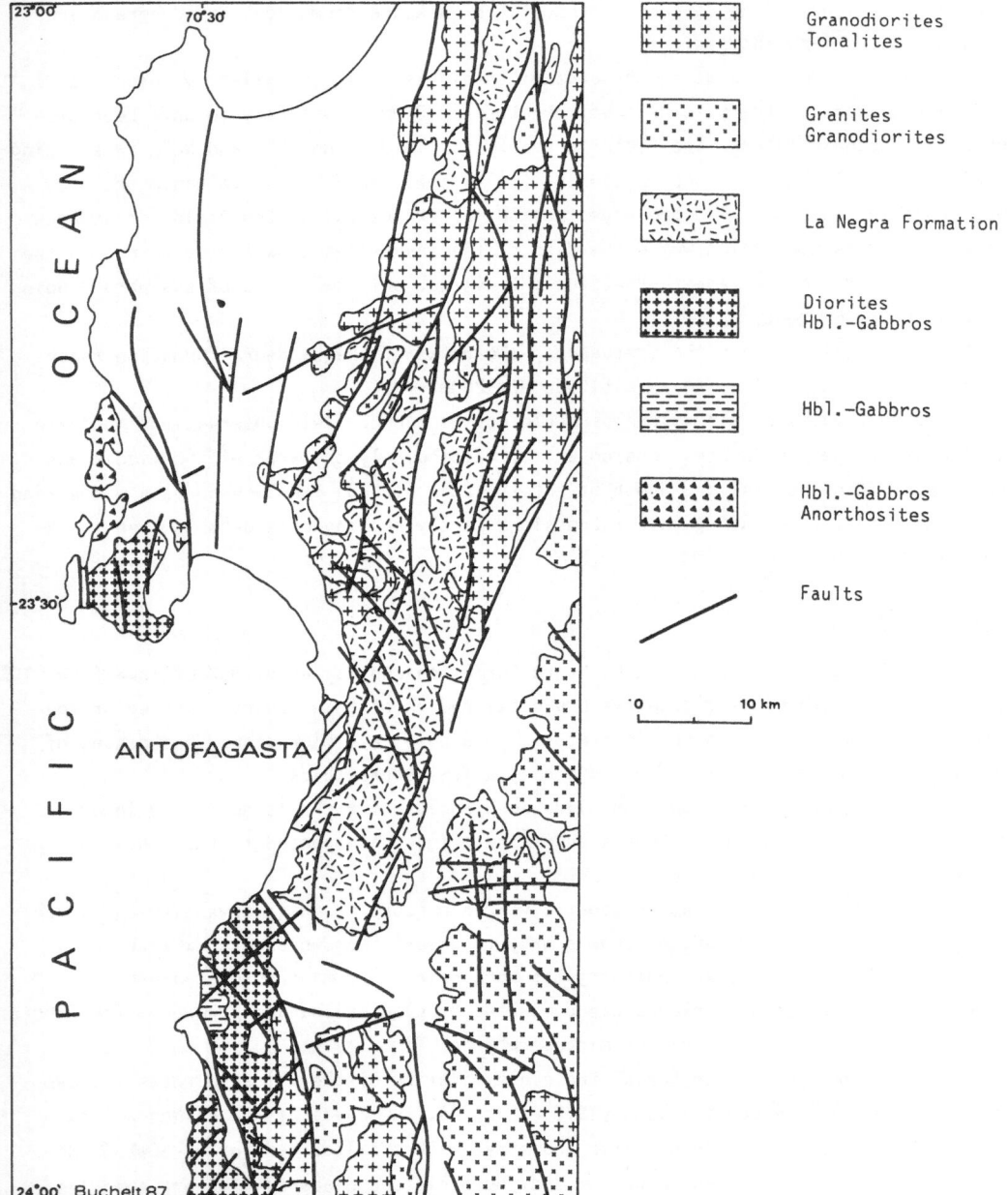

Figure 3: Generalized geological sketch-map of the Mesozoic Rocks in the
Antofagasta area.

sites. The extended euhedral prisms of the phenocrysts can reach up to 2cm. Often a homogeneous core is mantled by a more sodic rim and stages in the growth of the plagioclase show bands of small pyroxene inclusions (gradual or abrupt changes in solid-solution composition), randomly oriented.

The sanidine feldspar appears only sporadically in the shoshonites within the matrix.

Olivine is restricted to the basaltic rock-types and the earliest precipitant, forming single rounded grains of 2-3mm, that are normally surrounded by reaction rims. The basalts and the basaltic andesites contain rather Mg-rich olivine, predominantly in the range Fo_{70}-Fo_{80}. Corona fabrics with an iddingsitization of the rims and along fissures are due to secondary alteration. Olivine has only been observed in disequilibrium to enstatite and can therefore be interpretated as a "remnant" of a basaltic parental magma.

Enstatite appears as up to 1cm large euhedral phenocrysts in basaltic andesites. It is usually altered by bastitization and the oriented pseudomorphism is surrounded by small augite grains.

The normal Ca-poor pyroxene is orthopyroxene of bronzite-hypersthene composition and is restricted to basaltic andesites and andesites. The orthopyroxenes are not in equilibrium with the modal composition (matrix) in which they occur.

An equilibrium balance is demonstrated by the reaction trend of the clinopyroxenes and may be recognized as follows:

salite/diopside - salitic augite - augite - pigeonite.

The calcic pyroxenes are largely of augite composition, with a tendency to extend into the diopside and salite compositions. Diopside and salite occur as "independent" clinopyroxenes in basalts and basaltic andesites and more rarely in andesites. The dominant clinopyroxenes are augites with or without a salite composition (always oscillatory and sector zoned). The average composition of the salitic augites ranges from $Wo_{45}En_{50}Fs_5$ in the core to $Wo_{36}En_{46}Fs_{18}$ in the rim. Augite in basalts and basaltic andesites has the average composition of $Wo_{37}En_{45}Fs_{18}$, whereas augite of the more evolved andesites has an average composition of $Wo_{39}En_{32}Fs_{29}$.

Pigeonite occurs between glomeroporphyric plagioclases, following the precipitation of the feldspars.

The amphibole hornblende and occasional Fe-rich biotite occur in the common andesites.

Within the group Fe-Ti oxides the dominating mineral is titanomagnetite. Varying amounts of microphenocrysts are present in all rock types, with a greater quantity in the more acidic and potassic andesites. In addition to primary titanomagnetite, the widespread subsequent hydrothermal alteration may have raised the Fe- and Ti-content of many rocks.

GEOCHEMISTRY

The chemical data of the investigated volcanic rocks indicates that they are pre-dominantly basalts, basaltic andesites and andesites (25% basalts, 55% basaltic andesites, 20% andesites). Following PECCERILLO & TAYLOR (1976) the volcanites can be sub-divided further more in calc-alkaline series and high-K series, Fig. 4.

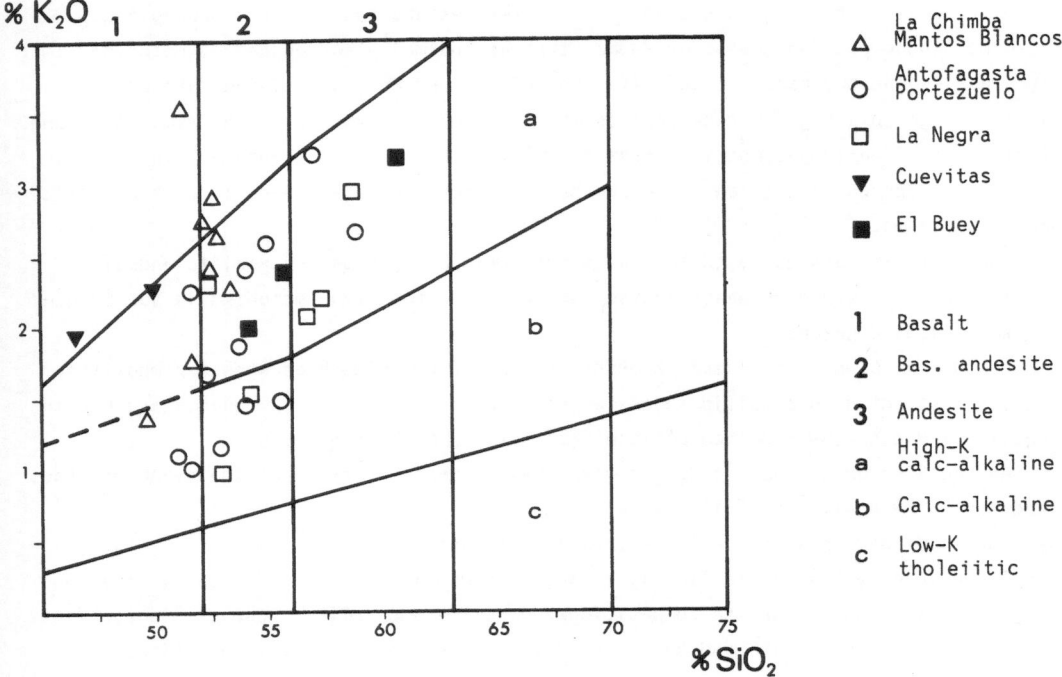

Figure 4: Classification scheme of the volcanic rocks. Note the abundance of the
basaltic andesites.

The scheme shows the distribution of the potassium content, which ranges in rocks with typical calc-alkaline- to high-K affinities, from 1 to 3wt.% K_2O. On the AFM-dia-gram of IRVINE & BARAGAR (1971) Fig. 5 the analysed rocks lie mainly in the calc-alkaline field. The rocks, which contain slight amounts of Fe, lie on the tholeiite/calc-alkaline boundary, however, most of them are placed along a typical calc-alkaline trend, contrary to the interpretation of LOSERT (1974) and PALACIOS (1978, 1982). A lack of tholeiitic affinities is recognizable.

In the AFM-diagram the analysed samples of the two parallel traverses of La Chimba-Mantos Blancos and Antofagasta-Portezuelo differ slightly in their alkali- and Mg-con-tent. Moreover, it could be established that the rocks of the Antofagasta traverse are richer in Al_2O_3 and belong to the high-Al_2O_3 field of KUNO (1966), in contrast to the rocks of the adjacent traverse (alkali field).

Using the geochemical parameters SiO_2 and K_2O a zonation of the Antofagasta tra-

verse was determined, which enables the separation of the lava successions into four
main groups:

Top   High-K-calc-alkaline basaltic andesites and andesites
            Calc-alkaline basaltic andesites
            Calc-alkaline basalts
Base   High-K-calc-alkaline basaltic andesites

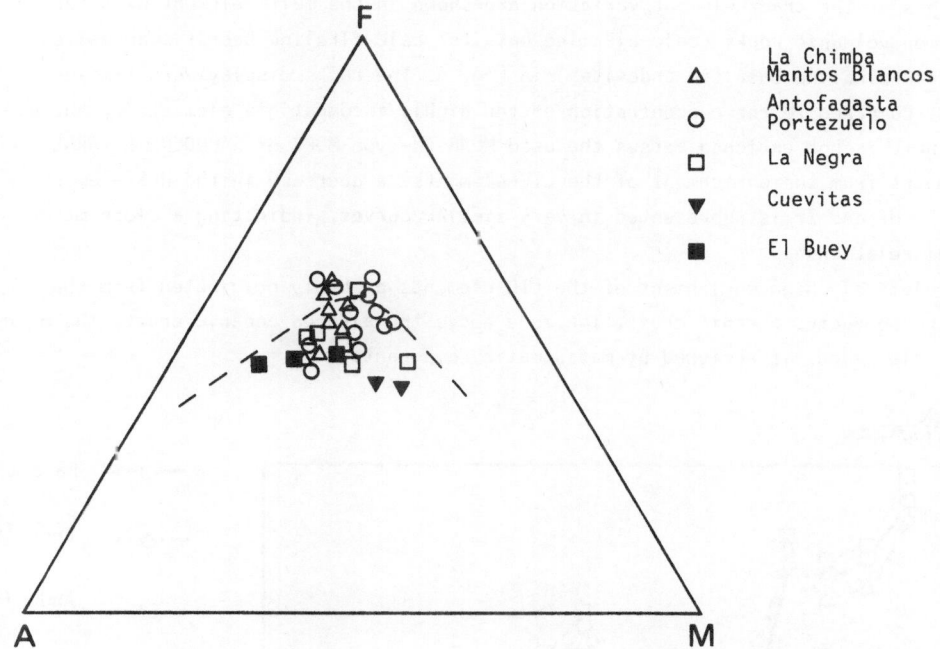

Figure 5: The AFM plot of the La Negra volcanites presents the dominant calc-alkaline
          character.

The K/Rb ratios show a progressive increase through the calc-alkaline series from
the volcanics at the base to the volcanics at the top. Ba increases from the basaltic
andesites to the andesitic compositional range. Sr exhibits a decrease from the basic
to the intermediate composition. While with an increasing differentiation a greater
amount of Zr can be observed, the contents of the elements Cr, Ni and V are depleted.

Averages of element ratios between compatible and incompatible elements in this
traverse show decreasing or increasing trends depending on their geographical position
and the degree of differentiation (Zr/P_2O_5 0,053-0,061, La/Y 0,22-0,26, Th/U 4,22-5,17,
Ni/Co 0,62-0,88, Cr/V 0,16-0,27, V/Ni 33.6-11,2).These trends can also be observed in
the more highly differentiated volcanics (high-K andesites) of the La Negra-group
and in the samples of the Quebrada El Buey.

When comparing the volcanites of the sampled traverses with each other a gradual

enrichment of LIL-elements between the lowest lavas of the Cuevitas-group and the lavas in the east part of La Chimba and Antofagasta-Portezuelo becomes apparent. Probably early stages of a subduction were less enriched with subduction related LIL-elements than later stages. Trace-element data for lavas from the island arcs and remnant arcs of the Mariana system indicate that the degree of LIL-element enrichment has
increased in successive epidodes of arc magmatism over the last 35 Ma (SAUNDERS &
TARNEY, 1984).

Averages of the trace element variation are shown in the multi-element plot for the
most common volcanic rocks (calc-alkaline basalts, calc-alkaline basaltic andesites,
high-K calc-alkaline basaltic andesites) in Fig. 6. The rocks display very similar
patterns. Compared to the concentration of the highly incompatible elements K, Rb, Ba
and Th the Y is not enriched versus the used MORB (N-type MORB of SAUNDERS & TARNEY,
1984). Apart from the enrichment of the LIL-elements, a decrease in the HFS-elements
Ta, Nd, P, Hf and Zr is represented in very similar curves, indicating a close magma-
genetical relationship.

The relatively high enrichment of the LIL-elements probably originated from the
dehydrated subducted oceanic crust. The zone above the altered oceanic crust, the overlying mantle wedge, is enriched by metasomatic components.

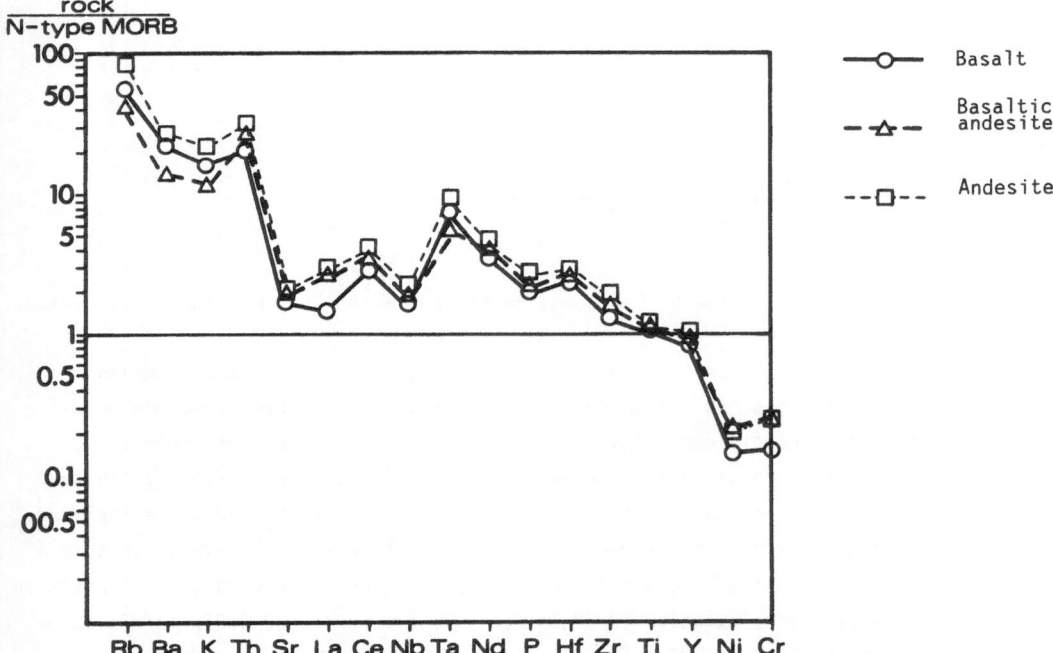

Figure 6: Multi-element plot (rock/MORB). Averages of the most common rocks (basalts,
basaltic andesites, andesites). Chemical data in BUCHELT & ZEIL, 1986.

The high concentrations of Ta relative to Zr and Zr relative to Y indicate an enrichment from the upper mantle as can be seen in within-plate basalts. The concentrations of the HFS-elements Ta, Nd, P, Hf and Zr show enrichments as they occur in back-arc volcanic settings with affinities to within-plate basalts.

CONCLUSIONS

The volcanic rocks of the La Negra Formation in the investigated area of Antofagasta belong mostly to the high-alumina basalt-andesite-dacite association, which is (together with rhyolite) the most characteristic volcanic association of the Andes. Basalts, basaltic andesites and andesites are predominantly lavas, the dacitic rocks pyroclastics.

Owing to the abundance of basic rocks (80%) it is justified to suppose an origin of basaltic parental magmas. The trend of reaction between the orthopyroxenes and the clinopyroxenes, the substitution of olivine with enstatite and the occurence of basic to intermediate (and rarely acidic) rocks without evidence of a vertical variation suggests more a polymagmatic origin. The mineralogical variations in the rocks could also have their origin in changes of subducted components or by crustal contamination.

Although some co-existing lava successions differ slightly in their chemical compositions (major elements), the distribution and variation of the trace elements in all rocks show a coherent character, which points to a close and similar magmagenesis and evolution.

During initial stages of back-arc spreading, the volcanism started with anomalous magmas (alkali basalt from an ensialic environment) in the Lower to Middle Sinemurian. The volcanic rocks overlie marine sediments of the Hettangian.

Episodically, the active arc migrated trenchward to the West and magmas erupted successively building up new arcs accompanied by basin openings. Intercalations of marine sediments indicate relatively shallow and near-shore deposits. The most intense volcanic activity was in the region of the present coast during the Middle- to Late Jurassic. While the basalts and andesites of the early arc at Cuevitas present a more alkaline trend, the later arcs (Mantos Blancos, La Chimba, Antofagasta and La Negra) have element compositions which are typical of calc-alkaline affinities with a minor trend to an alkaline affinity (shoshonites in the roof section of the La Negra-group). The area of Portezuelo and the east part of La Chimba seem to be transitional zones, where alkaline affinities change into calc-alkaline affinities, "more matured and evolved stages" of the volcanic arc. It is probable that during the Middle- to Late Jurassic many of neighbouring volcanoes produced immense successions of lava flows and tuff-layers, sporadically interrupted by deposits of terrestrial sediments.

The volcanics of the La Negra Formation in the investigated area of Antofagasta can be considered as volcanic rocks of a 40-50km long "retreating" back-arc of an active continental margin.

This work was carried out at the suggestion of Prof. Dr. W. Zeil within the research project "Mobilität aktiver Kontinentalränder", supported by the Deutsche Forschungsgemeinschaft (DFG). We wish to thank him for his review of the manuscript and for his critical comments.

REFERENCES

BUCHELT, M. & ZEIL, W. (1986): Petrographische und geochemische Untersuchungen an jurassischen Vulkaniten der Porphyrit-Formation in der Küstenkordillere Nord-Chiles.- In P. Giese (Ed.) " Forschungsberichte aus den zentralen Anden (21°-25°S)", - Berliner geowiss. Abh. (A), 66, 191-204, Berlin.

COIRA, B., DAVIDSON, J., MPODOZIS, C. & RAMOS, V. (1982): Tectonic and Magmatic Evolution of the Andes of Northern Argentina and Chile.- Earth Sci. Rev., 18, 303-332, Amsterdam.

DARWIN, Ch. (1838): Journal of Research.- Royal Geol. Soc., 448pp., London.

GARCIA, F. (1967): Geología del Norte Grande de Chile.- Simp. Geosincl. Andino, Soc. Geol. Chile Publ., 3, 138pp., Santiago de Chile.

HILLEBRANDT, A.v., GRÖSCHKE, M., PRINZ, P. & WILKE, H.-G. (1986): Marines Mesozoikum in Nordchile zwischen 21° und 26° S in P. Giese: Forschungsberichte aus den zentralen Anden (21°-25° S).- Berl. Geow. Abh., Reihe A, 66, 169-189, Berlin.

IRVINE, T.N. & BARAGAR, W.R.A. (1971): A Guide to the Chemical Classification of the Common Volcanic Rocks.- Canad. J. Earth Sci., 8, 523-548, Ottawa.

KUNO, J. (1966): Lateral variation of basalt magma type across continental margins and island arcs.- Bull. Volcanol., 29, 195-222, Napoli.

LOSERT, J. (1974): Alterations and associated copper mineralization in the Jurassic volcanic rocks of the Buena Esperanza mining area, Antofagasta Province, Northern Chile.- Publ. Depto. Geol. Univ., 41, 51-85, Santiago de Chile.

PALACIOS, C.M. (1978): The Jurassic paleovolcanism in northern Chile.- Unpubl. Thesis, 99pp., Tübingen.

    (1982): Volcanismo Jurasico en el Sector Sur de los Andes Centrales, (22°-26°S) Chile.- V Congr. Latinoameric. de Geol., Actas II, 83-96, Buenos Aires.

PECCERILLO, A. & TAYLOR,S.R. (1976): Geochemistry of Eocene Calc-Alkaline Volcanic Rocks from the Kastamonu Area, Northern Turkey.- Contrib. Mineral. Petrol., 58, 63-81, Berlin-Heidelberg-New York.

PICHOWIAK, S. & BREITKREUZ, Ch. (1984): Volcanic Dykes in the North Chilean Coast Range.- Geol. Rundsch., 73, 853-868, Stuttgart.

SAUNDERS, A.D. & TARNEY, S.R. (1984): Geochemical characteristics of basaltic volcanism within back-arc basins.- In B.P. Kokelaar & M.F. Howells (Eds.) "Marginal Basin Geology", 59-76, London.

ZEIL, W. (1979): The Andes- A geological Review.- Beitr. Regional. Geol. Erde, 13, 260pp., Gebr. Borntraeger, Berlin-Stuttgart.

CENOZOIC IGNIMBRITES OF THE CENTRAL ANDES: A NEW GENETIC MODEL

Cornelia Schmitt-Riegraf
Institut für Mineralogie, Universität Münster
Corrensstr. 24, D-4400 Münster

Hans Pichler
Mineralogisch-Petrographisches Institut, Universität Tübingen
Wilhelmstr. 56, D-7400 Tübingen

Abstract: New petrographical and geochemical studies of Cenozoic central Andean ignimbrites are the basis for a new discussion of their magma genesis. In combination with published isotope data a multi-stage-model can be developed:

1. The origin of the ignimbrites and intercalated lavas can be derived from the transition zone of the upper mantle/lower continental crust.

2. Heat producing factor is the mechanism of subduction. The existence of ^{10}Be in andesites (Harmon, 1986), which are closely connected in space and time with the ignimbrites, speaks for an oceanic crust participation.

3. Both ignimbritic and andesitic magmas suffered a multi-stage evolutionary process:
   - upper crustal contamination or "magma-mixing" in the upper crust, proved by mineral analyses and isotope data;
   - intensive plagioclase-dominant fractional crystallization and progressive segregation, demonstrated by the Ca-, Ba-, Sr-, and Eu^{2+}-data.

Introduction

Various hypotheses of the origin and evolution of the ignimbrites and andesites of the central Andes have been discussed in view of major element analyses (Zeil & Pichler, 1967; Pichler & Zeil, 1972; Kussmaul et al., 1977), trace elements (El-Hinnawi et al., 1969), REE analyses and isotopic data (McNutt et al., 1975; James et al., 1976; Klerkx et al., 1977; Thorpe et al., 1979; Hawkesworth et al., 1982; Déruelle et al., 1983; Francis et al., 1984; Harmon & Hoefs, 1984; Harmon, 1986).

New petrographical and geochemical data of the ignimbrites and their phenocrysts are presented. These data, combined with isotope data, are used to re-evaluate the petrogenesis and the magmatic evolution of these Cenozoic central Andean volcanic rock series.

The studied samples (table 1) were collected by Zeil, Pichler and Pichowiak during different field trips in 1962, 1970 and 1982.

Lecture Notes in Earth Sciences, Vol. 17
H. Bahlburg, Ch. Breitkreuz, P. Giese (Eds.),
The Southern Central Andes
© Springer-Verlag Berlin Heidelberg 1988

Geological setting

The "ignimbrite plateau" of the central Andes is located between 15° and 27°S latitudes covering an area of about 150,000 km². This corresponds to a flush production volume of about 70,000 km³ of acidic melt (Zeil & Pichler, 1967). The thickness of the individual ignimbrite eruptions ranges from about 15 m (Francis et al., 1974) to more than 1000 m (Zeil & Pichler, 1967) with an average thickness of about 500 m in the central part (James, 1971) and 150 m in the northern Chilean part (Francis & Rundle, 1976).

number	location/country	latitude	samples
29	transverse section across the Andean Cordillera of SW–Bolivia	19°45'--21°S	Z 1a to Z 24
5	transverse section across the Anaean Cordillera of NW–Argentina	23°30'--24°30'S	Z 31 to Z 63
23	transverse section across the Andean Cordillera of northern Chile	18°19'--24°19'45''S	18/62 to 320/62 HU 3, MQ 154

Table 1 Location of the analysed ignimbrite samples.

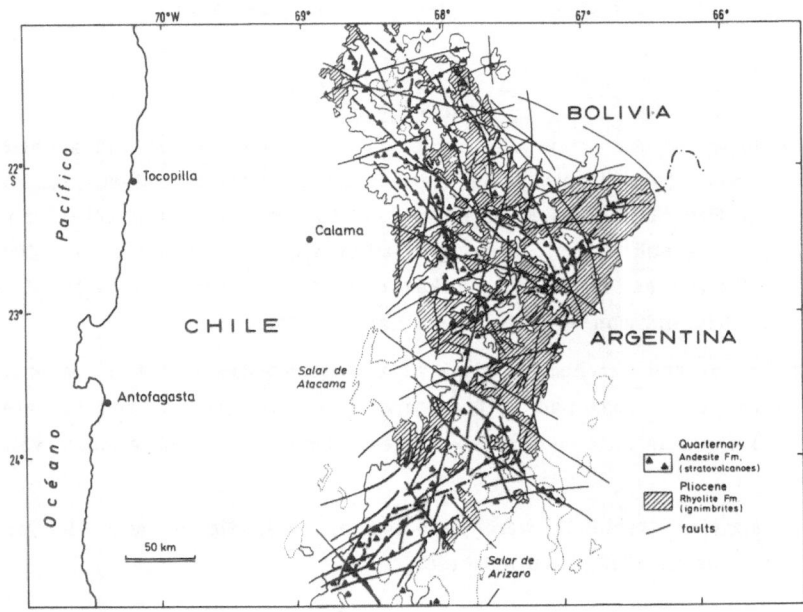

Fig. 1 Satellite image interpretation (courtesy of M. Buchelt, Berlin) of the central Andes illustrating the distribution of the Cenozoic 'Ignimbrite' and 'Andesite' Formations.

The interpretation of satellite images (Fig. 1) shows the tectonic environment of the main volcanic front. Mostly N--S or NW--SE striking fissures near the crest of the recent volcanic range are assumed to be the main eruption zone of the ignimbrites (Zeil & Pichler, 1967; James, 1971; Pichler & Zeil, 1972; Kussmaul et al., 1977). Less frequently the ignimbrite magmas are combined to calderas or shield complexes (Kussmaul et al., 1977; Baker & Francis, 1978; Baker, 1981; Hawkesworth et al., 1982; Francis et al., 1983; 1984; Sparks et al., 1985). The reported ages of the central Andean ignimbrites range from about 25 Ma to 1 Ma (Clark et al, 1967; Baker, 1977).

The existence of a low velocity layer beneath the recent and ancient volcanic front at depths of about 10 to 30 km is argued by Schwarz et al. (1984), Arenada et al. (1985), Wigger (1986) and Reutter et al. (1986) in favour of the still existing acidic melts in the crust beneath the Pre- and Western Cordillera.

Petrography

Detailed microscopical descriptions including investigations on minerals by universal stage methods have been published by Zeil & Pichler (1967), Pichler & Zeil (1972), and Kussmaul et al. (1977). Additional research has been done by the present authors.

Mineral	Alkalirhyolite	Rhyolite		Rhyodacite		Dacite	
alkalifeldspar	$Or_{7-2}Ab_{79-81}An_{1-13}$ $Or_{56-60}Ab_{39-43}An_{1-8}$	$Or_{56-68}Ab_{37-40}An_{1-2}$		$Or_{7}Ab_{73}An_{20}$		---	
		phenocryst	microlite	phenocryst	microlite	phenocryst	microlite
plagioclase	---	An_{45-48}	n.d.	An_{33-48} An_{22-34} An_{42-83} An_{38-46} An_{34-42} An_{38-51}	n.d. An_{44} n.d. An_{39-43} n.d. An_{41-62}	An_{37-44}	An_{56}
biotite	n.d.	n.d.		meroxene MgO:15.6-16.6 % and TiO_2: 2.8-3.6 % and MgO: 15.4-16.8 % and TiO_2: 4.4-5.2 %		n.d.	
hornblende	n.d.	hastingsite MgO:13-15 % and TiO_2: 1.5 %		hastingsite MgO: 14-15 % and TiO_2: 2-3 %		n.d.	
clinopyroxene	---	---		---		augite $Mg_{43-44}Fe_{14-16}Ca_{43-45}$	
orthopyroxene	---	hypersthene $Mg_{64}Fe_{84}Ca_1$		hypersthene $Mg_{64-66}Fe_{84-32}Ca_1$ $Mg_{63-66}Fe_{84-36}Ca_1$ $Mg_{66-69}Fe_{29-33}Ca_{1-2}$		hypersthene $Mg_{62}Fe_{36.8}Ca_{1.8}$	
iron-titanium oxides	n.d.	magnetite TiO_2: 3.7-9 %		magnetite-ilmenite TiO_2: 1-20 % TiO_2: 46.6 %		magnetite TiO_2: 3.5-11 %	

Table 2 Results of microprobe analyses of rock-forming minerals from alkalirhyolitic to dacitic ignimbrites (four groups).

The modal content of the phenocrysts is quite variable. Felsic components, such as quartz, sanidine (in part very inhomogeneous), and plagioclase, predominate. Mafic minerals, such as biotite, hornblende, augite, and bronzite-hypersthene, are very rare in ignimbrites (table 2).

Phenocrysts, microlitic lath-shaped feldspars and opaque components have been analysed by an ARL microprobe analyser. Variations of orthoclase contents within a single sanidine phenocryst of an alkalirhyolitic ignimbrite are noteworthy: Or_{7-9} with a Ba content of 358 ppm contrasting Or_{56-60} with a Ba content of 3256 ppm. The sanidines in the rhyolitic and rhyodacitic samples only show slight chemical inhomogeneities, which can be explained by submicroscopical antipertitic clusters of exsolution.

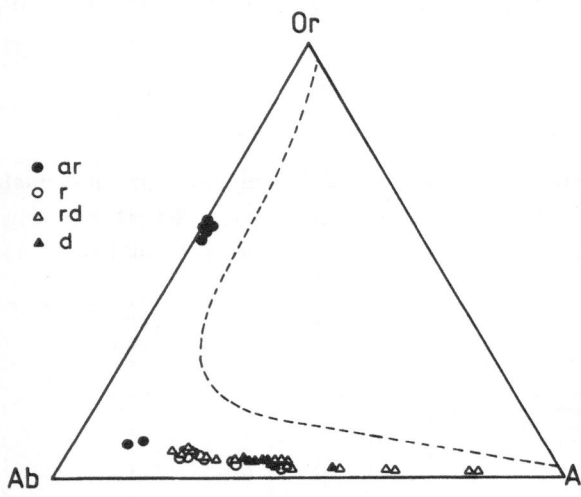

Fig. 2 Or-Ab-An compositions of feldspars from the four ignimbrite groups by way of microprobe analyses.

The anorthite content of the microlitic lath-shaped plagioclases in the pseudo-fluidal groundmass of most of the ignimbrites is often higher than in the phenocrysts (table 2). Plagioclase phenocrysts show a multiple zoning, or a combined discontinuous, oscillatory and convolute zoning (Schmitt-Riegraf and Pichler 1987), or a multiple zoning with partially oscillatory zones (fig. 3).

The anorthite content ranges from An_{23} to An_{83}. The biotites are oxidized meroxenes to lepidomelanes. The hornblendes are weakly to moderately oxidized. Pure green hastingsitic ones rarely exist. Clinopyroxenes of augitic composition have been recognized only in some dacitic ignimbrites. Orthopyroxene of bronzitic to hypersthenic composition have been found rarely in rhyolitic to dacitic ignimbrites (table 2, fig. 4).

The analyses of the iron-titanium oxides have been calculated after Carmichael (1967) on the basis of the ulvospinel or the ilmenite structure to estimate the Fe^{2+} and Fe^{3+} content. The ores are mostly magnetites with low TiO_2 values (table 2). The products of opacitization of biotites only show higher values of TiO_2. Based on the mineral calculations these alteration products are called "titanomagnetites" with ulvospinel-structure (TiO_2 about 20%) and ilmenite with ilmenite-structure (TiO_2 about 47%).

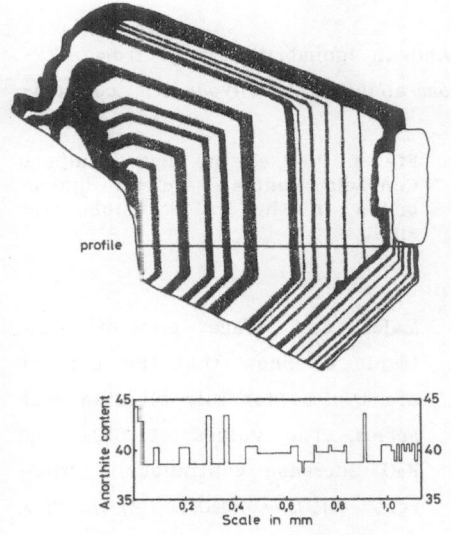

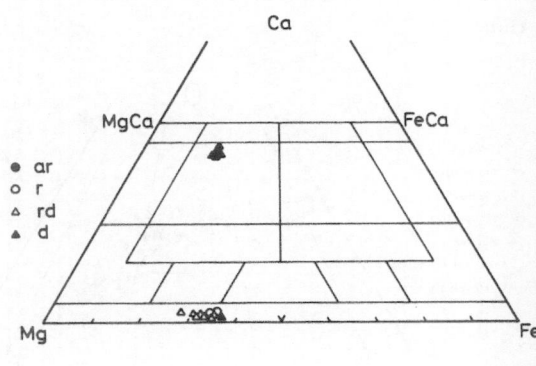

Fig. 3 A strongly corroded, euhedral plagioclase phenocryst with
a multiple zoning and small differences in the anorthite content of
the various, partially oscillatory
zones. Dacitic ignimbrite.

Fig. 4 Composition of pyroxenes from the
four ignimbrite groups by way of microprobe
analyses.

In some rhyodacitic ignimbrites the analyses of coexisting iron-titanium oxides could be
used for geothermometry. Using the $FeO-TiO_2-O_2$ system (Buddington & Lindsley, 1964;
Carmichael, 1967; Spencer & Lindsley, 1981) temperatures of 700±50°C have been calculated. The sensitive response of the Fe^{2+}/Fe^{3+} ratio towards oxidation during cooling
is problematic. A lateral oxidation of iron during eruption also influenced the TiO_2 distribution. Therefore most iron-containing minerals in the more or less welded ignimbrites have been oxidized during violent eruptions, former informations on temperature
and oxygen fugacity are destroyed or, at least, very uncertain.

The mineral analyses demonstrate that most of the investigated minerals were in chemical disequilibrium with the groundmass. The plagioclase phenocrysts with extremely
high anorthite content and the feldspar microlites with An content higher than the
phenocrysts could not have crystallized in the same acidic melt. This disequilibrium
points either to extensive contamination with upper crustal rocks or to "magma-mixing".

The glass shards in ignimbrites show sphaerolitic texture. Powder diffraction data showed that the devitrification-minerals are low-cristobalite and (K,Ca)-analbite. In
respect to the petrographic and geochemical characteristics the central Andean ignimbrites are very similar to those from the surroundings of Popayan, Central Cordillera of
southern Colombia (Schmitt, 1983).

Geochemistry

In the QAPF-double-triangle (fig. 5) the central Andean ignimbrites range from alka-lirhyolitic to dacitic composition and the lavas from andesitic to rhyodacitic composition.

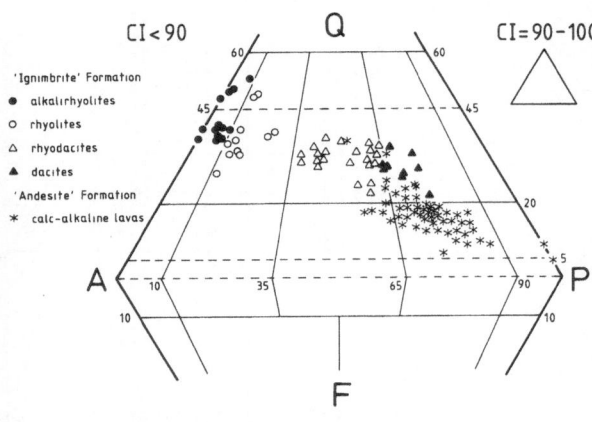

Fig. 5 Plot of the new analysed Cenozoic central Andean ignimbrites in the QAPF double triangle.

Major and trace element data (table 3) show that the samples are typical for calc-alkaline rock series. The values of SiO_2 and K_2O increase continuously, whereas Al_2O_3, Fe_2O_3, MgO, CaO, TiO_2 and P_2O_5 decrease. This decrease can be explained by plagioclase and augite segregation.

In the calc-alkaline rocks apatite occurs mostly as vermicular crystals within plagioclase phenocrysts. This plagioclase segregation also diminishes the P_2O_5 content. TiO_2 occurs in all mafic minerals. As much as those are segregated, TiO_2 decreases. Furthermore, most of the biotites and hornblendes host inclusions of iron-titanium oxides and zircon. This fact explains a decrease of the Zr and Th contents in the more fractionated rocks.

Major element analyses alone do not allow to differentiate between the following types of magma genesis: (i) fractional crystallization from mantle-derived melts, (ii) partial melting of subducted oceanic crust, or (iii) large scale bulk melting of lower continental crust producing massive volumes of intermediate composed magmas, followed by high-level plagioclase-dominated fractional crystallization. The combination of relative high amounts of SiO_2, K_2O and Al_2O_3, which result in normative cordierite, points to an anatectic sialic origin after Zeil & Pichler (1967).

Table 3 demonstrates the distribution of all determined major and trace elements. Rb increases with growing K_2O, whereas Ba is irregulary distributed. Sr diminishes with increasing SiO_2, whereas Zr increases from dacite to rhyodacite and decreases from rhyolite to alkalirhyolite. Ta increases and Cr decreases from dacite to alkalirhyolite. The total amounts of Th, Hf, and Cs show no characteristic developments. U and Sc decrease from dacite to rhyolite.

Interelement correlation in segregating minerals can be recognized by comparing normalized graphs: The values of twenty ions have been selected and are normalized to primitive mantle composition (system developed by P. Möller, personal communication, 1986). The elements are given in the order of their valency and ionic radii. All normalized contents are plotted in a logarithmic scale (fig. 6a–d). Large differences are only to be seen between the different rock types. Comparing the four ignimbrite types some differences in the element distribution patterns are obvious. The element distribution patterns of the monovalent cations are very similar. Although all patterns show a great variability in Rb and Cs, one observes a high increase from Na to K and a smaller increase from K to Rb and Cs. The general positive trend of curves of the divalent cations is nearly the same for dacitic to alkalirhyolitic ignimbrites. Their shift to lower values is caused by the increasing losses from Ba to Mg. The bend of the curves at iron may be caused by changes in oxidation state and the decrease in Ca. The impoverishment in Sc and the formation of an increasing Eu anomaly from dacite to alkalirhyolite is obvious. The contents of the other REE are very similar. The quadrivalent cations from Ti to Th show a trend behaviour similar to that of the divalent ions. From dacitic to alkalirhyolitic ignimbrites a continuous plagioclase-dominated segregation is observed. It causes the diminishing in Al_2O_3, Ca, Sr, Ba and Eu. A contemporary small pyroxene crystal fractionation leads to a slight impoverishment in Mg, Ca, Fe, Ti, Mn, Sc, Cr and Ni. Furthermore, with increasing plagioclase segregation, the acidity of the residual magma rises. The higher the SiO_2 content of the liquid, the higher are the amounts of the easily volatilized components (H_2O, CO_2, PO_4^{3-}). Thus the explosion index grows and the possibility of a violent eruption increases.

At a certain degree of acidity in the melt of nearly rhyolitic composition the trace elements Zr, Th, and U will be fixed in distinct mineral phases. Thus a further enrichment in the residual melt is impossible. Therefore, most of the rhyolitic to alkalirhyolitic ignimbrites show no continuous increase, but a slight decrease in these elements, in spite of the progressive differentiation.

Isotopic Data

For the discussion of the magma genesis isotopic analyses and K/Ar-ages from the literature were taken into consultation (Thorpe et al., 1979; Hawkesworth et al., 1982; Francis et al., 1984).

In fig. 7 the relation between geological ages (abscissa) and initial $^{87}Sr/^{86}Sr$ ratios (ordinate) is shown. The values within the figure are $^{87}Rb/^{86}Sr$ ratios. The dotted line shows the assumed $^{87}Sr/^{86}Sr$ development in the primitive andesitic magma as produced by the continuous subduction of oceanic crust or by partial melting of the lower crust.

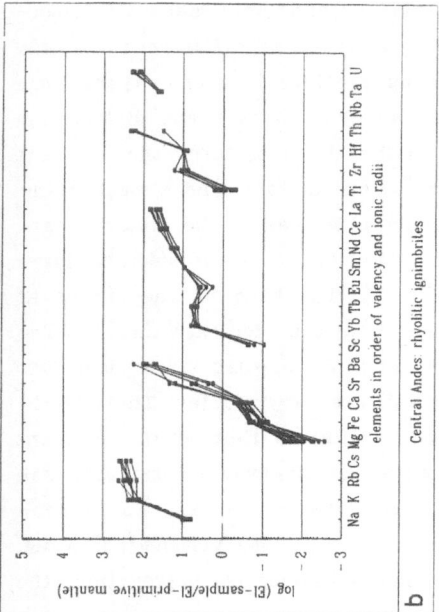

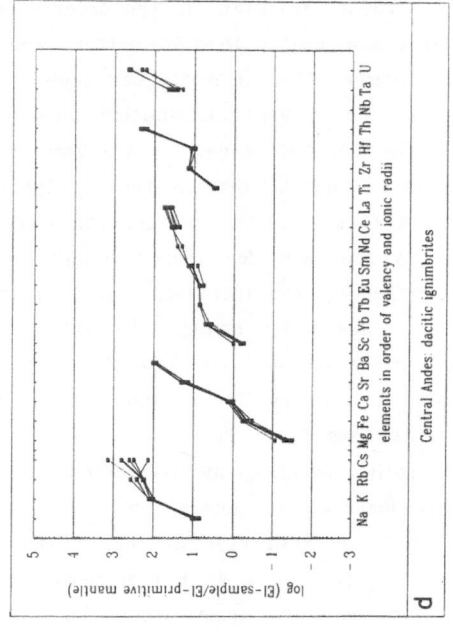

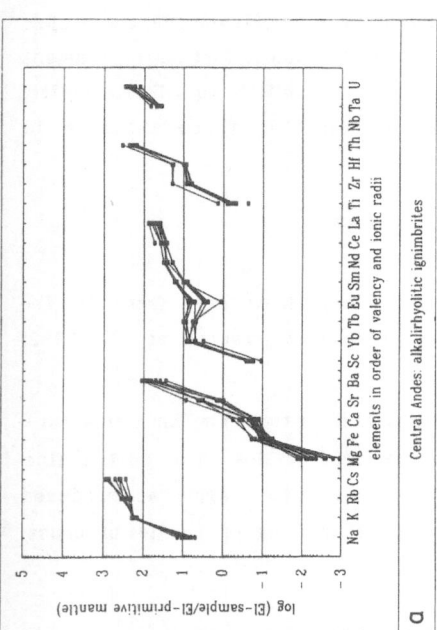

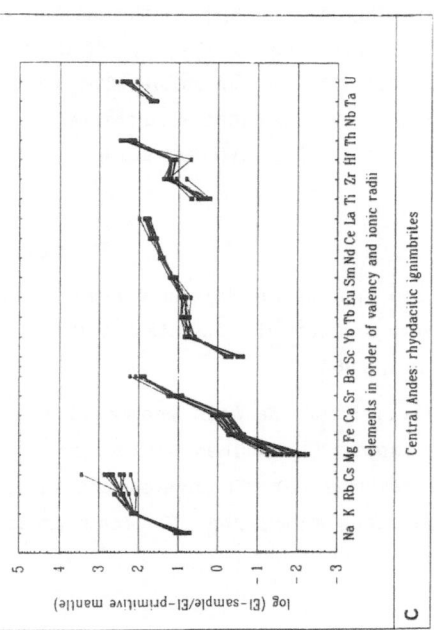

Fig. 6a–d Element distribution patterns of primitive mantle–normalized samples from the central Andes: **a)** alkalirhyolitic ignimbrites; **b)** rhyolitic ignimbrites; **c)** rhyodacitic ignimbrites; **d)** dacitic ignimbrites.

	Dacites		Rhyodacites		Rhyolites		Alkalirhyolites	
	average	range	average	range	average	range	average	range
SiO₂	64.6	62.9 – 66.3	68.4	62.9 – 73.4	74.4	71.7 – 84.1	76.4	72.3 – 80.4
Al₂O₃	15.8	15.3 – 16.4	15.1	12.5 – 16.1	12.9	12.2 – 14.4	12.3	10.3 – 14.2
Fe₂O₃	4.5	3.2 – 5.1	3.2	2.1 – 4.7	1.4	0.8 – 2.0	0.9	0.5 – 1.8
MnO	0.06	0.05 – 0.09	0.05	0.01 – 0.09	0.06	0.04 – 0.08	0.05	0.04 – 0.09
MgO	1.7	1.2 – 3.2	1.0	0.2 – 2.2	0.5	0.2 – 0.8	0.2	0.02 – 0.5
CaO	4.1	3.6 – 4.9	3.0	2.0 – 4.6	1.2	0.6 – 1.5	0.7	0.4 – 2.0
Na₂O	3.3	3.3 – 1.8	3.5	1.9 – 4.6	3.4	2.3 – 3.9	3.3	1.9 – 5.2
K₂O	3.3	3.3 – 3.6	3.7	3.0 – 4.3	4.4	3.6 – 5.4	4.7	4.0 – 5.5
TiO₂	0.63	0.5 – 0.7	0.52	0.3 – 1.0	0.21	0.2 – 0.3	0.13	0.1 – 0.3
P₂O₅	0.16	0.11 – 0.23	0.16	0.09 – 0.23	0.07	0.02 – 0.16	0.02	0.01 – 0.05
number	9		30		17		28	
Sc	11.5	9.2 – 17.3	6.7	4.2 – 11.2	3.0	1.6 – 4.2	3.6	1.9 – 4.4
Cr	38	14 – 112	9	6 – 13	5	4 – 7	3	2 – 4
As	14	6 – 27	9	5.4 – 23	10	6.8 – 13.8	9	6 – 14
Rb	148	110 – 245	187	76 – 250	160	95 – 273	186	135 – 186
Sb	1.47	0.6 – 2.6	0.64	0.35 – 1.0	0.8	0.58 – 1.2	0.87	0.56 – 2.56
Cs	6.8	3.6 – 16	5.0	1.9 – 7.1	3.5	2.4 – 4.7	6.7	3.3 – 10.7
Ba	720	630 – 803	732	526 – 999	556	339 – 755	544	205 – 544
Hf	3.9	3.29 – 4.28	4.8	1.7 – 5.9	3.2	2.67 – 3.45	3.9	2.96 – 6.77
Th	19.1	14.7 – 21.9	20.4	12.0 – 29.5	14.5	3.0 – 20.8	19.6	14.4 – 33
U	7.5	4.2 – 12	6.0	3.0 – 9.7	3.5	3.0 – 4.5	7.1	3.3 – 7.9
Zn	77	65 – 90	72	50 – 107	40	20 – 60	43	20 – 60
Sr	392	304 – 476	304	177 – 427	163	39 – 484	71	22 – 192
Y	19	13 – 21	21	14 – 27	20	16 – 24	24	8 – 31
Zr	149	133 – 160	174	163 – 270	120	82 – 181	84	67 – 210
La	29.8	22.3 – 36.3	42.2	34.4 – 62.8	32.9	25.5 – 45.1	32.3	26.6 – 49.1
Ce	60.3	52.9 – 72.8	80.7	66 – 104	61.4	52.1 – 74.1	64.7	49 – 103
Nd	27.6	23.9 – 31.2	32.2	28.5 – 36.5	21.1	16 – 24.2	29.6	22.3 – 38
Sm	4.9	3.26 – 5.66	5.6	4.52 – 6.87	3.6	3.54 – 3.78	4.8	3.57 – 6.77
Eu	1.04	0.85 – 1.12	1.09	0.72 – 1.29	0.46	0.27 – 0.59	0.62	0.42 – 1.15
Tb	0.7	0.68 – 0.89	0.7	0.51 – 0.89	0.51	0.42 – 0.62	0.69	0.05 – 1.0
Yb	1.93	1.70 – 2.08	2.19	1.63 – 3.0	0.20	0.17 – 0.26	0.26	0.13 – 0.35
Lu	0.31	0.24 – 0.34	0.34	0.23 – 0.42	0.32	0.28 – 0.39	0.37	0.17 – 0.49
number	6		11		5		9	

Table 3 Average major (in %) and trace element (in ppm) composition of the investigated ignimbrite samples.

The Cerro Purico Shield Complex (northern Chile) has produced both acidic ignimbrites and calc-alkaline lavas of basic andesitic to rhyolitic composition during the last 12 Ma in which at least five long-lasting inactive periods have been interrupted by episodes of violent eruptions.

Petrogenesis

Most authors relate the petrogenesis of the central Andean ignimbrites to anatexis of crustal rocks, or to the contamination of basaltic or andesitic magma with continental crustal material, or to the fractional crystallization of mantle-derived andesitic magma (Zeil & Pichler, 1967; Pichler & Zeil, 1972; Jakes & White, 1972; Lefèvre et al., 1976; Dostal et al., 1977a; Dostal et al., 1977b; Kussmaul et al., 1977; Thorpe et al., 1979; Francis et al., 1980; Stern et al. 1984; Sparks et al., 1985). These aspects will be discussed as follows:

(1) **Mantle-derived calc-alkaline magma**: The major and trace element patterns, including the REE of the dacitic ignimbrites show a similar distribution as the calc-alkaline lavas from the same area. The lavas show only slightly lower Th and U values. The ignimbrites probably have the same source as the andesites (Thorpe & Francis, 1979), but show a distinctive different development and eruption mechanism. James (1978) supposes that the isotopic and trace element composition of the central Andean volcanics can plausibly be explained by partial melting of subducted greywacke and altered oceanic basalt, and the subsequent reaction of that melt with the overlying mantle material in order to yield andesitic magma. Thus the common parental magma source could be the modified mantle wedge above the downgoing oceanic crust. The recycling hypothesis of older continental material by subduction and subsequent infiltration in the overlying mantle is also favoured by Harmon (1986) on the basis of the stable and radiogenic isotopes including ^{10}Be. Further reasons for a mantle-derived parental magma are the $^{87}Sr/^{86}Sr$ ratios. An exclusively anatectic upper crustal origin for the ignimbrites would have lead to much higher Sr initial ratios; in any case values higher than 0,71 have to be expected. The $^{87}Rb/^{86}Sr$ ratios >3 are in disagreement to the low $^{87}Sr/^{86}Sr$ ratios.

(2) **Contamination of mantle magma with lower or upper crustal anatectic melts**: In comparison with normal island-arc volcanics (Gill, 1981) the high SiO_2, Al_2O_3, K_2O, Rb, U, Th, and REE values (table 3), higher $^{87}Sr/^{86}Sr$, high $^{87}Rb/^{86}Sr$ and $^{18}O/^{16}O$ ratios (Déruelle et al., 1983; Harmon & Hoefs, 1984) of the magmas seem to be the result of contamination with crustal material. James (1981, 1984) postulated that Sr isotope ratios did not change during magmatic differentiation. Any variation in the isotopic ratios in the central Andean volcanic rock sequences must then be due to source inhomogeneity or to contamination. This argumentation can be re-

jected by the observed progressive plagioclase segregation in the magma chamber. The development in the interelement correlation graphs from dacitic to alkalirhyolitic and andesitic to rhyodacitic lavas suggests a plagioclase-dominated fractional crystallization and segregation or various stages of crustal partial melting. An exclusively anatectic origin is not likely for reason of the radiogenic isotopes. Plagioclase segregation, however, in the large magma chamber leads to the impoverishment of Ca, Ba, Sr and Eu^{2+} in the residual melt.

(3) **Fractional crystallization**: The rhyodacites to alkalirhyolites are characterized by the depletion of Ca, Sr and Ba in combination with the development of a negative Eu anomaly in contrast to the dacites. This phenomen can be explained by the segregation of calcic plagioclase. Both alkalifeldspar and plagioclase enrich Eu^{2+} relative to other REE (Möller & Muecke, 1984). Thus the subtraction of calcic plagioclase from the dacitic liquid promotes further element fractionation. Because feldspars are good hosts for Sr, Ba and Eu^{2+}, the residual melt fractions are impoverished in these elements by the crystallization of plagioclase (fig. 6a-d). Ca decreases relatively to Fe. Si increases relatively to Al_2O_3 (table 3). Both $^{87}Rb/^{86}Sr$ and $^{87}Sr/^{86}Sr$ ratios increase with progressive plagioclase-dominated fractional crystallization. The latter relation can be qualitatively seen in the example of the Cerro Purico Shield complex (fig. 7).

(4) **A new model of ignimbrite magma genesis**: A model of magma genesis is proposed for the Cerro Purico Shield complex (fig. 7). This hypothesis can probably be generalized for the central Andean Cenozoic ignimbrites:

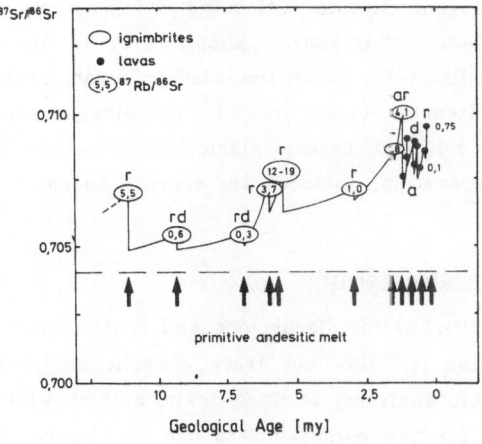

Fig. 7 Relation between geochronological results and initial $^{87}Sr/^{86}Sr$ and $^{87}Rb/^{86}Sr$ ratios in the volcanic rocks from the Cerro Purico Shield complex of northern Chile as an example for volcanic activity and magmatic evolution (a=andesite, ar=alkalirhyolite, d= dacite, r=rhyolite rd=rhyodacite).

(i)   The first erupted rhyolitic (r) ignimbrite has relatively low $^{87}Sr/^{86}Sr$ and high $^{87}Rb/^{86}Sr$ ratios. Major and trace element data point to an origin from a modified mantle-derived magma or from a lower crustal partial melt with both crustal contamination and plagioclase-dominated fractional crystallization.

(ii) Along with the eruption the magma chamber has been refilled with andesitic melt ($\approx$65%) which is proved by the lower isotope ratios. Subsequently, the mixed magma fractionated.

(iii) A new eruption, now of rhyodacitic composition, took place.

(iv) The refilling of the magma chamber followed ($\approx$20%). During a long period of inactivity the magma fractionated to become rhyodacitic in composition.

(v) Only a very small portion of andesitic melt has been mixed with the residual melt in the magma chamber.

(vi) Intensive plagioclase-dominated fractionation and segregation took place leading to the production of rhyolitic melts.

This sequence of volcanic rock production has been repeated several times and were followed by the outpouring of calc-alkaline lavas produced in the same way during the last 1,3 Ma. The high $^{87}Rb/^{86}Sr$ ratios (3,7-19) in contrast to low $^{87}Sr/^{86}Sr$ ratios can be caused by:

1. segregation of great quantities of plagioclase in the "last minute" before eruption;
2. contamination with great amounts of carbonatic rocks from the lower or upper crust;
3. high quantities of Rb solved by the agressive fluid phases in the ignimbritic melts out of micas originating from the surroundings of the magma chamber in the upper continental crust.

Conclusions

The central Andean dacitic to alkalirhyolitic ignimbrites are characterized by major, trace and REE patterns and initial radiogenic isotope ratios indicating a common source to the contemporaneous calc-alkaline lavas. This source, supposed to be the boundary zone upper mantle/lower crust, was modified by subducted oceanic crust, explained by the existence of ^{10}Be. Furthermore, intensive upper crustal contamination or magma-mixing with late-stage anatectic melts followed. Lateral plagioclase-dominant fractional crystallization including progressive segregation, produced the central Andean volcanics.

Acknowledgements

Gratefully thanks are due to P. Möller (Berlin) for discussions and constructive criticism of the manuscript; to P. Dulski (Berlin) for REE and trace element analyses; to H. Friedrichsen (Tübingen) for major element analyses; to W. Zeil and S. Pichowiak (Berlin) for providing samples; to M. Buchelt for the permission to use his interpretation of satellite images; to the Mineralogisch-Petrographisches Institut der Universität Tübingen for the permission to use the microprobe analyser; to the Deutsche Forschungsgemein-schaft for financial support of the studies in the one-year DFG project "Junge Vulka-nite der Mittleren Anden".

References

ARENADA, M., CHONG, G., GÖTZE, H.-J., LAHMAYER, B., SCHMIDT, S. & STRUNK, S. (1985): Gravimetric modelling of the northern Chilean lithosphere (20°-26° latitude south).- In: Actas IV Congreso Geologico Chileno, Antofagasta, 18-34; Antofagasta.

BAKER, M.C.W. (1977): Geochronology of Upper Tertiary volcanic activity in the Andes of North Chile.- Geol. Rdsch., 66, 455-465; Stuttgart.

-- (1981) The nature and distribution of Upper Cenozoic ignimbrite centres in the Central Andes.- J. Volcanol. geotherm. Res., 11, 293-315; Amsterdam.

BAKER, M.C.W. & FRANCIS, P.W. (1978): Upper Cenozoic volcanism in the central Andes - ages and volumes.- Earth and planet. Sci. Lett., 41, 175-187; Amsterdam.

BUDDINGTON, A.F. & LINDSLEY, D.H. (1984): Iron-titanium oxide minerals and their synthetic equivalents.- J. Petrol., 5, 310-357; Oxford.

CARMICHAEL, I.S.E. (1967): The iron-titanium oxides of salic volcanic rocks and their associated ferromagnesian silicates.- Contrib. Mineral. Petrol., 14, 36-64; Berlin/Heidelberg/New York.

CLARK, A.H., MAYER, A.E.S., MORTIMER, C.. SILLITOE, R.H., COOKE, R.V. & SNELLING, N.J. (1967): Implications of the isotopic ages of ignimbrite flows, southern Atacama desert, Chile.- Nature, 215, 723-724; London.

DÉRUELLE, B., HARMON, R.S. & MOORBATH, S. (1983): Combined Sr-O isotope relationship and petrogenesis of Andean volcanics of South America.- Nature, 302, 814-816; London.

DOSTAL, J., ZENTILLI, M., CAELLES, J.C. & CLARK, A.H. (1977): Geochemistry and origin of volcanic rocks of the Andes (26°-28°S).- Contrib. Mineral. Petrol., 63, 113-118; Berlin/Heidelberg/New York.

DOSTAL, J., DUPUY, C. & LEFEVRE, C. (1977): Rare earth element distribution in Plio-Quaternary volcanic rocks from Southern Peru.- Lithos, 10, 173-183; Amsterdam.

EL-HINNAWI, E.E., PICHLER, H. & ZEIL, W. (1969): Trace element distribution in Chilean ignimbrites.- Contrib. Mineral. Petrol., 24, 50-62; Berlin/Heidelberg/New York.

FRANCIS, P.W., MCDONOUGH, W.F., HAMMILL, M., O'CALLAGHAN, L.J. & THORPE, R.S. (1984): The Cerro Purico Shield complex, north Chile.- In: HARMON, R.S. & BARREIRO, B.A. (eds.), Andean magmatism, chemical and isotopic constraints, 106-123, Shiva Publ.; Nantwich.

FRANCIS, P.W., O'CALLAGHAN, L.J., KRETZSCHMAR, G.A., THORPE, R.S., SPARKS, R.S.J., PAGE, R.N., BARRIO, R.E. DE, GILLON, G. & GONZALEZ, O.E. (1983): The Cerro Galan ignimbrite.- Nature, 301, 51-53; London.

FRANCIS, P.W., ROOBOL, M.J., WALKER, G.P.L., COBBOLD, P.R. & COWARD, M. (1974): The San Pedro and San Pablo volcanoes of northern Chile and their hot avalanche deposits.- Geol. Rdsch., 63, 357-388; Stuttgart.

FRANCIS, P.W. & RUNDLE, C.C. (1976): Rates of production of the main magma types in the central Andes.- Bull. geol. Soc. Amer., 87, 474-480; New York.

FRANCIS, P.W., THORPE, R.S., MOORBATH, S., KRETZSCHMAR, G.A. & HAMMILL, M. (1980): Strontium isotope evidence for crustal contamination of calcalkaline volcanic rocks from Cerro Galan, Northwest Argentina.- Earth and planet. Sci. Lett., 48, 257-267; Amsterdam.

GILL, J. (1981): Orogenic andesites and plate tectonics.- Springer; Berlin, 390 pp.

HARMON, R.S. (1986): Overview of stable and radiogenic isotope variation along the Andean Cordillera between 5°N and 46°S.- Lecture at the Workshop 'Orogeny on Continental Margins', February 19th-22nd, Berlin, 1986.

HARMON, R.S. & HOEFS, J. (1984): Oxygen isotope ratios in late Cenozoic Andean volcanics.- In: HARMON, R.S. & BARREIRO, B.A. (eds.), Andean magmatism, chemical and isotopic constraints, 9-20, Shiva Publ.; Nantwich.

HAWKESWORTH, C.J., HAMMILL, M., GLEDHILL, A.R., CALSTEREN, P. van & ROGERS, G. (1982): Isotope and trace element evidence for late-stage intra-crustal melting in the High Andes.- Earth and planet. Sci. Lett., 58, 240-254; Amsterdam.

JAKES, P. & WHITE, A.J.R. (1972): Major and trace-element abundances in volcanic rocks of orogenic areas.- Bull. geol. Soc. Amer., 83, 29-40; New York.

JAMES, D.E. (1971): Plate tectonic model for the evolution of the Central Andes.- Bull. geol. Soc. Amer., 82, 3325-3346; New York.

-- (1978): On the origin of the calc-alkaline volcanics of the central Andes: a revised interpretation.- Yb. Carnegie Instn. Washington, 77, 562-590; Washington.

-- (1981): Role of subducted continental material in the genesis of calc-alkaline volcanics of the central Andes.- In: KULM, L.D., DYMOND, J., DASCH, E.J.,& HUSSONG, D.M. (eds)., Nazca Plate: Crustal formation and Andean convergence.- Mem. geol. Soc. Amer., 154, 769-790; New York.

-- (1984): Quantitative models for crustal contamination in the Central and Northern Andes.- In: HARMON, R.S. & BARREIRO, B.A. (eds.), Andean magmatism, chemical and isotopic constraints, 124-138, Shiva Publ.; Nantwich.

JAMES, D.E., BROOKS, C. & CUYUBAMBA, A. (1976): Andean Cenozoic volcanism: Magma genesis in the light of strontium isotopic composition and trace-element geochemistry.- Bull. geol. Soc. Amer. 87, 592-600; New York.

KLERKX, J., DEUTSCH, S., PICHLER, H. & ZEIL, W. (1977): Strontium isotopic composition and trace element data bearing on the origin of Cenozoic volcanic rocks of the central and southern Andes.- J. Volcanol. geotherm. Res., 2, 49-71; Amsterdam.

KUSSMAUL, S., HÖRMANN, P.K., PLOSKONKA, E. & SUBIETA, T. (1977): Volcanism and structure of southwestern Bolivia.- J. Volcanol. geotherm. Res., 2, 73-111; Amsterdam.

LEFEVRE, C., HAMEL, J. & DUPUY, C. (1976): Rapports $^{87}Sr/^{86}Sr$ dans les andésites et les shoshonites du Pérou.- In: 4ème Réunion annuaire des Sciences de la Terre, Paris 1976, p. 259.

MCNUTT, R.H., CROCKET, J.H., CLARK, A.H., CAELLES, J.C., FARRAR, E., HAYNES, S.J. & ZENTILLI, M. (1975): Initial $^{87}Sr/^{86}Sr$ ratios of plutonic and volcanic rocks of the central Andes between latitudes 26° and 29° south.- Earth and planet. Sci. Lett., 27, 305-313; Amsterdam.

MÖLLER, P. & MUECKE, G.K. (1984): Significance of the europium anomalies in silicate melts and crystal-melt equilibria: a re-evaluation.- Contrib. Mineral. Petrol., 87, 242-250; Berlin/Heidelberg/ New York.

PICHLER, H. & ZEIL, W. (1972): The Cenozoic rhyolite-andesite association of the Chilean Andes.- Bull. volcanol., 35 (1970), 424-452; Napoli.

PITCHER, W.S., ATHERTON, M.P., COBBING, E.J. & BECKINSALE, R.D. (1985) (eds.): Magmatism at a plate edge. The Peruvian Andes.- Blackie; Glasgow, 328 p.

REUTTER, K.-J., SCHWAB, K. & GIESE, P. (1986): Oberflächen- und Tiefenstrukturen in den Zentralen Anden.- Berliner geowiss. Abh. A, 66, 247-264; Berlin.

SCHMITT, C. (1983): Petrologische Untersuchungen junger Vulkanite in Südkolumbien.- Doctoral Thesis Univ. Tübingen, 207 pp.; Tübingen.

SCHMITT-RIEGRAF, C. & PICHLER, H. (1987): Petrogenesis and evolution of the Cenozoic "Ignimbrite" and "Andesite" Formation in the Central Andes as indicated by their trace and major element data.- Zbl. Geol. Paläont. Teil I, **1987** (7/8), 937-953; Stuttgart.

SCHWARZ, G., HAAK, V., MARTINEZ, E. & BANNISTER, J. (1984): The electrical conductivity of the Andean crust in northern Chile and southern Bolivia as inferred from magnetotelluric measurements.- J. Geophys., **55**, 169-178; Berlin/Heidelberg/New York.

SPARKS, R.S.J., FRANCIS, P.W., HAMER, R.D., PANKHURST, R.J., O'CALLAGHAN, L.O., THORPE, R.S. & PAGE, R. (1985): Ignimbrites of the Cerro Galán caldera, NW Argentina.- J. Volcanol. geotherm. Res., **24**, 205-248; Amsterdam.

SPENCER, K.J. & LINDSLEY, D.H. (1981): A solution model for coexisting iron-titanium oxides.- Amer. Mineralogist, **66**, 1189-1201; Menasha.

STERN, C.R., FUTA, K., MUEHLENBACHS, K., DOBBS, F.M., MUÑOZ, J., GODOY, E. & CHARRIER, R. (1984): Sr, Nd, Pb and O isotope composition of Late Cenozoic volcanics, northernmost SVZ (33-34°S).- In: HARMON, R.S. & BARREIRO, B.A. (eds.), Andean magmatism, chemical and isotopic constraints, 96-105, Shiva Publ.; Nantwich.

THORPE, R.S. & FRANCIS, P.W. (1979): Variations in Andean andesite composition and their petrogenetic significance.- In: Uyeda, S. (ed.), Processes at subduction zones.- Tectonophysics, **57**, 53-70; Amsterdam.

THORPE, R.S., FRANCIS, P.W. & MOORBATH, S. (1979): Rare earth and strontium isotope evidence concerning the petrogenesis of north Chilean ignimbrites.- Earth and planet. Sci. Lett., **42**, 359-367; Amsterdam.

WIGGER, P. (1986): Krustenseismische Untersuchungen in Nord-Chile and Süd-Bolivien.- Berliner geowiss. Abh. A, **66**, 31-48; Berlin.

ZEIL, W. & PICHLER, H. (1967): Die känozoische Rhyolith-Formation im mittleren Abschnitt der Anden.- Geol. Rdsch., **57**, 48-81; Stuttgart.

CENTRAL ANDEAN GRAVITY FIELD AND ITS RELATION TO CRUSTAL STRUCTURES

Götze, H.-J., S. Schmidt and S. Strunk
Institut für Geophysikalische Wissenschaften, FU Berlin
Malteserstr. 74-100, D-1000 Berlin 46

ABSTRACT

The gravity anomaly map of the Central Andes was prepared on the base of nearly 3500 gravity observations. The map consists of the terrain-corrected station complete Bouguer gravity anomalies onshore and free-air anomalies in adjacent marine areas, using the sea-level datum and a reduction density of 2.67 g/cm³. Contour intervals of 10 mGal are used. The most obvious characteristic feature of this map is the strong negative anomaly beneath the Central Andean belt. However, the data include also much information which may be directly related to geological features. In order to emphasize this fact, a regional gravity field has been calculated using a bipolynomial trend surface of 5th degree and eliminating it from the Bouguer gravity. Correlations of regional structures, represented by geological and tectonic maps, with the corresponding residual gravity map provide essential information regarding the structural provinces, the configuration of basement rocks, buried mesozoic basins as well as present and older magmatic belts.

INTRODUCTION

This article is concerned with an interpretation of a regional gravity survey carried out since 1982 by the gravity working group of the FU Berlin (Götze, 1986; Götze et al., 1987b). The study area (Fig. 1) covers the Central Andes between 20° and 26°S, extending from the Pacific Ocean coast to the western border of the Argentinean Chaco. Hence the investigated strip crosses the entire complex of the Andean orogen with its different geological units, changing from West to East in structure, composition and genesis.

The purpose of the survey was to investigate regional structural features as indicated by gravity anomalies and to relate geological and further geophysical information to these anomalies in order to support and improve ideas about the crustal behaviour of the Central Andes.

Lecture Notes in Earth Sciences, Vol. 17
H. Bahlburg, Ch. Breitkreuz, P. Giese (Eds.),
The Southern Central Andes
© Springer-Verlag Berlin Heidelberg 1988

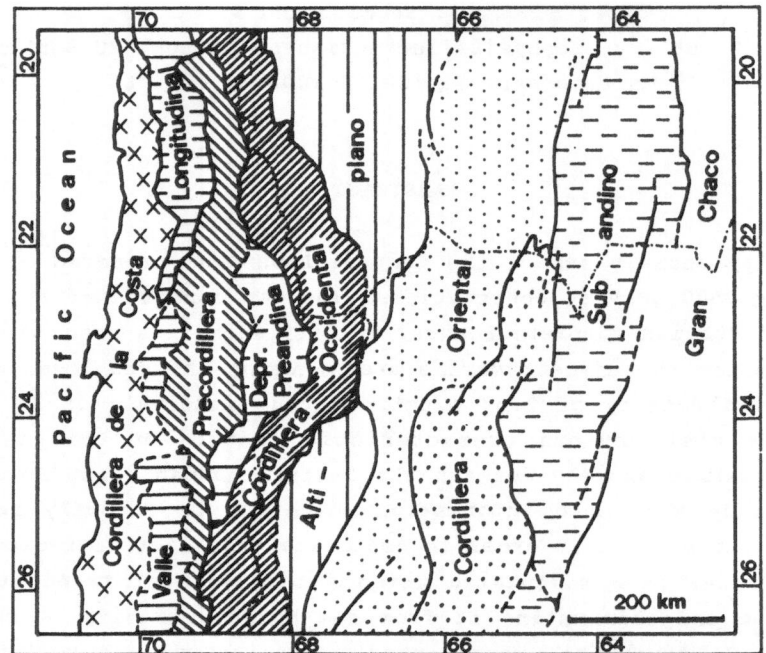

Fig. 1: Morphostructural units of the Central Andes, redrawn after Reutter et al., 1986.

DATA BASE AND BOUGUER ANOMALY

Previously existing gravity data (Dragicevic, 1970; IGM, 1974; Cerrato, 1975) within the surveyed area followed only some of the principal roads and merely were able to give a general view of the regional anomaly field. To supplement these data, gravity observations at about 2500 additional stations were made along all accessible roads and railways at roughly a 5-km interval (Araneda et al., 1985; Götze et al., 1987b). Finally the data base was completed by offshore gravity data, made available by the Bureau Gravimetrique International (Toulouse).

The resulting file of more than 3500 gravity measurements tied to the IGSN-71 gravity datum contains reduced Bouguer anomaly values using a sea-level datum and a reduction density of 2.67 g/cm³. Each station (including the older ones) is terrain-corrected using a digital topographic model of approximately 5 km x 5 km grid spacing.

Figure 2 shows the compiled Bouguer gravity map with a contour interval of 10 mGal. As usual, in the offshore area the Bouguer anomaly is replaced by the free air anomaly so that variations of the seafloor-depths are represented by directly associated anomalies: The elongated minimum of -200 mGal following the 71.5° longitude marks the strong gravimetric anomaly of the Peru Chile trench in front of the coast.

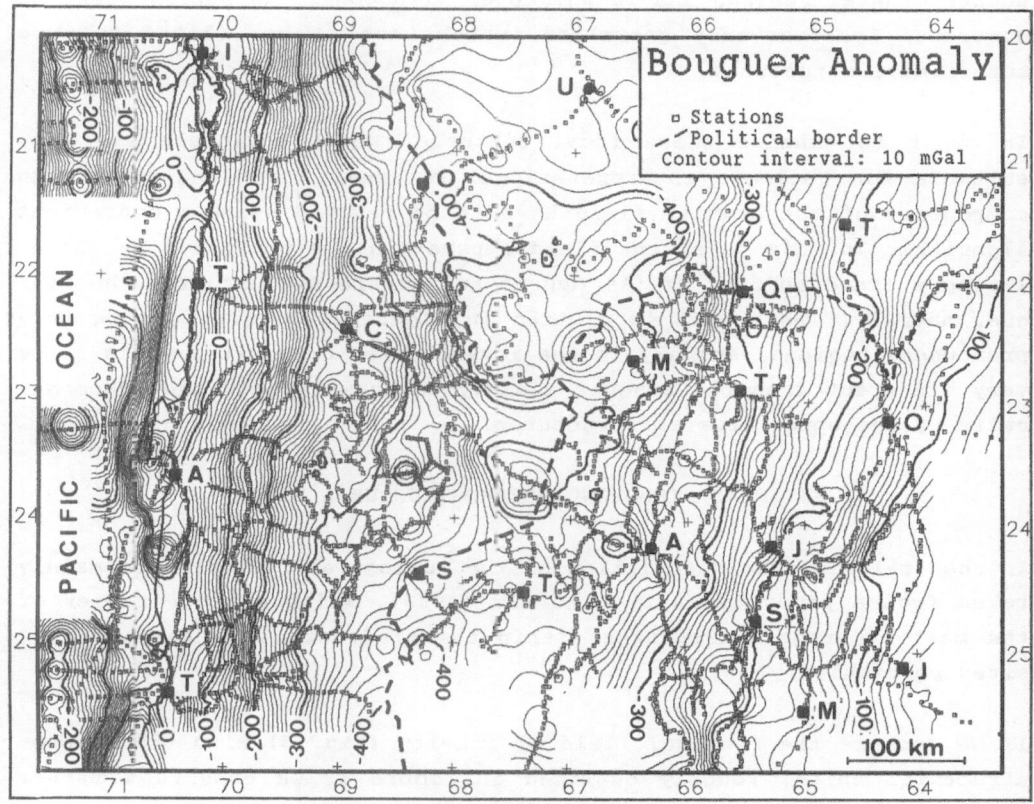

Fig. 2: Station complete Bouguer anomaly of the Central Andes. Offshore: Free-air anomaly. Contors are computer drawn and not smoothed. Contour interval: 10 mGal; datum: sea level; reduction density: 2.67 g/cm³.

Onshore the well-known regional minimum of about -400 mGal dominates, mostly related to the crustal thickening beneath the enormous excess of topographic masses. However, it has clearly been shown by model calculations, that the effect of gravity of the subducted oceanic lithosphere has to be considered too, because of its strong long-wave anomaly of up to 150 mGal (Strunk, 1985). In accordance with seismic refraction investigations (Wigger, this issue), the regional gravity field is satisfied assuming a crustal thickness increasing from about 30 km beneath the coastal line up to 70 km beneath the Central Ridge, and then decreasing further to the east to about 40 km in the Brazilian shield area.

This regional gravity anomaly caused by lateral density variations in deeper crustal regions and in the upper lithosphere is superimposed by numerous high-frequency anomalies, which have to be related to more superficial structures.

In order to enhance these local features, which obviously are more suitable for geologic interpretational purposes, the long-wavelength components of the gravity field have to be removed using a convenient algorithm. In this case, the field separation was performed by calculating a trend surface of 5th degree and subtracting it from the Bouguer anomaly. The result is plotted in Figure 3. Because of the very pronounced regional anomaly in the investigated area, it appears to be very different from the Bouguer anomaly emphasizing now the inhomogeneity of the upper crustal structures.

RESIDUAL FIELD AND MORPHOSTRUCTURAL UNITS

In the following the description of the geological features are mainly taken from a paper by Reutter et al. (1986), which gives a review of the most important structures within the study area. So they are not cited particularly.

If we compare the residual field of gravity (Fig. 3) with the morphostructural units, roughly outlined in Figure 1, we observe a rather complicated and sometimes surprising picture, which nevertheless in most cases shows a good correspondence.

The general strike direction (see Fig. 1), indicating N/S in the western part of N-Chile, NE/SW in the Argentinean area and NW/SE near the Chilean/Bolivean border is also reflected in the gravity field. One important exception is to be observed south of 22°S between 69° and 67°W, where the main gravity trend of the residual gravity anomalies is NW/SE, but the geological striking tends toward NE/SW (see below).

The Coastal Cordillera is characterized by a chain of positive anomalies, which clearly are caused by the granodioritic to basic intrusives (Jurassic to Lower Cretaceous) predominating in this structural unit together with high dense Paleozoic sediments and granitoides. This Jurassic "Magmatic Arc" morphostructurally is limited by the Longitudinal Valley, a young tectonical depression zone with sediments up to 4000 m thickness (Zeil, 1979). However, they are not all accompanied by negative gravity values. On the contrary, the positive anomalies also extend over the western part of the Longitudinal Valley. This gravimetric result is another evidence of the rather complicated behaviour of its basement, which seems to be a structural high and not a rift valley (Buchelt und Reutter, 1986).

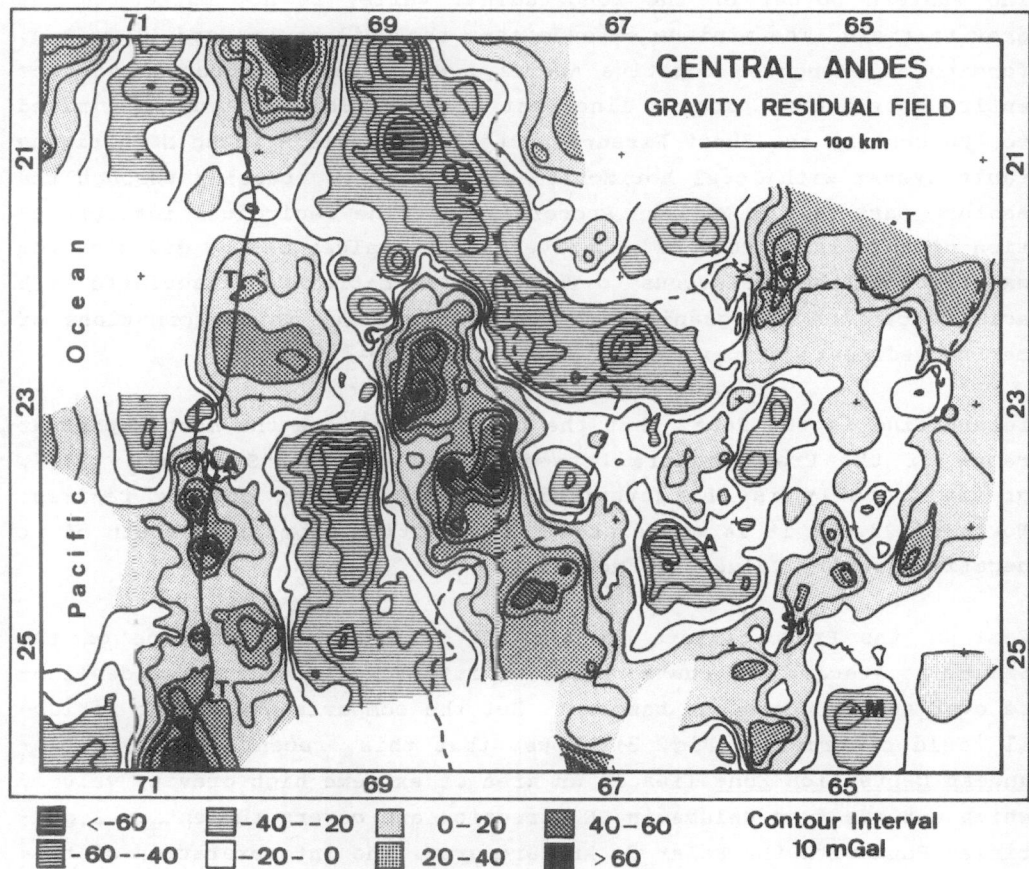

Fig. 3:   Residual Bouguer anomaly of the Central Andes, using a bipolynom 5th degree as a regional field. Hand contoured, using the computer-contoured map as a guide. Contour interval: 10 mGal.

South of 23°S the existence of the continuation of the Longitudinal Valley is not yet clear, but Reutter et al. (1986) consider a large basin south of 24.5°S, filled with vulcanites of Upper Tertiary age to be an equivalent structure. This unit is accompanied by considerable low residual gravity values, which are separated from the positive anomalies of the Coastal Cordillera by extreme high horizontal gradients northwest of Sta. Catalina (25°S, 69.8°W). The interpretation of this feature may be given by a pronounced, deep reaching fault line with NS trend, marking the western border of a depression structure ("graben"?).

The eastern border of the Longitudinal Valley is not marked in the gravity field. The minimum extends over the Chilean Precordillera too, forming an elongated negative anomaly, which can be pursued over the entire geotraverse along a line west of 69°W. This minimum is limited to the east by the "West Fissure System", which is a young NS-striking fault system with local horizontal displacements, passing through the eastern part of the Chilean Precordillera. The geological interpretation of this large zone of negative gravity values may be given by the existence of the Cretaceous to Tertiary "Magmatic Arc" associated with acid vulcanites and granitic plutons, as well as thick formations of marine sediments (Upper Triassic and Jurassic).

Interesting is the fact, that the Domeyko Range as the most important range of the Precordillera between 22.3° and 25.5°S (Chong, 1977), gravimetrically is characterized by two different gravity regimes: North of 23.5°S it is associated with positive anomalies, south of it negative gravity values are dominating.

East of the Precordillera follows the young depression zone of the Salars de Atacama and Punta Negra, an interior-draining basin developed over an inhomogeneous basement. But the comparison with the residual field of gravity (Fig. 3) shows, that this -recently active- Preandean Depression Zone lies in an area of extreme high gravity values, which extends from Calama in SE direction and covers the entire Argentinean Puna with the Salar de Arizaro area. The interpretation of this prominent high is not yet finished, but in accordance with geological observations, e.g. a compilation of age determinations by Palma et al. (1986), a high dense Paleozoic-Precambrian crustal fragment (terrane?) in the upper and lower crust is discussed by Götze et al. (1987a).

Worth mentioning is the fact, that the area of the South Bolivian Altiplano is characterized by high residual gravity values too. This structure, a high mountainous plateau (3500-4000 m), is a basin region

with a sedimentary cover (Cretaceous and Tertiary) locally reaching 15
km. Hence it may be stated, that all depressional units of larger
extent are accompanied by high gravity regimes! It seems, that the
negative gravity effect of these depressions, which all are tied to
the development of salt lakes (interior-draining) (Fig. 4), is superimposed by positive gravity effects of an underlying, very high dense
basement or basic intrusives.

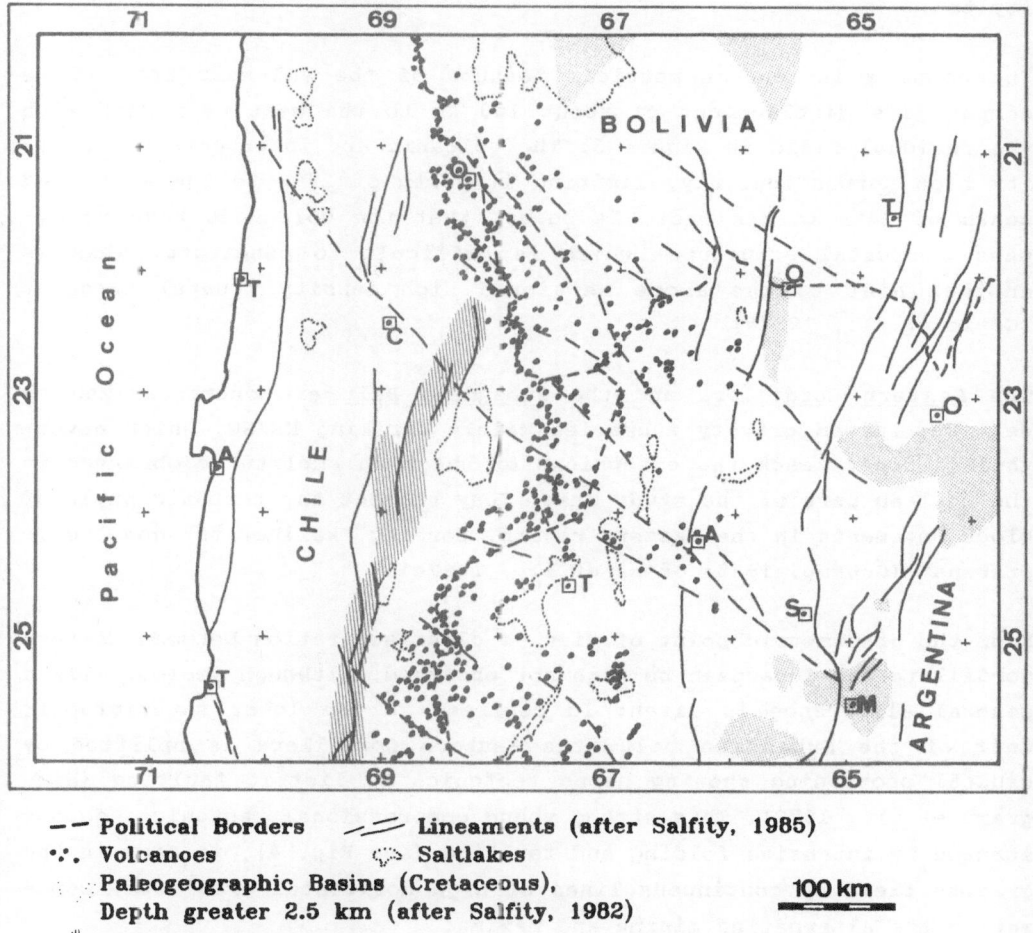

-- Political Borders   Lineaments (after Salfity, 1985)
∵. Volcanoes   Saltlakes
Paleogeographic Basins (Cretaceous),
Depth greater 2.5 km (after Salfity, 1982)   100 km
⫴ Domeyko Range

Fig. 4:   Some geologic/tectonic elements within the study area.

Morphologically the study area is dominated by the <u>Western</u> <u>Cordillera</u>
with its chain of impressive volcanoes (Upper Miocene to recent),
reaching altitudes up to 7000 m. The location of these volcanoes is
also indicated in Figure 4. This morphological unit represents the
"Volcanic Arc" of the recent subduction of the oceanic Nazca plate.

The residual gravity field marks this arc as an elongated, clearly limited area of very low values. It indicates the low dense rhyolitic nappes and ignimbrites with densities down to 2.0 g/cm³ (Zeil, pers. comm.). Magnetotelluric measurements in this region (Schwarz et al., this issue) yielded extreme low electrical resistivity (0.5 - 1 Ωm) in depths greater than 10 km, so that we have to assume recent magma chambers with molten material (very low densities) effecting the gravity too.

Interesting is one geometrical feature of the volcanic arc, if we compare its displacement of about 100 km to the west near 24°S with the residual field in Figure 3: The volcanic arc intersects the gravity high perpendicularly, limiting it north of 24°S to the east, and south of 24°S to the west. It seems, that the volcanoes have to by-pass a crustal structure, which is difficult to penetrate. This is another hint to the above mentioned high density crustal fragment (Götze et al., 1987a).

The Eastern Cordillera and the Subandean Belt are characterized by well determined gravity anomalies mainly striking NE/SW, which nevertheless don't reach the extensions or the high amplitudes observed in the Chilean part of the study area. They reflect the tectonic style of block movements in the western region, forming "keilhorsts" and "keilgrabens" (Schwab, 1985; Götze et al., 1987c).

From the gravimetric point of view, a clear separation between Eastern Cordillera and Subandean belt is not observed, although geologically a general difference is given: In contrast to the lower stratigraphic unit of the Subandean belt, the Eastern Cordillera is uplifted by crustal shortening showing nappe tectonics by listric faulting (Mingramm et al., 1979). This strong young compressional tectonics, demonstrated by intensive folding and faulting (see Fig. 4), is seen in the gravity field by continuous lines of high horizontal gradients, separating the alternating minima and maxima.

Furthermore it should be mentioned, that the residual gravity field shows a good correlation with paleogeographic elements too. The dotted areas in Figure 4 indicate the distribution of Cretaceous taphrogenic basins compiled by Salfity (1982). Comparing this map with the residual gravity (Fig. 3) it is evident, that most of the Cretaceous basins are associated with pronounced positive anomalies. This supports the observation e.g. of Gallisky and Viramonte (1987), that the Cretaceous

basin formation has been accompanied by intensive magmatism of basic
chemism. The high densities of these intrusions today produce positive
gravity values. The observed high coincidence of positive gravity
values and Cretaceous basins leads to the presumption, that the basic
magmatism has been widespread within these large basins.

CONCLUSIONS

The gravimetrical data, as one part of the geophysical data base, aid
in a better definition of the regional structural framework of the
Central Andes. Several important features, not clearly identified by
geological mapping, are recognized. So the pronounced gravity high
between Calama and the Salar de Arizaro is linked to an enormous
crustal fragment (a terrane?) which is characterized by palaeozoic
ages. From the gravimetric viewpoint one cannot tell apart the limits
of the Eastern Cordillera and Altiplano/Puna, although major fault
zones are mapped by geologists. These results will at least shed new
light on the interpretation of the geological evolution of this
subduction related orogenic area.

ACKNOWLEDGEMENTS

The gravity survey was carried out within the scope of a geoscientific
research program of Argentinean, Chilean and German Universities and
first we have to thank all our colleagues. Many institutions from
South America gave us support and our fieldwork could not be done
without their assistance. We are grateful to all of them. Particulary
we appreciate B. Lahmeyer, who joined the field campaigns in South
America and supported our interpretation work as well. The financial
support was gratefully given by the "Deutsche Forschungsgemeinschaft".

REFERENCES

Araneda, M., Chong, G., Götze, H.-J., Lahmeyer, B., Schmidt, S. &
    Strunk, S. (1985): Gravimetric modelling of the Northern Chilean
    lithosphere (20°-26° Lat. South).- 4. Congr. Geol. Chileno Actas,
    1: 2/18 - 2/34; Antofagasta.

Buchelt, M. & Reutter, K.-J. (1986): Photogeologische Untersuchungen
    an einer LANDSAT-Szene aus dem Anden Segment zwischen 20° 30' und
    22°S.- Berliner geowiss. Abh. (A), 66: 205-208; Berlin.

Cerrato, A.A. (1975): Contribuciones a la Geodesia aplicada.- 16. Gen-
    eral Assembly of the IUGG (Grenoble), Inst. de Geodesia, Univ. de
    Buenos Aires; Buenos Aires.

Chong, G. (1977): Contribution to the knowledge of the Domeyko Range in the Andes of Northern Chile.- Geol. Rdsch., 66: 374-403; Stuttgart.

Dragicevic, M. (1970): Carta gravimétrica de los Andes meridionales e interpretación de las anomalías de gravedad de Chile central.- Publ. 93, Dep. de Geofis. y Geod., Univ. de Chile; Santiago.

Gallisky, M.A. & Viramonte, J.G. (1987): Cretaceous Paleorift in Northwestern Argentina. Petrological approach.- Journal of South American Earth Sciences; Oxford. In press.

Götze, H.-J. (1986): Schweremessungen und deren Interpretation im mittleren und östlichen Teil der Anden-Geotraverse.- Final Report, DFG Project Go 380/1; Bonn. Unpublished.

Götze, H.-J., Chong, G., Lahmeyer, B., Omarini, R.H., Salfity, J.A., Schmidt, S., Strunk, S. & Viramonte, J.G. (1987a): The gravity field and its relation to the geological structures between 20° and 26°S.- In preparation.

Götze, H.-J., Lahmeyer, B., Schmidt, S., Strunk, S. & Araneda, M. (1987b): A new gravity data base in the Central Andes (20°-26°S).- EOS; under review.

Götze, H.-J., Lahmeyer, B., Schmidt, S. & Strunk, S. (1987c): Gravity field and magafault-system of the Central Andes (20°-26° L.S).- Geol. Vereinigung, 77th Ann. Meeting Basel - Abstracts. Terra Cognita, 7(1): 57; Cambridge.

Instituto Geográfico Militar (IGM) de Bolivia (1974): Mediciones Gravimétricas en Bolivia.- Departamento Geofísico; La Paz.

Mingramm, A., Russo, A., Pozzo, A. & Cazau, L. (1979): Sierras Suban-dinas.- 2. Simp. Geol. reg. Argent., Acad. nac. Cienc., 1: 95-137; Córdoba.

Palma, M.A., Parica, P.D. & Ramos, V.A. (1986): El granito Archibarca: su edad y significado tectónico.- Revista de la Asociación Geoló-gica Argentina, 41(3-4); Buenos Aires.

Reutter, K.-J., Schwab, K. & Giese, P. (1986): Oberflächen- und Tie-fenstrukturen in den Zentralen Anden.- Berliner geowiss. Abh. (A), 66: 247-264; Berlin.

Salfity, J.A. (1982): Evolución paleogeográfica del Grupo Salta (Cretácico-Eogénico), Argentina.- 5. Congreso Latinoamerica de Geología, 1: 11-26; Buenos Aires.

Salfity, J.A. (1985): Lineamientos Transversales al rumbo Andino en el Noroeste Argentino.- 4. Congr. Geol. Chileno Actas, 1, 2/119-2/1-37; Antofagasta.

Schwab, K. (1985): Basin Formation in a Thickening Crust - the Intra-montane Basins in the Puna and the Eastern Cordillera of NW-Ar-gentina (Central Andes).- 4. Congr. Geol. Chileno Actas, 1: 2/138 - 2/158, Antofagasta.

Strunk, S. (1985): Auswertung gravimetrischer Messungen und deren 3-D Interpretation im Bereich der andinen Subduktionszone Nordchiles. Diplomarbeit.- Institut f. Geophysik; Clausthal.

Zeil, W. (1979): Zur Geodynamik des Anden-Orogens.- Geologie en Mijn-bouw, 58(2): 187-192; Amsterdam.

SEISMICITY AND CRUSTAL STRUCTURE OF THE CENTRAL ANDES

Peter J. Wigger

Institut für Geophysikalische Wissenschaften der
Freien Universität Berlin
Rheinbabenallee 49, D-1000 Berlin 33

ABSTRACT

The earthquakes of the central Andes belong to the Circum-Pacific earthquake-belt and provide about 5% of its seismic energy. The central Andes are one of the most active seismic areas in the world.
As the analysis of earthquake catalogue data revealed, the energy budget of this area is dominated by a few distinct earthquake accumulations. The strongest cluster of earthquakes is located under the Pre- and Western Cordillera at a depth of about 100 km . Earthquake hypocentres define a Benioff-plane dipping with 20°-30° to the east. Between depths of 300 km and 500 km exists a seismic gap. Very deep events occur between 500 km and 700 km. A north-south-segmentation of the downgoing Nasca-plate is a well known fact: while there is a dip-angle of 20°-30° under the central Andes, it is just 10° further north and south. Energy estimations support the possibility of thermodynamic phase transitions as a cause of earthquakes within the descending Nasca plate.
Four crustal seismic profiles had been measured in 1982 and 1984. As signal source blasts of the Chuquicamata coppermine were used. Additionally seismic results from the coastal range of N-Chile, Peru and the Altiplano were used to describe the crustal structure of the Central Andes. The correlations of S- and P-wave lead to a well structured upper crust. A discontinuity in 30 km was derived where P-wave velocity increases from 6.3 to 7.2-7.4 km/s. Below that discontinuity a very thick lower crust is indicated characterized by a LVZ with a mean velocity of 6.4-6.6 km/s. Total crustal thickness reaches more than 70 km already beneath the Pre-Cordillera. The discontinuity at a depth of 30 km is interpreted as a paleo-Moho of the Jurassic crust, which was a sedimentation area of marine sediments. A hypothesis of how the anomalous thick lower crust is originated is provided by an accretion of material from below: the subducted oceanic coast partially melts at depths of greater than 100 km, rises and accumulates under the old Jurassic crust.

INTRODUCTION

The Central Andes, part of the Circum-Pacific earthquake belt, are belonging to the most active seismic areas of the world. Many publications describe the earthquakes and their spatial distribution in South America especially in Peru and Chile, e.g. STAUDER (1973, 1975), BARAZANGI & ISACKS (1976, 1979), ISACKS & BARAZANGI (1977), and HASEGAWA & SACKS (1981). As contribution to the study of the

Lecture Notes in Earth Sciences, Vol. 17
H. Bahlburg, Ch. Breitkreuz, P. Giese (Eds.),
The Southern Central Andes
© Springer-Verlag Berlin Heidelberg 1988

geodynamics of the South American active continental margin different seismological studies have been carried out by the interdisciplinary geoscientific research group working at the Central Andean Geotraverse between 21° and 25° S:

- basing on the international earthquake catalogues and sources of local networks a statistical examination of the earthquake data was made to describe seismicity and energy release in the area between 15° and 30° S in detail.
- local seismic networks with mobile recording stations had been operating in northern Chile and southern Bolivia for four periods between two and five months in the years 1982 and 1984 to get data about possible time delays and different attenuation of the seismic waves along the geotraverse as well as information about local seismicity and
- refraction seismic measurements - using the blasts of the Chuquicamata (N-Chile) copper mine as signal source - at four profiles running to different directions.

The results of earthquake catalogue data analysis and crustal seismic observations will be introduced whereas treating of the mobile stations' data has not yet been finished (see also BUNESS, 1984; BUNESS et al., 1986, and WIGGER, 1986).

SEISMICITY AND ENERGY RELEASE

The earthquake data file for a period from 1906 to 1980 covering the area 60°-75° W and 15° and 30°S contains 8696 events. This file was compiled by the Seismological Central Observatory, Erlangen (W-Germany) by use of mainly ISS, ISC, USCGS, ERL, PDE and GUT data. The catalogue shows two distinct inhomogenities. First there is a drastic increase of events after 1960. Only 12% are dated previous to this year but 88% after 1960. This inhomogenity is a result of the installation of the World-Wide Standardized Seismograph Network (WWSSN). On the other hand up to 1960 99% of the seismic energy after the data file was released. This second inhomogenity is an expression of a problem in earthquake statistic in the beginning of instrumental seismology where magnitudes have been calculated too high (ABE, 1981). For this reason the interpretation of the data refer mainly to the period 1961-1980.

The epicenters of the mentioned region are plotted in two maps (fig. 1) devided for magnitudes (m_b) greater and smaller than 5.0.

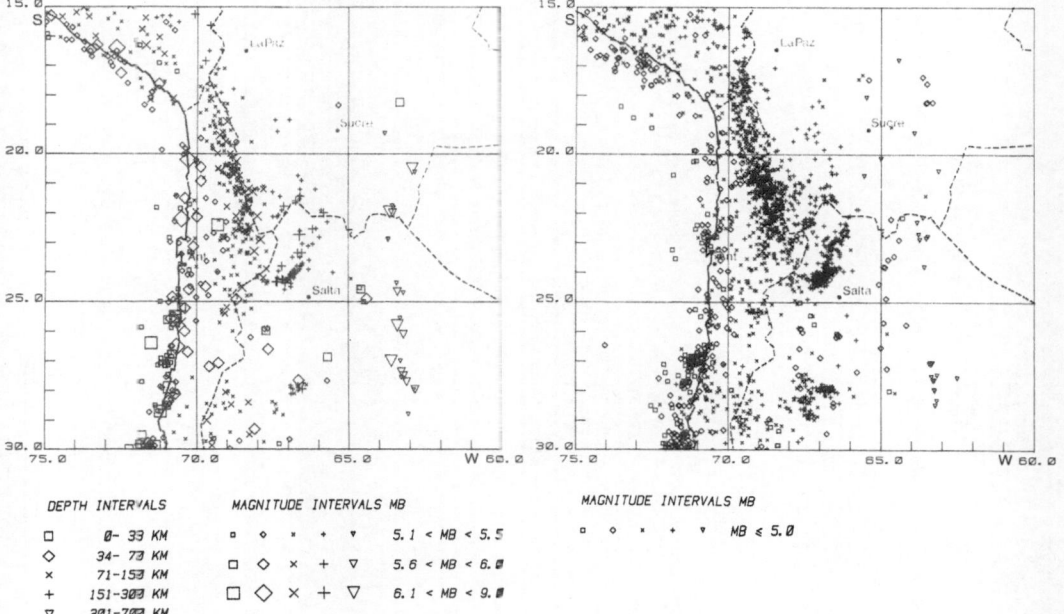

DEPTH INTERVALS		MAGNITUDE INTERVALS MB					
□	0- 33 KM	□	◇	×	+	▽	5.1 < MB < 5.5
◇	34- 73 KM	□	◇	×	+	▽	5.6 < MB < 6.0
×	71-153 KM	□	◇	×	+	▽	6.1 < MB < 9.0
+	151-303 KM						
▽	301-703 KM						

MAGNITUDE INTERVALS MB

□ ◇ × + ▽   MB ≤ 5.0

Fig. 1: Distribution of epicenters in the Central Andes for the period 1961-1980 displayed in two maps. Left with magnitude m_b > 5, right m_b < 5

Different depths are displayed by different symbols. The main features of the epicentre distribution are here pointed out shortly:
- According to the subduction zone the depths of foci describe a Wadati-Benioff zone dipping to the east.
- There is a nearly beltlike arrangement of epicenters. Shallow quakes are mainly bounded to the coastal range, intermediate events can be met under the Pre- and Western Cordillera, another concentration of intermediate events is beneath the southern Altiplano and Puna, and the deep-focus earthquakes are situated under a strip between 63° and 64° W of longitude.
- Between 25° and 27° S there exists an area with a strong reduced seismicity for intermediate events.

If we look to the three E-W sections (19°-21° S, 21°-23° S and 23°-25° S) in fig. 2 further characteristics of this subduction zone become obvious. The dip angle of the Wadati-Benioff zone at these latitudes is 25-30°. There is a very strong concentration of foci in depths

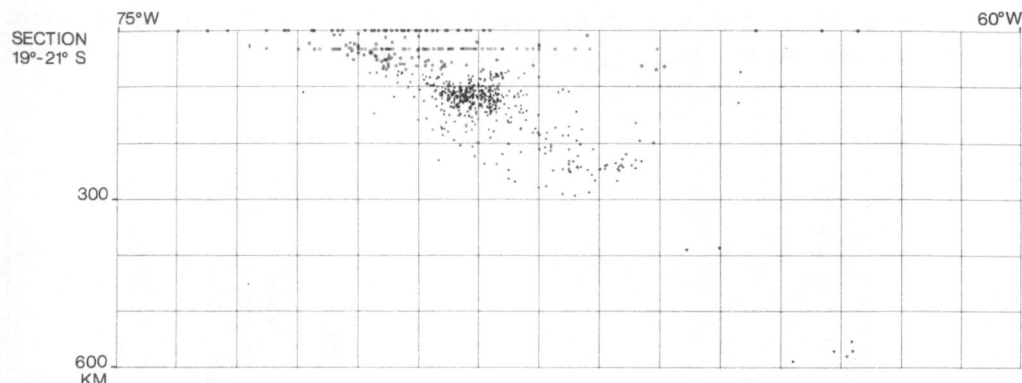

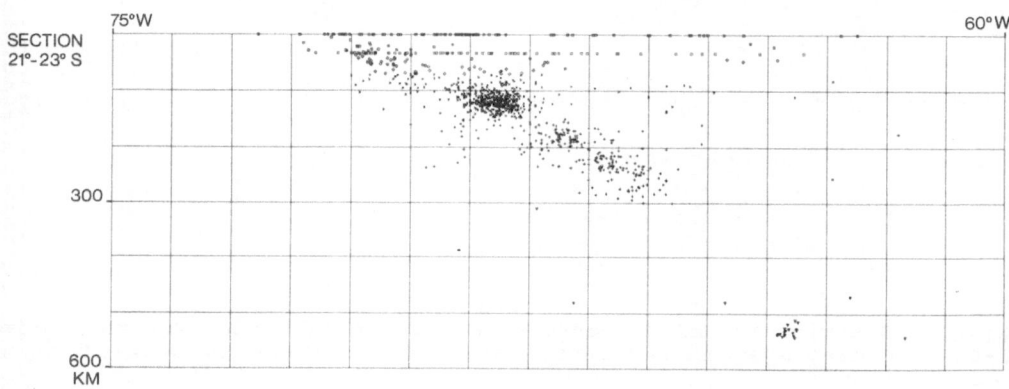

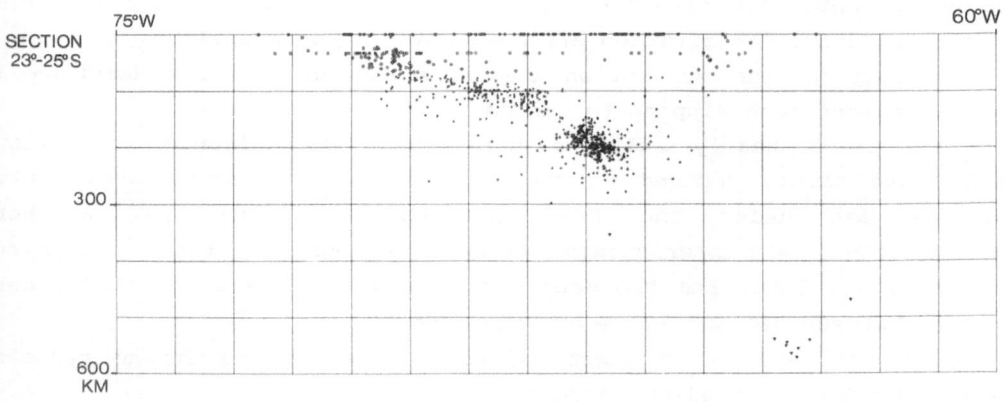

Fig. 2: Distribution of hypocenters along three E-W sections between 19° and 25° S, period 1961-1980. Notice the change of depth for the concentration of intermediate earthquakes and the pronounce seismic gap

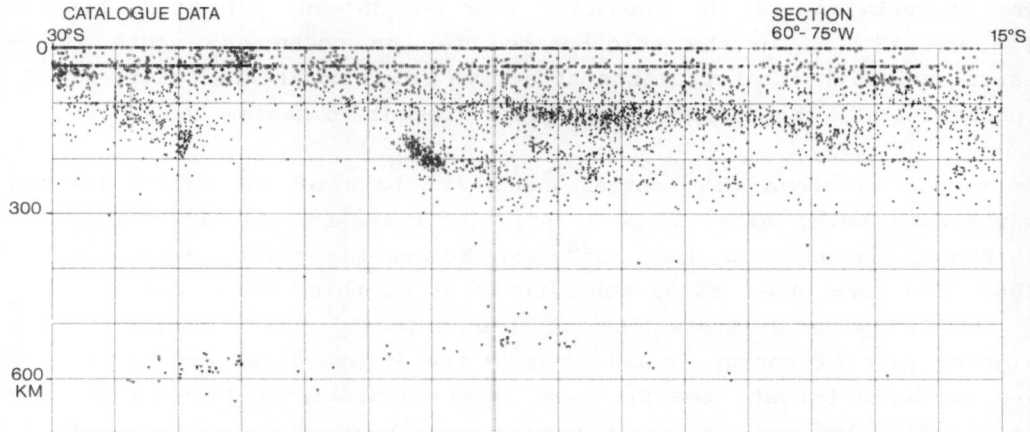

Fig. 3: All hypocenters between 15° and 30° S projected to a meridian. The earthquake nest in about 200 km depth beneath the southern Altiplano and Puna has a sharp border to an aseismic zone

between 100 and 140 km under the Pre- and Western Cordillera for the sections between 19° and 23° S. But farther to the south this concentration disappears and another earthquake accumulation in about 200 km depth dominates the scene. In depths from 300-500 km a seismic gap has to be stated. Deep earthquakes under the Central Andes are concentrated in the depth range of 500-600 km. Shallow events at 63°-66° W beneath the Eastern-Cordillera and the Subandean belt document recent tectonic activity. The question arises whether the broad scattering of hypocenters of about 150 km describing the downgoing slab is real or the result of an inaccurate localization. For southern Peru BOYD et al. (1982) showed that the distribution of hypocenters lays within a narrow zone of some tens of kilometers. It may be concluded that the situation beneath Northern Chile is similar. Foci determination could have higher precision, if there are more stations established in that area.

Fig. 3 shows all hypocenters of the mentioned region projected to a meridian in a N-S section. The concentration of intermediate events in depths between 100 and 200 km (S-N) is slightly ondulated and can be followed up to 23° S. Around 24° S we meet a nest of earthquakes in 200 km depth with a sharp border to the south and followed by an aseismic zone for intermediate depths. At this level seismicity starts again at 27° S and the dip angle of the subducted slab changes from 25° to a very flat subduction of 10°.

The segmentation of the subducted slab – 25°-30° dip angle between 17° and 27° S, in the neighborhood to the north and south 10° – correlates with the appearance of active volcanos in the sector of 30° subduction. In the region of 10° dip no recent volcanism can be met.

As mentioned above the Central Andes are part of the Circum-Pacific earthquake belt. After DUDA (1965) the released seismic energy for this belt amounts to 2.95 10^{26} erg during the period from 1897 to 1964. The area under study contributes 5% of this energy and only 21% of this region delivers 90% of the released seismic energy. The highest part of energy is released by the intermediate events in 100-140 km depth beneath the Pre- and Western-Cordillera between 20° and 24° S. This epicentral region covers only 3.6% of the whole area but contributes 42% of the energy. A histogram which shows the released energy of the period from 1961 to 1980 for a 100 km broad strip (fig. 4) is dominated by the maximum at the Pre- and Western Cordillera. Remarkable is the second maximum caused by the deep-focus events. A comparison of the cumulative seismic energy with the electrical conductivity integrated from the earth's surface down to 25 km depth (SCHWARZ et al., 1986) shows the highest conductivity maximum also at the Pre- and Western Cordillera.

Isolines of the cumulative magnitude, a value of energy release in logarithmic scale calculated by the moving-block method which was introduced by BATH (1981 and 1982) for the Central Andes are shown in fig. 5. Highs are exposed for the coastal range, the zone of deep earthquakes and the strongest high for the Pre- and Western Cordillera caused by the extensive cluster of events in about 100 km depth.

A parameter which characterizes the behavior of a region in respect of its ability to store deformation energy is the b-value. The dependance between the frequency of earthquakes N and its magnitudes M

$$\log N = a - b M$$

was shown by GUTENBERG & RICHTER (1944). Typical b-values reach from 0.6 for consolidated shield regions to 1.0-1.8 for Circum-Pacific subduction zones (KARNIK, 1969). A low value means , that the deformation energy can be strongly accumulated, a high b-value indicates that the energy may be released early in a great number of small events.

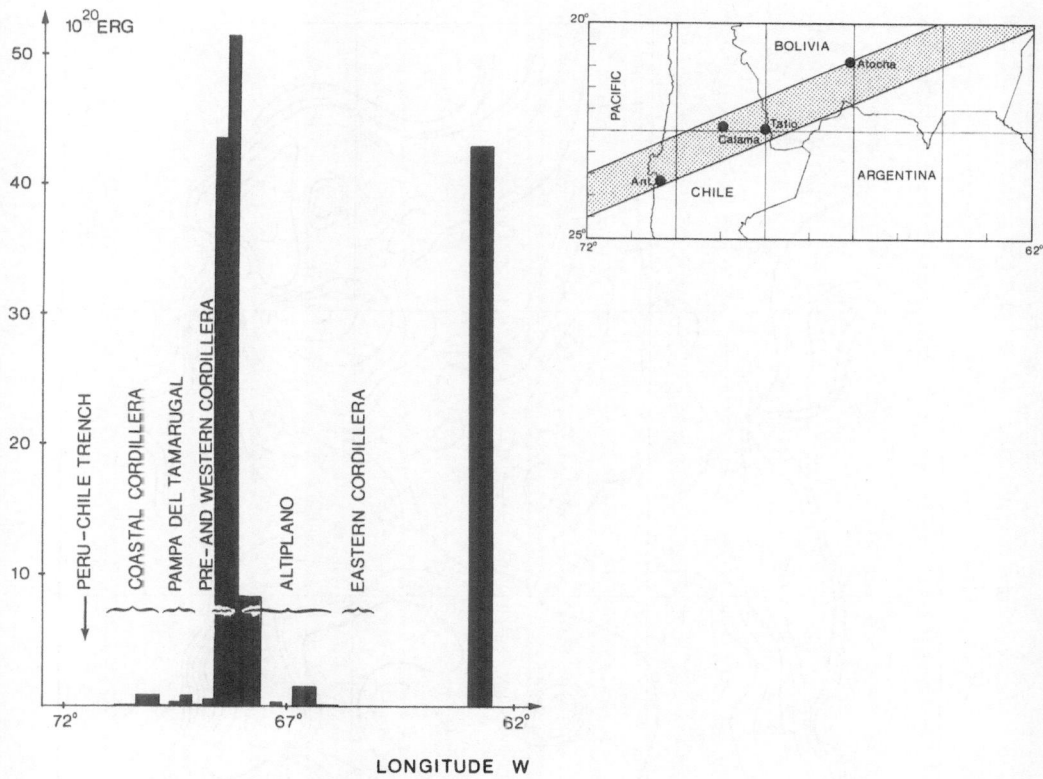

Fig. 4: Released energy of the period 1961-1980 along the indicated strip. Intermediate and deep earthquake dominates the energy balance

The mean value for the Central Andes reaches to 1.24, a typical value for subduction zones, but there are regional differences. The distribution for the coastal range is given in fig. 6. Inside the sector where the dip angle of the downgoing slab is 25°-30° the b-value is significant lower (~0.8) than in the adjacent areas with 10° dip angle (~1.1), another indication of the north-south segmentation of the subduction zone.

The intermediate events result in a b-value of 1.07-1.10. These relatively high values are in accordance with the numerous small events of that region although after the magnitude-frequency relation events up to a magnitude of 7.0 cannot be excluded. The very low b-value for the deep quakes (0.48) may be explained by unusually high stress in the upper mantle. ACHARYA (1971) reports very low values for deep earthquakes in other subduction zones, too.

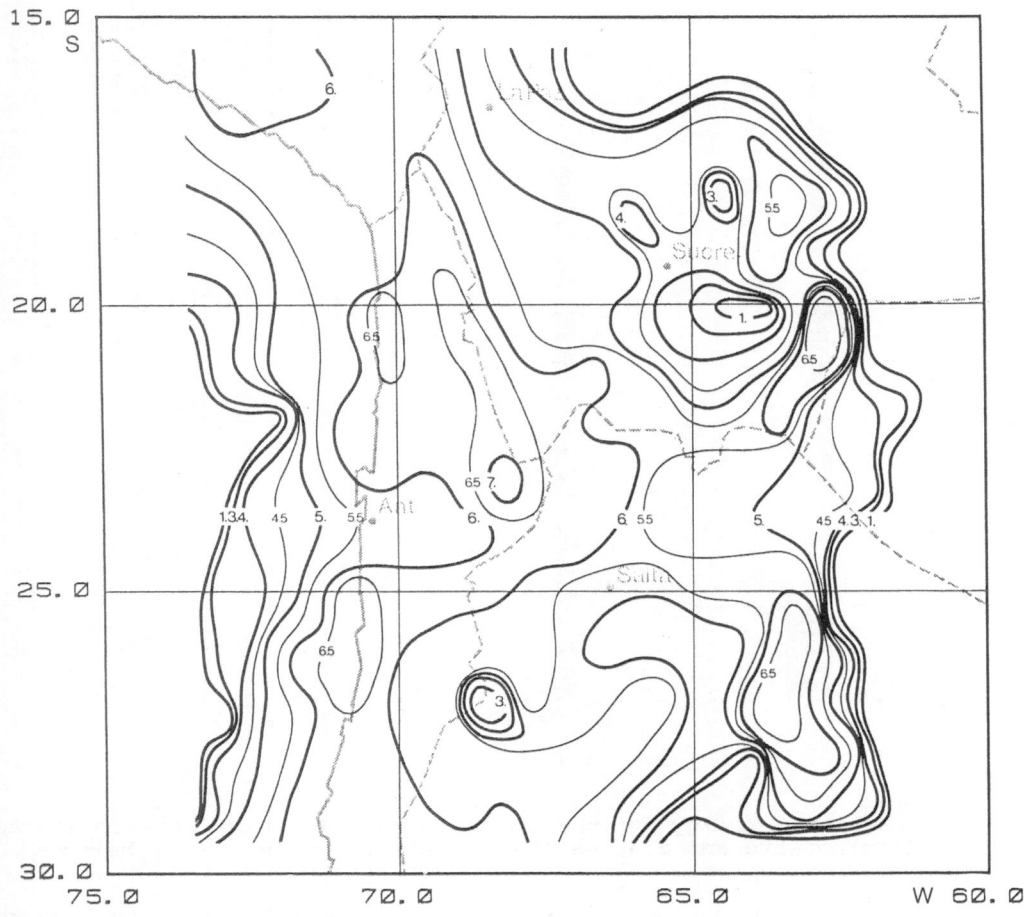

Fig. 5: Contour lines of cumulative magnitudes for the Central Andes, calculated with the moving-block method after BATH (1981 and 1982). The highest maximum covers the Pre- and Western Cordillera between 20° and 24° S

Focal mechanism of the deep earthquakes shows compression in direction of the subduction (STAUDER 1973). This may be explained either by a detached fragment of the downgoing slab or more probably by an isolated rest of a paleoplate subducted during an earlier epoch, now colliding with more rigid material at the "700 km-discontinuity". The intermediate events show tension parallel to the direction of the dip of the plate, and for the shallow quakes of the 30-60 km range STAUDER determined underthrusting of the oceanic plate.

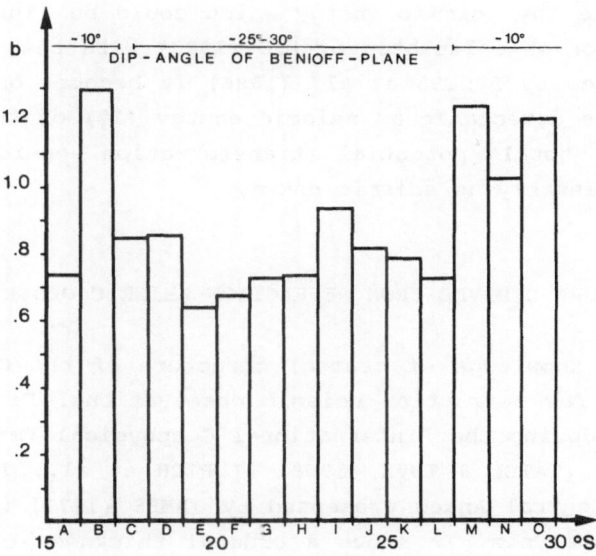

Fig. 6: Distribution of b-values along the coastal range between 15° and 30° S

STAUDER's solutions of focal mechanism hint to a mechanic process as origin of the earthquakes, the double-couple mechanism. For this mechanism also a second solution is possible. STAUDER derived for the intermediate events tension in direction of the dip. An equivalent solution is compression perpendicular to this direction. Furthermore his study was restricted to earthquakes with magnitudes greater than 6.0. In an investigation of a large number of intermediate events with depths more than 100 km, including also the small ones, DZIEWONSKI & WOODHOUSE (1983) found remarkable deviations from the double-couple mechanism. HAAK & GIESE (1986) assume that earthquakes with this deviation may be caused by phase transformations in the subducted oceanic plate at depths of about 100 km. This is the depth of the earthquake accumulation under the Pre- and Western Cordillera. During the subduction process the oceanic crust will be dehydrated and metastable phase transitions of amphibolite to eclogite according to the pressure and temperature condition will take place. Near the critical temperature of 800° C the phase change will proceed more or less instantly in the stability field of ecoglite. LIU (1983) and many other authors have discussed this metastable phase transition as a potential source of energy for earthquakes.

When we compare the seismic energy which could be liberated from this phase transition as calculated by LIU (1983) with the actual liberated energy estimated by BUNESS et al. (1986) it becomes obvious that only a small part is detectable as seismic energy (1%) or that only a small part of the total potential transformation energy is actually dissipated metastably as seismic energy.

CRUSTAL STRUCTURE DERIVED FROM REFRACTION SEISMIC OBSERVATIONS

Up to now our knowledge of crustal structure of the Central Andes is based on very few refraction seismic observations. First measurements were realized during the "International Geophysical Year", 1957 by the Carnegie group (TATEL & TUVE, 1958; ALDRICH et al., 1959). A crustal model of the Central Andes presented by JAMES (1971) was derived from surface wave analysis. It shows a crustal thickness of 70 km beneath the Altiplano. There is furthermore a profile running at the Altiplano with a shot point in Peru and Bolivia (OCOLA et al., 1971; OCOLA & MEYER, 1972), and for the Pacific range four profiles parallel to the coast off Antofagasta are described by FISHER & RAITT (1962).

Within this project four profiles have been observed during four combined field campaigns, together with earthquake recordings and magnetotellurics in Northern Chile and southern Bolivia in the years 1982 and 1984. The profiles are running from shotpoint Chuquicamata (N-Chile) to the east, south-east, south and south-west (fig. 7).

As signal source for the recordings the blasts of the open pit copper mine Chuquicamata were used. Total charge of these blasts can amount up to 250 tons of explosive material, but the charge is distributed in some tens of holes which are arranged in rows. These rows are fired with a time delay of 20-35 ms in between and the total duration of one explosion may reach 1.5 seconds. That is the cause why every blast produces a differing seismic signal with its own characteristics, especially for the frequency content. Since the observations were made successively with only few recording stations during longer time periods it is nearly impossible to compare the seismograms relating to their amplitudes and frequency content. Nevertheless, it could be proved that the energy is sufficient to produce seismic signals up to 300 km distance from Chuquicamata.

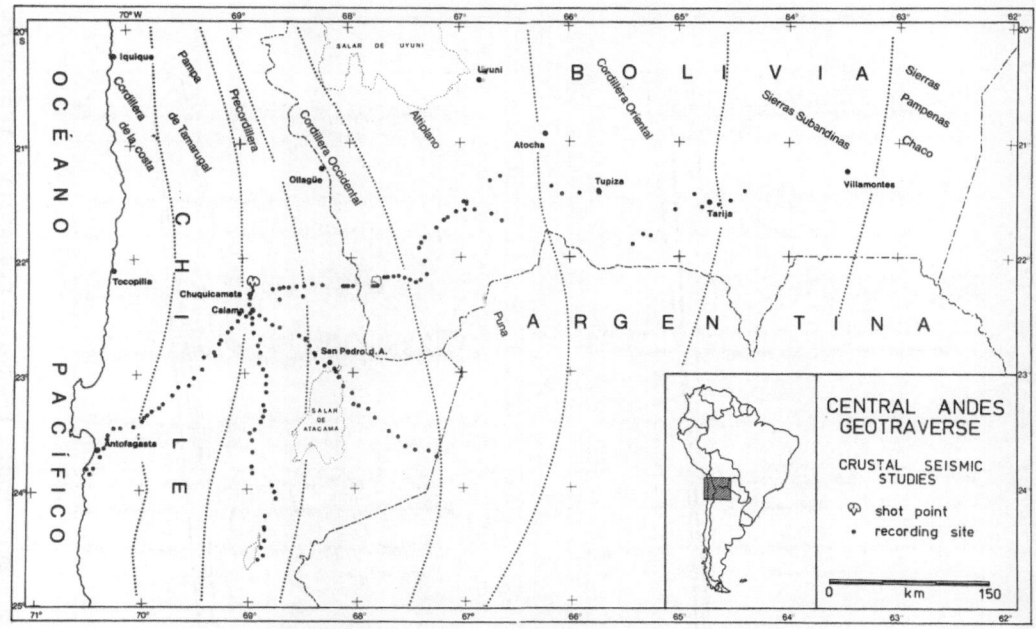

Fig. 7: Location map showing seismic profiles and shot point Chuquicamata

For the field recordings MARK L4 seismometers were used. The output signal was modulated (FM) and analogously recorded on magnetic tape. Time base as well for time break recording as in the field were quartz clocks with daily adjustment. The resulting time error is smaller than 10 ms. The data of the field tapes were digitized with a sample rate of 100 Hz/channel. Further processing was made at a CYBER 170-835 computer.

A special grave problem for refraction seismic recording in that region is the very high seismicity. Fig. 8 shows the record section for the profile Chuquicamata-SW. The traces which are indicated by a dot show the seismic signal caused by the blast of the mine superposed upon the signal of a local earthquake. At top of fig. 8, the record section is displayed with a reduction velocity of 6 km/s and the P-wave correlation is drawn. The record section below with a reduction velocity of 3.46 km/s shows the S-wave correlation. The 240 km long profile shows clear arrivals up to the Pacific ocean and a set of pro- and retrograde phases belonging to discontinuities which lay beneath the Pampa de Tamarugal can be correlated.

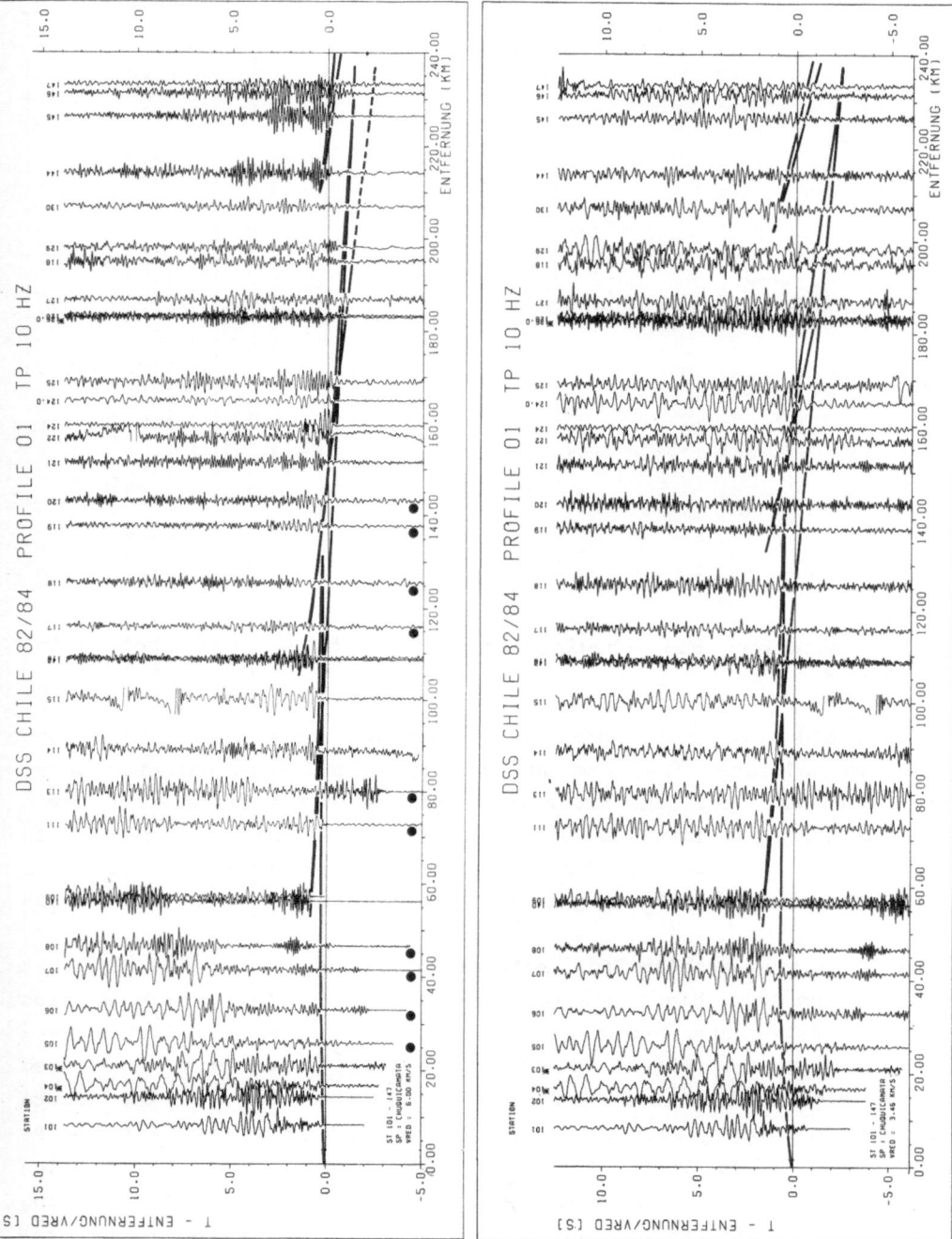

Fig. 8: Record sections profile Chuquicamata to SW. Top: P-wave correlation, reduction velocity 6.0 km/s. Below: S-wave correlation, reduction velocity 3.46 km/s

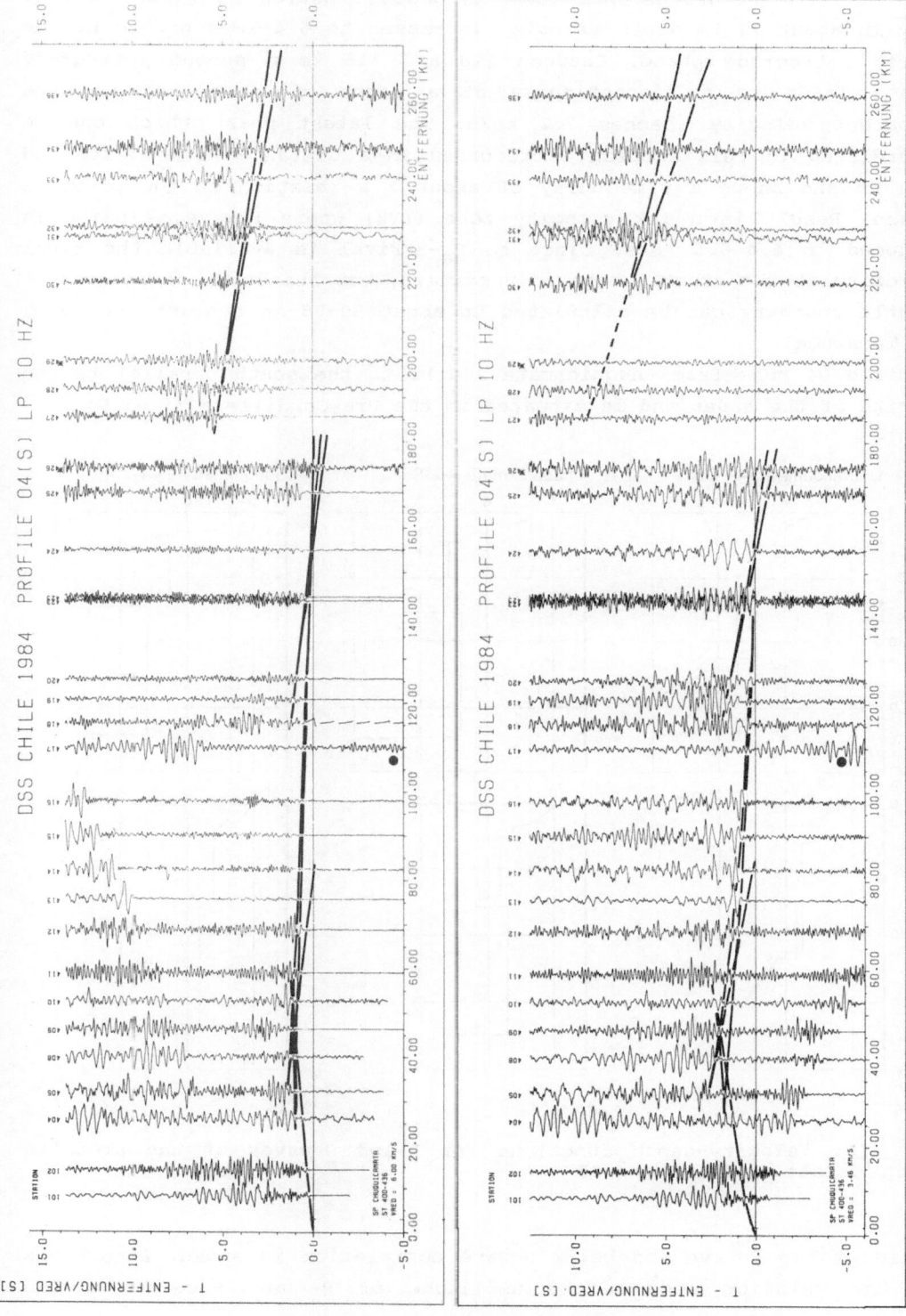

Fig. 9: Record sections profile Chuquicamata to south. Top: P-wave correlation, reduction velocity 6.0 km/s. Below: S-wave correlation, reduction velocity 3.46 km/s

The derived one-dimensional velocity-depth function is given in fig.
10. In about 11 km depth velocity increases to 6.4 km/s proved by the
first retrograde phase. Between 210 and 115 km a second retrograde
phase leads to a discontinuity at a depth of 28-29 km, and the
apparent velocity reaches 7.2 km/s. The latest phase which can be
correlated in this section is recorded in a distance between 240 and
210 km and shows a time delay of about 2 s relating to the previous
phase. Result is a low velocity zone (LVZ) where P-wave velocity is
reduced to 6.4-6.6 km/s. Since no P_n-arrival is available the final
velocity at the crustal base is uncertain, but the depth of the crust-
mantle boundary can be calculated to about 50-55 km beneath the Pampa
de Tamarugal.

Profile 04 runs from Chuquicamata 260 km to the south parallel to the
strike of the Andes and is situated in the Pre-Cordillera (fig. 9).

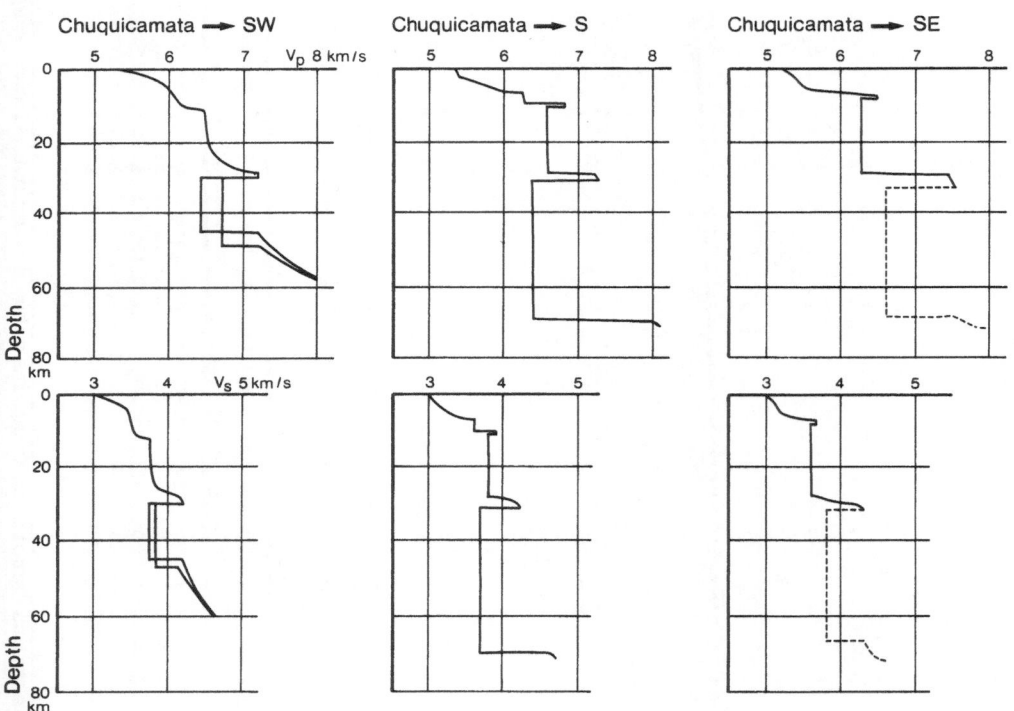

Fig. 10: Velocity-depth functions for P-and S-waves of the profiles
Chuquicamata to SW, S and SE

Again at top P-wave and below S-wave correlation is shown. Remarkable
is the relation between the amplitude of P- and S-waves in the

distance range of 70-170 km and 185-200 km. In the first case P-arrivals nearly disappear, whereas S-arrivals become very large. In the second case the situation is just in opposite to that. Is this phenomenon only caused by different shot characteristics as mentioned above, or is it originated in the crustal structure. That is an open question.

Alternating pro- and retrograde phases can be correlated as well for P- and S-waves up to 180 km which lead to a strong structured upper crust. Discontinuities are indicated in 6 and 10 km where the apparent velocities (v_a) reach 6.2, respectively 6.8 km/s and in 29 km depth where the velocity increases to 7.3 km/s (fig. 10). With a time delay for P-arrivals of about 7 s to the previous phase a retrograde wave group indicates a very thick lower crust. v_a of this group amounts to 8.1 km/s and the depth of the belonging discontinuity was calculated to 70 km. The mean velocity in the about 40 km thick lower crust amounts to 6.4 km/s.

The record section of profile Chuquicamata to SE which is entering the Western-Cordillera is very similar to the previous one. Its velocity-depth relation ($v(z)$) is given in fig. 10. The deeper part of the $v(z)$-relation is drawn with a dashed line because the far distant recordings are noisy and so the correlation is not as certain as the other ones. Profile Chuquicamata to east was recorded up to a distance of 470 km, crossing the Western Cordillera, Altiplano and the Eastern Cordillera. Highest recording station was set up in the Western Cordillera at a height of 4800 m above sea level. Very clear P_g-arrivals could be recorded up to 100 km. For greater distances no further onset could be correlated clearly. There are almost some weak indications for S-wave arrivals between 180 and 210 km shot point distance. The traveltime is 67s at 180 km. Basing on 4.6 km/s S-wave velocity maximum depth is calculated to 76 km beneath the Altiplano, but there are no indications for this discontinuity with P-waves. The observed P_g-group penetrates the crust at most 10 km. From this depth downwards we have to assume a LVZ. The thickness of this LVZ cannot be determined by this observations.

We can assume different reasons that there are no clear arrivals at distances greater than 100 km. The LVZ produces a shadow zone where no seismic waves emerge. For greater distances the energy of the blast might be insufficient for the far distant recordings. In this relation some other observations support this idea. CHINN et al. (1980) describe this part of the Western Cordillera and southern Altiplano as a region that attenuates/scatters high frequency seismic waves. Passing this region, the energy of the blast may be attenuated strongly. For depths greater than 10 km down to 30 km beneath the Pre- and western Cordillera and between 30 and 40 km beneath the Altiplano SCHWARZ et al. (1984 and 1986) derived an extremely high electrical conductivity and postulate partial melt as origin. The same conclusion found ARANEDA et al. (1985) interpreting the minimum in the residual field of Bouguer gravity of this region. In spite of these different informations the structure of the lower crust of this Andean part cannot be described in more detail.

Before continuing the discussion on the data a general remark must be made. Being not reversed the profiles allow only restricted discussion. Relatively high error bars remain for velocities and depths. But the existence of the derived discontinuities are beyond any doubt. The question is now how to interprete these discontinuities in the depth of 30 and 70 km and the enormous thickness of the lower crust in a geological frame. The velocity-depth distributions (fig. 10) down to 30 km look very similar to one of a normal crust. Taking in account that during Jurassic times the crust between the Coastal and Western Cordillera had been base of marine sedimentation (v. HILLEBRANDT et al., 1986) it has to be stated that the maximum thickness of the Jurassic crust could not have been of more than 30-35 km. For this reason the discontinuity in about 30 km depth is interpreted as paleo Moho of Jurassic age. The deep reflector - about 50-55 km under the Pampa de Tamarugal and 70 km under the Pre-Cordillera - is considered as base of the actual crust.

The continuation of crustal structure to the west and east basing on data cannot yet be described. As mentioned above a segmentation of the subduction zone is existing, we also have to expect varying crustal structure in N-S direction. Nevertheless, it may be helpful to cite some results of adjacent regions. A profile for the Peruvian and Bolivian Altiplano was introduced by OCOLA & MEYER (1972). They have calculated a total crustal thickness of 72 km and found intracrustal discontinuities in about 30 and 40 km depth with a velocity increase to 6.8-6.9 km (fig. 11). They also describe a very thick lower crust of about 30 km, but the velocity here, some 100 km farer to the north, is a little higher (6.9 km/s). If we want to have an idea of crustal thickness of the most eastern part of the Central Andes geotraverse adjoining to the Brasilian shield we may refer to the result of GIESE

& SCHÜTTE (1980) who derived 40 km for the Minas Gerais region in Brasil.

In the Coastal Cordillera of northern Chile as well as in southern Peru precambrian rocks are exposed. These two coastal ranges may be compared. LUETGERT & MEYER (1981) found 36 km crustal thickness and a velocity increase to 7.2 in 24 km (fig. 11). Furthermore the results of refraction seismic measurements from the Pacific ocean off Antofagasta (FISHER & RAITT, 1962) can be applied to construct a section from the Pacific ocean to the Brasilian shield.

Considering the mentioned results a section of reference latitude 22.5° S was drawn (fig.4 12). At the top the Bouguer gravity, topography and structural units of the Andes are displayed. The plotted earthquakes of the period 1961-1980 belong to 21°-23° S. The thick line indicates the top of the subducted Nazca plate approximately. Actual crustal thickness increases from about 35-40 km beneath the Coastal Cordillera to 70 km already under the Pre-Cordillera. GÖTZE & SCHMIDT (1984) give for this range a Bouguer gravity of -200 to -320 mgal. The minimal anomaly with values lower than -400 mgal is reached under the Altiplano and Puna. Following the gravity values we have to expect a maximal crustal thickness beneath this area of more than 70 km.

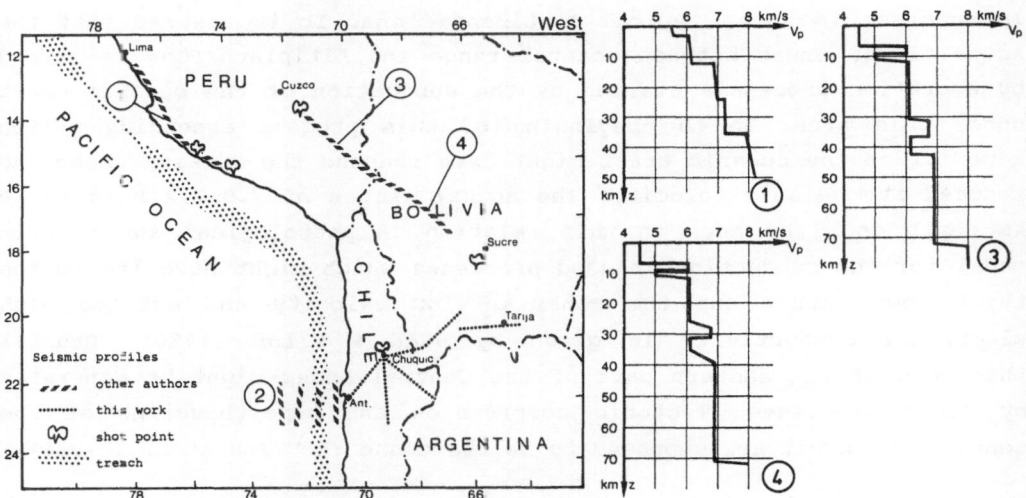

Fig. 11: Location map of crustal seismic profiles of the Central Andes and belonging velocity-depth functions. (1) LUETGERT & MEYER (1981), (2) FISHER & RAITT (1962), (3) and (4) OCOLA & MEYER (1972)

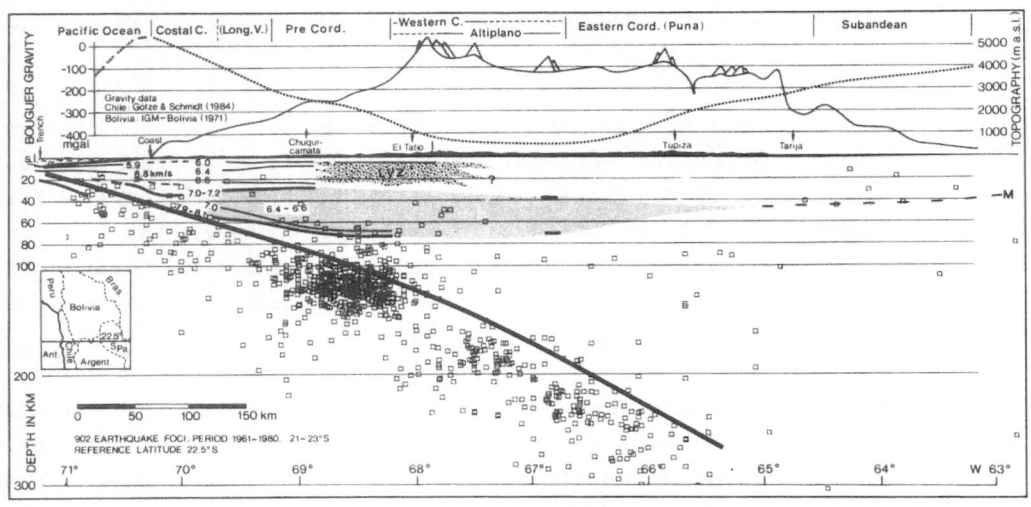

Fig. 12: West-east section of the Central Andes referring to the strip 21°-23° S. At the top Bouguer gravity and topography are drawn. Thickness of the LVZ and its prolongation to the east is unknown. Crustal thickness in the eastern part of this section bases on reasonable assumptions discussed in the text. The thick line dipping to the east should indicate the top of the downgoing Nazca plate and is drawn after the distribution of the hypocenters.

The thick crust in the western part is mainly resulting from the very thick lower crust. Velocity values of 6.4-6.6 km/s indicate that the lower crust is not a typical continental one. It is assumed that the actual lower crust between coastal range and Altiplano/Puna is formed by accretion processes started by the subduction of the oceanic crust under this area. In the beginning of this process ascending molten material of the oceanic crust might have reached the Jurassic Moho and reduced its seismic velocity. The actual values of 7.0-7.2 km/s in 30 km depth could be seen in this relation. A petrological and dynamic model for the subduction induced processes which might have led to the thick lower crust and the zones of low velocity and extreme high electrical conductivity is given by HAAK & GIESE (1986). Crustal thickness of the eastern part of the Central Andes might be generated by other processes. Tectonic compression and underthrusting of the continental crust are supposed to be the cause (REUTTER et al., 1986).

The introduced model can be seen as representative for the Central Andes only in a first approximation. A look to the earthquake data and the residual field of the Bouguer anomaly (ARANEDA et al., 1985) shows strong variations for the Central Andes. For the description of the crustal structure connected with these variations and having reversed profiles for different ranges in the Central Andes extended crustal seismic measurements were carried out in late 1987. The data are now treated and first results will be available soon.

Acknowledgements

Fieldwork in such remote areas - like the Central Andes - can only be carried out with many helpful hands: logistical support has been given by Programa Sismologico, Antofagasta; Universidad del Norte and CORFO, Antofagasta; CODELCO, Chuquicamata; MOP-Dir. de Riego and Urdangarin y Hnos., Calama; Universidad de Chile, Santiago; Embassy of the Federal Republic of Germany, Santiago; Universidad Mayor de San Andres, COMIBOL, GEOBOL and the Embassy of the Federal Republic of Germany, the latter all in La Paz, YPFB , Santa Cruz and YPFB, Tarija. Our colleagues Mario Aramayo, Eloy Martinez, Salvator del Pozo, Edgar Ricaldi and Dr. Gerhard Schwarz took part in the field survey. The Seismological Central Observatory, Erlangen (W-Germany) provides the earthquake data file. We are very grateful to all mentioned persons and institutions. We wish to express our gratitude to the Deutsche Forschungsgemeinschaft and Freie Universität Berlin for financial support.

REFERENCES

ABE, K. (1981): Magnitudes of large shallow earthquakes from 1904-
   1980. - Phys. Earth Planet. Int., 27, 72-92; Amsterdam.
ACHARYA, H. (1971): Magnitude-frequency relation and deep-focus
   earthquakes. - Bull. Seis. Soc. Am., 61, 1345-1350. Berkeley.
ALDRICH, L.T., TATEL, H.E., TUVE, M.A. & WETHERILL, G.W. (1959):
   The Earth's crust. - Carnegie Institution of Washington: Year-
   book 57, 104-111; Washington D.C.
ARANEDA, M., CHONG, G., GÖTZE, H.-J. LAHMAYER, B., SCHMIDT, S. &
   STRUNK, S. (1985): Gravimetric modelling of the northern
   Chilean lithosphere (20°-26° latitude south). - IV Congreso
   Geologico Chileno, Actas, II, 18-34; Antofagasta.
BARAZANGI, M & ISACKS, B. (1976): Spatial distribution of earth-
   quakes and subduction of the Nazca plate beneath South Ameri-
   ca. - Geology, 4, 686-692; Boulder.
BARAZANGI, M. & ISACKS, B. (1979): Subduction of the Nazca-plate:
   Evidence from spatial distribution of earthquakes. - Geophys.
   J.R. Astron.Soc., 57, 537-555; Oxford.
BATH, M., (1981): Seismic energy mapping applied to Sweden. -
   Tectonophysics, 81, 85-98; Amsterdam.
BATH, M. (1982): Seismic energy mapping applied to Turkey. -
   Tectonophysics, 82, 69-87; Amsterdam.
BOYD, T., SNOKE, S., SACKS, I. & RODRIGUEZ, A. (1982): High-
   resolution determination of the Benioff zone geometry beneath
   southern Peru. - Annual report of the Director - Department of
   terrestrial Magnetism, Carnegie Inst. of Washington, Yearbook
   82, 500 p.; Washington D.C..
BUNESS, F. (1984): Die Seismizität der Zentralen Anden. -
   Diploma Thesis, Freie Universität; Berlin
BUNESS, F., WETZIG, E. & WIGGER, P. (1986): Seismologische Unter-
   suchungen in den Zentralen Anden. - Berliner Geowiss. Abh.,
   (A), 66, 5-30; Berlin.
CHINN, D.S., ISACKS, B.L., BARAZANGI, M. (1980): High frequency
   seismic wave propagation in western South America along
   the continental margin, in the Nazca plate and across the
   Altiplano. - Geophys. J.R. Astron. Soc., 60, 209-244; Oxford.
DUDA, S. (1965): Secular energy release in the Circum Pacific
   belt. - Tectonophysics, 2, 409-452; Amsterdam.
DZIEWONSKI, A.M. & WOODHOUSE, J.H. (1983): An experiment in
   systematic study of global seismicity: Centroid-Moment
   tensor solutions for 201 moderate and large earthquakes of
   1981. - J. Geophys. Res., 88, 3247-3271; Washington D.C.
FISHER, R.L. & RAITT, R.W. (1962): Topography and structure of the
   Peru-Chile trench. - Deep-Sea Res., 9, 423-443; Oxford.
GIESE, P., SCHÜTTE, K.-G. (1980): Resultados das medidas de
   sismica de refracao a leste da Serra do Espinhaco, M.G.
   Brasil. - In: Nuevos Resultados de la Investigacion Geo-
   cientifica Alemana en Latinoamerica: Proyectos de la Dt. Forschungsgemeinschaft. Ed. por la Dt. Forschungsgemeinschaft
   y el Inst. de la Colab. Cientifica, Tübingen, RFA.
GÖTZE, H.-J. & SCHMIDT, S. (1984): Gravimetrische Messungen
   im chilenischen Teil der Anden - Geotraverse (21-25° südl.
   Breite). - 9. Geowissenschaftliches Lateinamerika-Kolloquium,
   21.-23.11.1984, Tagungsheft; Marburg.
GUTENBERG, B. & RICHTER, M. (1944): Frequency of earthquakes
   in California. - Bull. Seis. Soc. Am., 34, 185-188; El
   Cerrito, Cal.
HAAK, V. & GIESE, P. (1986): Subduction induced petrological

processes as inferred from magnetotelluric, seismological
and seismic observations in N-Chile and S-Bolivia. - Berliner
Geowiss. Abh., (A), 66, 231-246; Berlin.
HASEGAWA, A. & SACKS, I.S. (1981): Subduction of the
Nazca Plate beneath Peru as determined from seismic
observations. - J. Geophys. Res., 86, 4971-4980. Washington D.C
HILLEBRANDT, v., A., GRÖSCHKE, M., PRINZ, P. & WILKE, A.-G.
(1986): Marines Mesozoikum in Nordchile zwischen 21 ° und 26°
S. - Berliner Geowiss. Abh., (A), 66, 169-190; Berlin.
IGM-BOLIVIA (1971): Mapa gravimetrico de Bolivia. - La Paz.
ISACKS, B. & BARAZANGI, M. (1977): Geometry of Benioff zones:
Lateral segmentation and downwards bending of the subducted
lithosphere. - In: TALWANI, M. & PITMAN, W., eds.: Island
arcs, deep-sea trenches, and back arc basins. AGU, Ewing Series
1, 99-114; Washington D.C.
JAMES, D.E. (1971): Andean crustal and upper mantle structure. -
J. Geophys. Res., 76, No. 14, 3246-3271; Washington D.C.
KARNIK, V. (1969): Seismicity of the European area. Part 1. -
D. Reidel Publishing Co.; Dordrecht, Holland.
LIU, L. (1983): Phase transformations, earthquakes and the des-
cending lithosphere. - Phys. Earth Planet. Int., 32, 226-240.
LUETGERT, J.H. & MEYER, R P. (1981): Crustal structure of coastal
Peru, 12° S to 16.2° S latitude. - 21st General Assembly of the
IASPEI, July 21-30, London, Canada.
OCOLA, L.C., MEYER, R.P. & ALDRICH, L.T. (1971): Gross crustal
structure under Peru-Bolivia Altiplano. - Earthquake notes,
XLII, Nos. 3-4, 33-48; Atlanta, USA.
OCOLA, L.C. & MEYER, R.P. (1972): Crustal low-velocity zones
under the Peru-Bolivia Altiplano. - Geophys. J.R. Astron. Soc.,
30, 199-209; Oxford.
REUTTER, K.-J. SCHWAB, K. & GIESE, P. (1986): Oberflächen- und
Tiefenstrukturen in den Zentralen Anden. - Berliner Geowiss.
Abh., (A), 66, 247-264; Berlin.
SCHWARZ, G., HAAK, V., MARTINEZ, E. & BANNISTER, J. (1984): The
electrical conductivity of the Andean crust in northern Chile
and southern Bolivia as inferred from magnetotelluric measure-
ments. - J. Geophys., 55, 169-178; Heidelberg.
SCHWARZ, G., MARTINEZ, E. & BANNISTER, J. (1986): Untersuchungen
zur elektrischen Leitfähigkeit in den Zentralen Anden. Berliner
Geowiss. Abh., (A) 66, 49-72; Berlin.
STAUDER, W. (1973): Mechanism and spatial distribution of Chilean
earthquakes with relation to subduction of the oceanic plate. -
J. Geophys. Res., 78, 5055-5061; Washington D.C..
STAUDER, W. (1975): Subduction of the Nazca-Plate under Peru as
evidenced by focal mechanism and by seismicity. J. Geophys.
Res., 80, 1053-1064; Washington D.C.
TATEL, H.E. & TUVE, M.A. (1958): Seismic studies in the Andes. -
AGU., Transactions, 39, 580-582; Washington D.C.
WIGGER, P. (1986): Krustenseismische Untersuchungen in Nord-Chile
und Süd-Bolivien. - Berliner Geowiss. Abh., (A), 66, 31-48;
Berlin.

STRUCTURES AND CRUSTAL DEVELOPMENT OF THE CENTRAL ANDES BETWEEN 21^0 AND 25^0S

K.-J. REUTTER[*], P. GIESE[**], H.-J. GÖTZE[**], E. SCHEUBER[*],
K. SCHWAB[***], G. SCHWARZ[**] & P. WIGGER[**]

[*]Institut für Geologie der Freien Universität Berlin
Altensteinstr. 34 a, 1000 Berlin 33

[**]Institut für Geophysik der Freien Universität Berlin
Rheinbabenallee 49, 1000 Berlin 33

[***]Institut für Geologie und Paläontologie der Technischen Universität
Leibnizstr. 10, D-3392 Clausthal-Zellerfeld

ABSTRACT

The tectonics of the morphostructural units of the Central Andean segment between 21^0S and 25^0S are reviewed and their relation to the deep crustal structures, as far as known from geophysical research, is discussed. Special regard is given to the superposition of structures due to the stepwise eastward displacement of four arc systems subsequently developing on the continental margin during the Andean Cycle: (1) Lias - Early Cretaceous, (2) Mid-Cretaceous, (3) Latest Cretaceous - Eocene, and (4) Miocene - Holocene. Within these arc systems, three areas of main tectonic activity can be distinguished: The subduction zone and subduction complex, the magmatic arc, and the backarc region. The subduction complexes of the fossil stages are not preserved and the tectonic activity of the present subduction complex affects the continent only locally along the coast. The structures of the four magmatic arcs are exposed respectively in the Coastal Range, the Longitudnal Valley, the Chilean Precordillera, and in the broad area extending from the Preandean Depression to the western part of the Eastern Cordillera. Notwithstanding great structural differences between the individual arcs, there are common features such as the close reationship between deformation and magmatism, the incorporation of basement into fold and horst structures, and conjugate reverse fault systems. In the case of oblique subduction, longitudinal strike slip faulting, which may be left-handed (Atacama Fault of the Coastal Range) or right-handed (West Fissure of the Chilean Precordillera), follows the magmatic arc. In the backarc region, east vergent fold and thrust belts developed only in the stages 3 and 4. All the different arc stages seem to have contributed gradually to crustal thickening as there is a general development in these systems from ensialic marine conditions over an environment of continental lowlands to the present high plateau situation.

Lecture Notes in Earth Sciences, Vol. 17
H. Bahlburg, Ch. Breitkreuz, P. Giese (Eds.),
The Southern Central Andes
© Springer-Verlag Berlin Heidelberg 1988

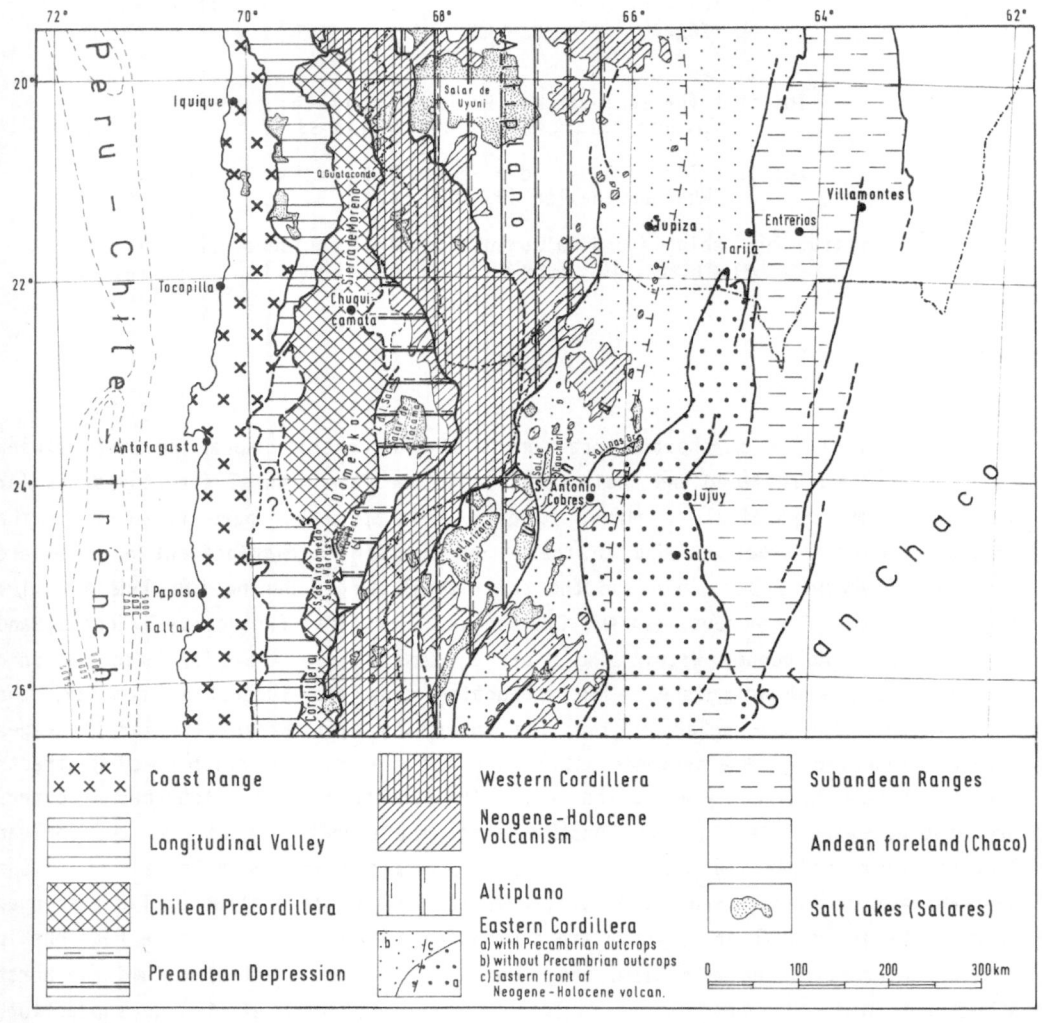

fig. 1: The morphostructural units of the Central Andean segment under consideration.

INTRODUCTION

The distinct morphostructural units of the Central Andes (fig. 1) are tectonic elements of an arc geotectonic setting generated at the western rim of the South American continent caused by its convergence with the oceanic plate of the Pacific. Within this setting, the Western Cordillera represents the magmatic arc, the backarc area comprises the Altiplano, the Eastern Cordillera, the Subandean Ranges as well as the Chaco lowlands, while the inner slope of the trench, the Coastal Range, the Chilean Central Valley, the Chilean Precordillera, and the Preandean Depression belong to the forcarc system (fig. 2). Morphology, tectonic behaviour, and internal structures of these units are not only determined by the stresses active in the upper plate as a consequence of plate convergence, but are controlled by physical parameters such as e.g. lithology, temperature distribution, partial pressure of fluids and the presence of liquid magmatic bodies within the different crustal levels, as well as by geological history, as the preexisting structures and inhomogeneities may control the younger deformation.

In this respect, the geotectonic development at least during the Mesozoic and the Cainozoic, i.e. during the Andean Cycle, should be considered. In the segment of the Andes studied the trend of displacement of the arc system towards the interior of the continent is well developed, better than in other parts of the Central Andes (fig. 7). Thus, it is known that the chain of the orogenic andesitic volcanism shifted from the Coastal Range, where it was situated during the Early Jurassic, to its present position in the Western Cordillera with intermediate positions in the Chilean Precordillera during Cretaceous and Early Tertiary times (COIRA et al., 1982). Evidence of a similar migration of arc plutonism was published by RIVANO et al., 1985. Together with the magmatic arc the continental arc system as a whole, i.e. stress fields and deformations characteristic for the forearc, arc and backarc region must have moved eastwards during that time. This is clearly shown in the eastern parts of the Central Andes where the backarc tectonics, i.e. the Subandean thrust belt progressively extends into the Subandean foreland not yet affected by Andean deformation (fig. 7). This also means that the present arc and forearc tectonics affect areas which some time before had been deformed, respectively, in the backarc and arc regimes. A structural comparison of the Central Andean morphostructural units has therefore to take tectonic superpositions into account which, as a consequence, become more intense· from east to west. This consideration should not only be valid for the surface structures and morphological features but also for deep crustal structures detected by geophysical methods. Seismological studies, seismic refraction work (WIGGER, this issue) gravity measurements (GÖTZE et al., this issue) and magnetotelluric surveys (Schwarz, et al, 1986) are available to probe the deeper crustal structures of the Central Andes.

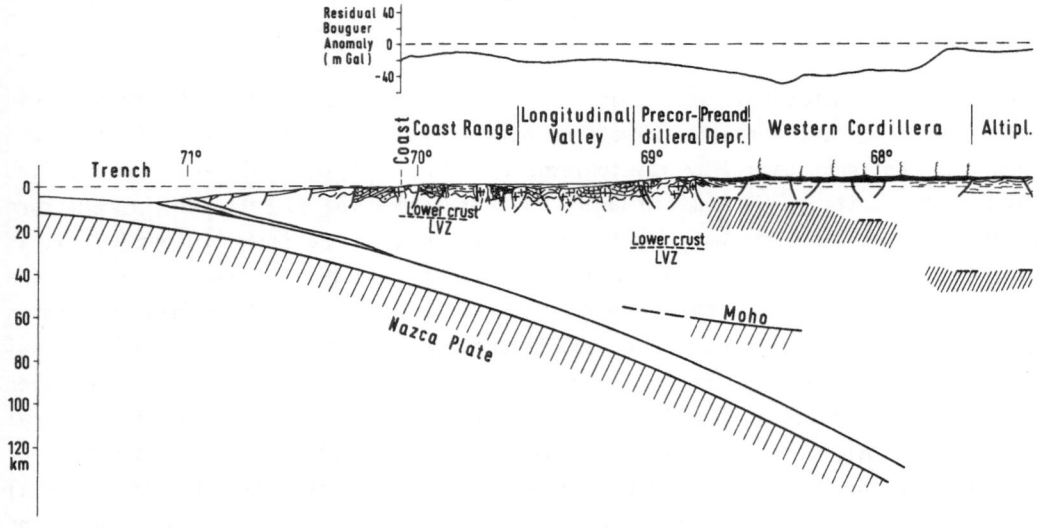

<u>fig. 2</u>: Cross section through the Central Andes at 21°25'S showing crustal structures according to geological and geophysical data. The subsurface structures of the fold and thrust belt of the Subandean Ranges are hypothetical.

In the following, the surface structures and, as far as they are known, the deep crustal structures of the morphostructural units of the Central Andean segment between 21° and 25°S shall be reviewed briefly in order to elucidate the structural development during the Andean Cycle.

TECTONICS OF THE MORPHOSTRUCTURAL UNITS

THE ANDEAN FORELAND (CHACO)

The Subandean lowlands of the Chaco are an active depositional basin whose continental deposits are supplied by rivers that drain the Eastern Cordillera and the Subandean Ranges. The Late Tertiary and Pleistocene sediments (Chaco Fm.) show growing thicknesses from E to W exceeding 2000 m near the Subandean foothills. They overlie either notably thinner Early Tertiary and Cretaceous sediments, or alternatively directly Triassic or Carboniferous sediments at the top of a more or less complete and thick Palaeozoic sedimentary sequence. There is no pronounced

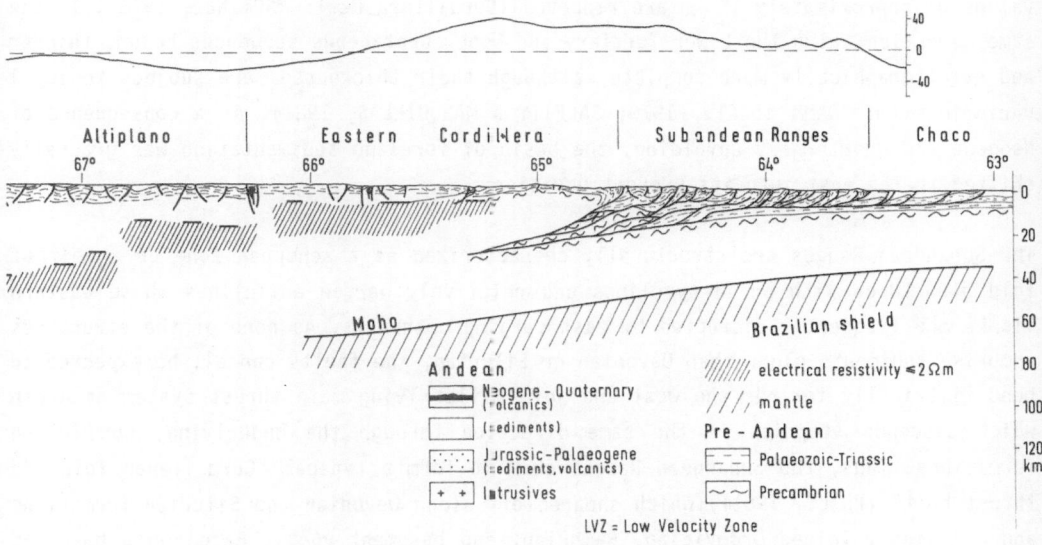

angular unconformity between the Neogene sediments and their substrate, but as in the vicinity of the Subandean Ranges, the underlying formations form young and still active structures (oil production) beneath a flat surface of recent deposition. Internal unconformities should exist within the Chaco Fm. itself.

In conformity with the increasing sedimentary thicknesses and incipient structures, the top of the Precambrian basement is inclined towards the Subandean Ranges and is situated at a depth of about 7 to 10 km (ALLMENDINGER et al., 1983) in the zone of the foothills (MARTINEZ et al., 1973). Based on the view that the crust of the Brazilian shield starts to point downwards underneath the Cordillera (LYON-CAEN et al., 1987), crustal thickness is accordingly supposed to increase in that direction and may reach 35 km. Except for the absence of young marine sediments, all these sedimentary and structural features are characteristic of a Molasse basin and a typical foreland. The Bouguer gravity field of the Chaco area is characterized by quite normal values of gravity fluctuating around -50 mGals. This is nearly the normal gravity of consilidated shield areas.

SUBANDEAN RANGES

The boundary of the Subandean Ranges with the foreland is determined by the appearance of young Andean structures at the surface. The stratigraphic column is very similar to that of the Chaco. The thicknesses of the Lower Palaeozoic formations gradually increase from the east (roughly 4 km) towards the Eastern Cordillera, where values of approximately 15 km are reported (Cordillera Real: MARTINEZ, 1978). In the same direction, also the Lower Tertiary and Upper Cretaceous sequences become thicker and stratigraphically more complete, although their thicknesses are subject to local variations (MINGRAMM et al., 1979; SALFITY & MARQUILLAS, 1981). As a consequence of Neogene and Quarternary upfolding, the basin of foreland sedimentation was gradually shifted to the east into its present position.

The Subandean Ranges are structurally characterized as a schuppen zone or a belt of fold structures with broad synclines and relatively narrow anticlines whose eastern flanks are frequently disrupted by east verging upthrusts. As none of the structures comprise sediments older than Devonian or Silurian, the faults can all be expected to bend listrically towards the west and join a flat lying main thrust system at depth which descends stepwise in the same direction through the underlying, not folded formations. Thus, the Subandean Ranges seem to form a typical "Cordilleran fold and thrust belt" (PRICE, 1981), which sheared off along Devonian and Silurian formations and overlies unfolded Ordovician, Cambrian, and basement rocks. Based on a balanced cross section near the Bolivian border (MINGRAMM et al., 1979), ALLMENDINGER et al. (1983) calculated a minimum shortening of the Subandean zone of 60 km since the Late Miocene.

The development of the Subandean zone during the Cretaceous and Early Tertiary has been interpreted by GALLISKI & VIRAMONTE (in press) as a foreland paleorift of "low volcanicity type". It corresponds to a continental backarc basin. This situation is underlined by the extrusion of alkali-basaltic lavas during the Late Cretaceous. The Neogene and Quarternary tectonics can be considered as an intracrustal thrust or A-Subduction system (BALLY, 1975) which came into being in the previous backarc basin.

There are only few data available about the rocks underlying the Subandean thrust plane. It is now supposed that the Ordovician, Cambrian and the Precambrian rocks have not been affected by the young Andean compressional tectonics (ALLMENDINGER et al., 1983; ROEDER, 1986). As a consequence of the increasing thicknesses of Palaeozoic and younger sediments, the top of the Precambrian had already dipped towards the west before thrust tectonics began. The piling up of thrust sheets must have increased this disposition. The crust can therefore also be expected to thicken from E to W. According to gravity models by STRUNK (1985) and GÖTZE et al. (1986), the Moho descends to about 50 km near the boundary to the Eastern Cordillera.

The magnetotelluric profile (SCHWARZ et al., 1986, fig. 3) does not show any significant features in this respect. Low resistivities are found in an upper layer about 5 km thick representing non-metamorphic sediments (Tertiary, Mesozoic and Upper Palaeozoic). In the section stretching from Villamontes to Tarija and Tupiza (Bolivia, fig. 1), at least in the eastern part of the Subandean Ranges no essential tectonic stacking of sediments seems to be indicated by magnetotelluric sounding. In the border zone between the Subandean Ranges and the Eastern Cordillera, uniform resistivities of about 100 Ohm m may be interpreted as a thick (2 km) stacking of Palaeozoic and Mesozoic metamorphic sediments.

EASTERN CORDILLERA, ALTIPLANO, PUNA

The eastern boundary of the Eastern Cordillera is characterized by the appearance of stratigraphically deeper levels, e.g. Lower Palaeozoic and Precambrian, than those exposed in the Subandean Ranges. Precambrian rocks are widely distributed in the Eastern Cordillera of NW-Argentina; they disappear as a consequence of a structural plunge - that at least in part had existed since pre-Cretaceous times - towards Bolivia, under Cambrian quartzites and thick (>5 km) Ordovician sequences of shales and sandstones. For the same reason, the easternmost structures of the Eastern Cordillera in Argentina merge along strike into the westernmost part of the Subandean thrust belt east of Tarija. From the highly uplifted eastern border of the Eastern Cordillera, the structural level also gradually descends towards the Altiplano. In Bolivia, the Altiplano, characterized by its mainly Cainozoic sediments, is clearly separated by faults from the mostly Palaeozoic sediments of the Eastern Cordillera. No geologically well defined limit between the Puna and the Eastern Cordillera exists in Argentina. The structures in both areas are very similar. Therefore the whole area is treated together in this paper. As a result of the possibly fault controlled southeastward extension of the young volcanism (SALFITY, 1985), the western boundary of the Puna and the Altiplano, as drawn by the present volcanic arc of the Western Cordillera, is very irregular. Both areas form an endorheic high plateau with several active intramontane depositional basins between the two Cordilleras.

In the Puna and the Altiplano, the structure of the crust underlying the sedimentary sequences of the Andean cycle changes in such a way that mechanical influences on the Andean tectonics can be expected. While the foreland, the Subandean Ranges, and the Eastern Cordillera of Bolivia did not suffer important deformations prior to the Andean Cycle and, hence, their thick mainly pelitic sediments are supposed to overlie a relatively undisturbed basement, the situation changes to the west and southwest with what is called the "Faja Eruptiva de la Puna". Volcanic intercalations in the Ordovician sediments, granitic intrusions, folding, and some metamorphism indicate the effects of an orogeny ascribed to the "Oclóyic Phase" of the Famatinian Cycle (Ordovician-Silurian boundary; ACEÑOLAZA & TOSELLI, 1976, 1981; MENDEZ et al., 1973; PALMA et al., 1986; for more details see BAHLBURG et al., this issue). This orogeny was interpreted as the result of a collision of a hypothetical "Arequipa Massif" (DALMAYRAC et al., 1977; MARTINEZ, 1978) with the Brazilian Shield. Precambrian rocks in the Chilean Precordillera and in the western part of the Altiplano were considered as indications of this massif (LEHMANN, 1978). During the Silurian and Devonian the "Oclóyic Orogen" formed a broad uplift (=Arco Puneño), at the western side of which thin continental and shallow marine sediments, contrasting with the contemporaneous facies to the east of the uplift, have been described (DONATO & VERGANI, 1985) which represent the marginal facies of sedimentary sequences in Chile (BREITKREUZ, 1986). The different condition of the area to the west of the Puna and the Altiplano is also underlined by the development from the Late Carboniferous to the Triassic when a broad magmatic belt developed with mainly acid and intermediate volcanics and shallow marine as well as continental sedimentary intercalations (COIRA et al., 1982).

As a consequence of the Palaeozoic tectonic events and possible pre-Cretaceous movements, Cretaceous and Early Tertiary sediments, as first deposits of the Andean Cycle in the area of the Eastern Cordillera, Altiplano and Puna (Salta and Oran Groups), unconformably overlie Palaeozoic and Precambrian rocks. In contrast to the Subandean Ranges, these sediments were distributed throughout the area and filled up depressions with several 1000 m of mostly continental deposits (SALFITY & MARQUILLAS, 1981; MINGRAMM et al., 1979; MARQUILLAS & SALFITY, this issue). Enormous thicknesses of more than 5000 m locally (MARTINEZ, 1978; SALFITY, 1985; SCHWAB, 1985) are a special feature of the Altiplano and the Puna. According to ARANIBAR (Lecture, 4th Chil. Geol. Congr., Antofagasta, 1985), blocks with reduced sedimentary thicknesses may border on other blocks with very great thicknesses, so that these blocks are assumed to be separated from each other by faults such as the San Andreas and Coniri faults of the northern part of the Altiplano (MARTINEZ, 1978) which might possibly be reactivated palaeo-faults of the basement (SALFITY, 1985). Some marine intercalations (e.g. Yacoraite Fm.) show that sedimentation took place in lowlands. A rather tensional regime during sedimentation is indicated by some basaltic flows of Early and Late Cretaceous age and, in the Puna, local basic plutonic rocks. (GALLISKI & VIRAMONTE, in press).

After intense compressional tectonics at the end of the Eocene and during the Oligocene ("Incaic Phase"), Neogene sedimentation took place in several separate intramontane longitudinal basins in the Puna and the Eastern Cordillera (JORDAN & ALONSO, 1987) under the influence of graben-like tectonics (SCHWAB, 1985, fig. 3). Internal unconformities, generally attributed to the "Quechua and Diaguita Phases", underline the synorogenic character of Neogene sedimentation. Another special feature of the Neogene and Quarternary sedimentation is its frequent association with volcanism. Miocene to Holocene volcanics and subvolcanic bodies are found in the Altiplano, Puna and in the western part of the Eastern Cordillera (COIRA et al., 1982). These eastern volcanics show backarc affinities as their potassium content is relatively high (KUSSMAUL et al., 1977). As this Neogene magmatism is partly older than the installation of the A-Subduction of the Subandean Zone during the Late Miocene (ALLMENDINGER et al., 1983), no causal relationships can be established.

239

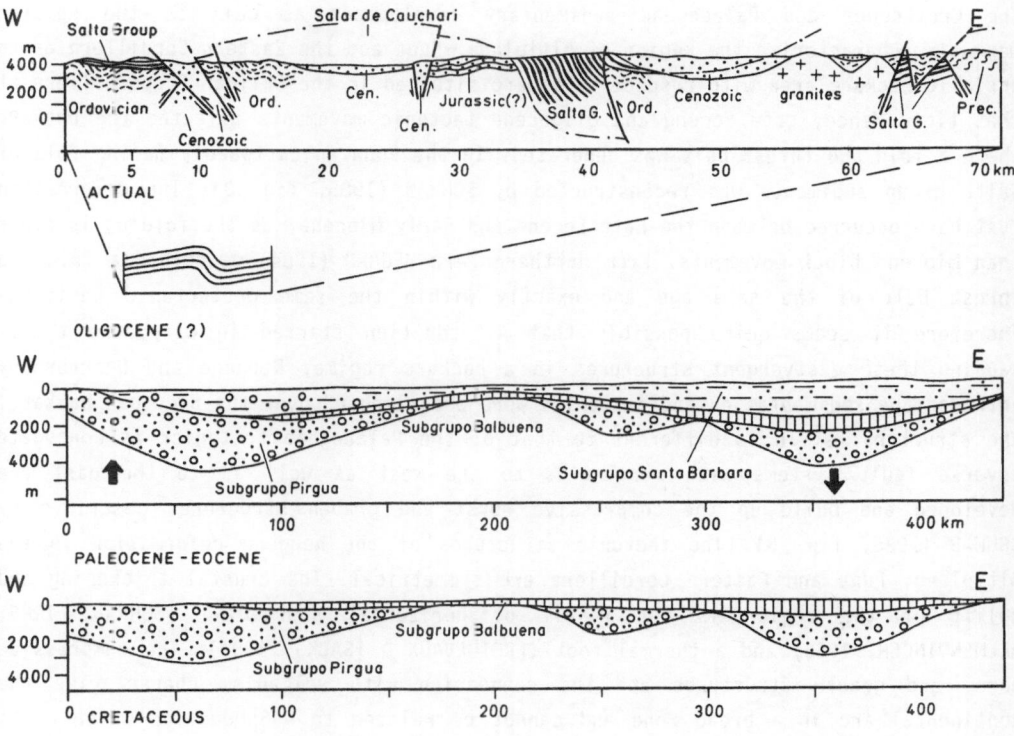

<u>fig. 3</u>: The subsequent stages of the tectonic development of the Puna, along a section at 23°40'S lat. in the Salar de Cauchari area (SCHWAB, 1985, modified). (1) Formation of backarc basins during the Cretaceous and Early Tertiary (lower two sections), (2) development of a backarc fold and thrust belt (small inserted section), and (3) superposition of conjugate reverse faulting in the course of arc tectonics (upper section).

The Cretaceous and Palaeogene sedimentary development as well as the related magmatism characterize the region of Altiplano, Puna and the Eastern Cordillera as an ensialic backarc area with respect to an arc situated in the Chilean Precordillera at that time. Hence, Late Eocene and Oligocene tectonic movements affected the backarc where a fold and thrust belt was generated. In the Puna an eastwardly facing fold of Salta group sediments was reconstructed by SCHWAB (1985, fig. 3). The deformation must have occurred between the Late Eocene and Early Miocene, as the folding is older than Miocene block movements. From northern Peru MEGARD (1984) described a fold and thrust belt of the same age and exactly within the same geotectonic position. Therefore it seems quite possible that A-Subduction started in Oligocene times, causing these eastvergent structures in a backarc regime. Neogene and Quarternary intramontane sedimentation and tectonics were superimposed over the previous backarc. The structural style is different to that of the Palaeogene tectonics as conjugate reverse fault systems with vergencies to the west as well as to the east are developed and build up the compressive horst and graben structures described by SCHWAB (1985, fig. 3). The tectonic structures of the Neogene deformation in the Altiplano, Puna and Eastern Cordillera are symmetrical. The crustal thickening and uplift of the region is the result of horizontal shortening (SCHWAB, 1985; ALLMENDINGER, 1986) and a thermal root (FROIDEVAUX & ISACKS, 1984). The compressive horst and graben structures and the connection with volcanism characterize the continental arc in a broad zone and cannot be related to A-Subduction which might have caused the east-facing backarc structures during the Oligocene. The superposition of these different tectonic styles shows eastward migration of the arc configuration.

The A-Subduction of the Subandean zone is contemporaneous to the structures of the Altiplano, Puna and Eastern Cordillera. Its sole thrust (ROEDER, 1986: "Main Andean Thrust") dips beneath the Eastern Cordillera where it is supposed to enter the basement at a major ramp. Whereas in Argentina the frontal Eastern Cordillera is characterized by a series of steep upthrusts, in Bolivia the huge Sama anticline near Tarija can be considered a frontal ramp anticline. The high crustal uplift of this frontal structure is reflected by positive isostatic and residual Bouguer anomalies (GÖTZE et al., 1987). According to the estimates of ALLMENDINGER et al. (1983), the Eastern Cordillera moved at least 60 km to the east with respect to the foreland in the course of Subandean thrust tectonics. ROEDER (1986) estimated 100 km of Neogene tectonic transport for the Andes of Northern Bolivia. As during that time the Eastern Cordillera was compressed and shortened, it was displaced as a whole with respect to the Chaco plain or the Subandean basement.

The magnetotelluric measurements near Tarija by SCHWARZ et al., (1986) (fig. 4) do not reveal this important thrust fault, but further west, near Tupiza, very low resistivities were found in a depth of about 10 km and deeper, which may be related to that thrust and/or the young volcanic manifestations of the western part of the Eastern Cordillera. Towards the west, beneath the Altiplano, the depth of the layer of very low resistivity in the crust descends to 20 and 40 km. These anomalies may be interpreted as magmatic impregnations or shear zones which possibly separate a rigid upper crust from a more ductile lower crust. It also is possible that the main shear zone of A-subduction passes through this layer.

This megathrust must have caused a considerable crustal thickening in addition to the internal thickening of the crust of the Eastern Cordillera, Altiplano and Puna region. The residual gravity of the Altiplano and Puna is also mainly characterized by NE-SW oriented highs and lows which indicate deeper sedimentary basins and belts of Palaeozoic rocks (e.g. the Faja Eruptiva Oriental). Strong local gradients of gravity mark the main faults or systems of faults in the picture of gravity. Based on gravity mesurements GÖTZE, (1986) calculated crustal thicknesses which increase beneath that region from about 50 km (E) to 70 km (W). Thus, it may be supposed that two crustal elements of normal thickness override each other (ROEDER, 1986).

WESTERN CORDILLERA

The Western Cordillera represents the Miocene-Holocene volcanic arc which consists of rhyolitic ignimbrites and andesitic volcanoes. While its eastern border is very irregular due to volcanoes, volcanic intrusives and ignimbrites extending (according to SALFITY, 1985) along reactivated transverse fault systems into the Altiplano and Eastern Cordillera, its western border is easier to define. It shows that, within the segment considered here, the axis of the Western Cordillera is not a straight N-S structure but that the portion north of $23^{\circ}30'$ trends NNW, the middle portion SSW and only the part to the south of 25°S trends N-S.

The parts that do not run N-S intersect with the neighbouring morphostructural units at a low angle. Thus, from $23^{\circ}30'$S towards NNW, the Western Cordillera is successively superposed on the northern prolongations of the Salar de Atacama and the Upper Loa Valley, both elements of the Preandean Depression, and, north of 21°S, on the Chilean Precordillera. The southern extensions of the Salar de Atacama are intersected by the middle portion of the Western Cordillera in the segment. These contrasting directions are due to the fact that the tectonic structures are determined by intracrustal stresses whereas the position of the volcanic arc is controlled by the subcrustal subduction of the Nazca Plate beneath the critical depth of about 110 km (GILL, 1981).

The volcanic arc of the Western Cordillera came into being during the Miocene east of an extinct Latest Cretaceous - Eocene arc situated in the ambit of the Chilean

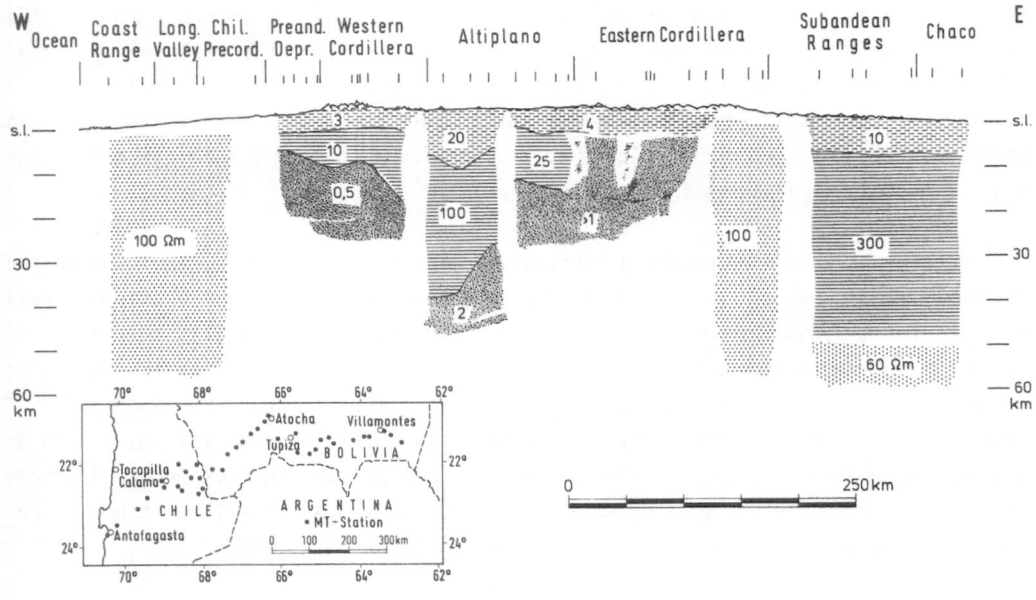

fig. 4: Electric resistivity profile through northern Chile and southern Bolivia based on magnetotelluric data (Resistivity distribution calculated from 1D models, SCHWARZ et al., 1986).

Precordillera and, hence, in the former backarc area. Due to prior backarc tectonics and subsequent erosion, the substrate of the Miocene-Holocene volcanics may consist of Palaeogene and Cretaceous rocks as well as of Palaeozoic volcanics, sediments and intrusives. Palaeozoic rocks as a substrate prevail in the southern part of the segment along the Argentinian-Chilean border, thus showing stronger uplift and erosion prior to Neogene volcanic activity.

The effects of the Miocene-Holocene arc tectonics can better be observed in the border zone of the Western Cordillera than within it. The growing structures were possibly buried beneath the extruding volcanic products whose syntectonic nature is also revealed by internal unconformities (LAHSEN, 1982). It has been suggested that the volcanoes formed along fracture zones and that block faulting occurred within the Western Cordillera (LAHSEN, 1982), but evidently there are no important normal faults or even graben structures. Generally, there seems to be no fundamental difference between the tectonics of the Western Cordillera and those of the neighbouring Altiplano except for the special mechanical conditions imposed by the large quantities of intruded and extruded volcanic material.

The minimum of the Bouguer gravity field of about -450 mGals was observed in the area of the Western Cordillera (GÖTZE, 1986, GÖTZE et al. 1987). Although the regional gravity field also attains its minimum we learnt from potential field separation techniques that the negative values mentioned are partly caused by gravity sources of the upper crust (down to 5 km). This points to the anomalies of electrical conductivity and their interpretation. The level of low resistivity in the western part of the Altiplano shallows towards W and reaches minimum values of less than 1 Ohm m below the western part of the Western Cordillera at a depth of merely 8-10 km. This extremely high conductivity may be interpreted by not or not completely solidified magmatic intrusions at that level or by circulating thermal waters. The first alternative seems to be confirmed by refraction seismics along a line from Chuquicamata into the Altiplano and the Eastern Cordillera, as WIGGER (1986) noted a strong attenuation of seismic waves beneath the Western Cordillera.

PREANDEAN DEPRESSION

In the northern part of Chile considered here, an important morphologic and tectonic depression between the Western Cordillera and the Chilean Precordillera is developed, most spectacular expression of what is the basin of the Salar de Atacama. To the south it is succeeded by the Salar de Punta Negra, and still further south endorheic basins along the western border of the Western Cordillera indicate the persistence of this morphostructural element. To the north of the Salar de Atacama the depression of the upper Loa Valley can be regarded as a structure of the same type.

As mentioned above, the Salar de Atacama region was affected during the Late Eocene and the Oligocene by strong compressive movements which caused folding and faulting of the Purilactis Group (essentially Latest Cretaceous - Eocene according to CHARRIER & REUTTER, in prep.), a partial equivalent to the Salta Group in Argentina (MARQUILAS & SALFITY, this issue). During the Miocene and Pliocene further compression also affected the sediments which had been accumulating with great thicknesses in the depression since the Late Oligocene. The similarity in tectonic and sedimentary developments between the Altiplano and the Salar de Atacama depression suggests that the latter was tectonically a part of the Altiplano until the installation of the Western Cordillera oblique to the tectonic grain separated both areas in the course of the Miocene.

There are, however, also differences. The area of the present Preandean Depression was covered by Jurassic marine and Lower Cretaceous continental sediments forming the eastern parts of an ensialic Jurassic backarc basin. These sediments were completely eroded prior to the sedimentation of the Purilactis Group. Furthermore, andesitic lavas in the upper part of this group (CHARRIER & REUTTER, in prep.) indicate proximity to an Eocene (and older) volcanic arc situated in the Chilean Precordillera.

The Preandean Depression came into being contemporaneously to the Western Cordillera. The Depression only roughly followed preexisting tectonic structures. Therefore, if the straight north-south trending axis of the Precordillera is taken as a reference line, the Salar de Punta Negra depression developed farther west than the Salar de Atacama, so that it would cut into the Precordillera. Furthermore, the Upper Loa Depression lies farther west structurally than the Salar de Atacama. All these individual tectonic depressions are not directly connected but separated from each other by ranges. It may be supposed that these depressions developed due to compression between the isostatically uprising blocks of the Precordillera and the Western Cordillera within a crustal portion weakened by magmatic processes. This assumption is supported by the interpretation of magnetotelluric measurements in the upper Loa Valley by SCHWARZ et al. (1986: station ARL) who found resisitivities of about 2 Ohm m down to 15 km and of less than 1 Ohm m below 20 km (fig. 4). The absorption of seismic signals from Chuquicamata towards the east (WIGGER, 1986) also points in the same direction. Residual gravity in the Preandean Depression zone is controlled by an enormous gravity high of 60-100 mGals. This anomaly covers the area between Calama in the NW and the Argentinan Puna crossing the Salar de Atacama and the Western Cordillera (GÖTZE et al., this issue). The extension, width and striking of this hitherto unknown anomaly corresponds perfectly with the "Faja Eruptiva Occidental" proposed by PALMA et al. (1986) The local negative anomalies caused by salt deposits in the Salar de Atacama are completely masked by this gravity high.

Other local positive anomalies here are related to deep seated intrusions of basic magmas. These authors indicate a crustal thickness of about 70 km, and almost the same thickness was interpreted by WIGGER (1986) from refraction seismic data for the Precordillera to the south of Chuquicamata.

CHILEAN PRECORDILLERA

The mountain ranges to the east of the Chilean Longitudinal Valley (Sierra de Domeyko, Sierra de Moreno) rise to heights of about 4.000 m and are morphologically clearly separated from the Western Cordillera by the Preandean Depression. From a structural point of view, the Precordillera presents a good example of what may be called 'arc tectonics'.

The pre-Jurassic development of the Precordillera is similar to that of the Preandean Depression and the Western Cordillera. Palaeozoic sediments and plutonic rocks of different ages are mostly overlain by Late Carboniferous to Triassic volcanics and sediments. During the Late Triassic, the Lias or locally the Dogger, a marine transgression occurred which led to the deposition of thick carbonatic sequences. The eastward extension of the sea is not known, as in the Preandean Depression and farther east, erosion preceded the deposition of Cretaceous sediments; to the west the basin was limited by the Jurassic volcanic arc, which was then located in the Chilean Coastal Range (v.HILLEBRANDT et al., 1986). The Jurassic palaeogeographic configuration is generally interpreted as an ensialic backarc basin.

At the approximate time of the Jurassic-Cretaceous boundary, marine sediments were gradually replaced by transitional and continental clastics, locally several km thick. They are conformably or unconformably overlain by volcanic formations of intermediate and acid composition. These volcanics indicate that a new magmatic arc was built up within the former Jurassic backarc basin after the extinction of the corresponding magmatic arc in the present Coastal Range during the Early Cretaceous.

From the southwestern part of the Sierra de Moreno ROGERS (1985) has dated "Mid" Cretaceous volcanic rocks (Rb/Sr-isochron: 104,7±19 Ma). Plutonic rocks of about the same age are reported by MARINOVIC & LAHSEN (1984) from the southern part of the Sierra de Moreno (K/Ar in biotite: 103±4 Ma). The "Mid" Cretaceous volcanics underlie with angular unconformity (MUÑOZ, 1986) a younger, only slightly warped volcanic sequence (Chile-Alemania Fm., CHONG (1973), or equivalents) of Latest Cretaceous to Eocene age (HERVE et al., 1985). In the southern part this latter essentially Palaeogene formation has a great areal extension and overlies, unconformably and without intervening continental clastic sediments, the folded Palaeozoic to Cretaceous rocks of the Precordillera and the Longitudinal Valley. As Palaeogene volcanics can also be found in the Preandean Depression too, the Palaeogene magmatic arc may have had a slightly more westerly position than the ill defined Cretaceous arc, i. e. both arc systems overlap each other.

The Precordillera is structurally characterized as a belt of strong compressional tectonics with intense folding and faulting. However, it is not a fold and thrust belt of the foreland type as can be seen from the fact that Palaezoic sedimentary and plutonic rocks and locally also Precambrian metamorphic rocks are involved in the fold structures. These rocks appear in the cores of two or three anticlines whose limbs consist of Mesozoic sequences. This implies that the structures are rather broad (10-15 km) although, normally, they are strongly compressed to such a degree that the limbs are steep or partly overturned and that the core is upthrusted with respect to limbs. In these anticlines vergencies to the W and to the E are developed, it seems, however, that the westward vergencies are slightly more widespread.

The flanks and especially the cores are frequently intruded by granodioritic or dacitic stocks which can in part be considered as synkinematic. CHONG & REUTTER (1985) therefore proposed a model in which the deformation was triggered by intrusions that destabilized the stressed crust and allowed shear movements in the upper rigid level of the crust with respect to the lower weak and viscous level (fig. 5). The wavelength of the folds would imply a detachment at an original depth of about 8-10 km. This value corresponds to the depth of the present highly conductive layer beneath the Preandean Depression. It may thus be concluded that a similar anomaly existed in the Precordillera at the end of the Eocene.

Another kind of deformation can occasionally be observed in the steep limbs of the Precordilleran anticlines. The Mesozoic strata are folded around almost vertical axes. The Z-array of the folds suggests that they were formed by dextral transcurrent faults along the strike of the limbs. The sense of shear along these faults corresponds to an oblique northeastward subduction of the Farallon Plate during the Palaeogene (WHITMAN et al., 1983, , fig. 6). In the cases mentioned, folding is somewhat older than the transcurrent movement, but, as in less inclined rocks wrenching cannot be easily recognized, it is supposed that compression and shear acted contemporaneously.

The exact age of the deformation is not known. In some places it is evident that there were two tectonic events, possibly due to tectonics of the Mid Cretaceous and the latest Cretaceous - Eocene arc. Thus, in the Sierra de Argomedo (southern part of the segment) the Palaeogene volcanics of the Chile-Alemania Formation, which unconformably overlie Jurassic sediments in the western flank of a Precordilleran anticline, were upfolded in a steep position, that is to say that here structures of Cretaceous (probably Late Cretaceous) age were reactivated during the Palaeogene

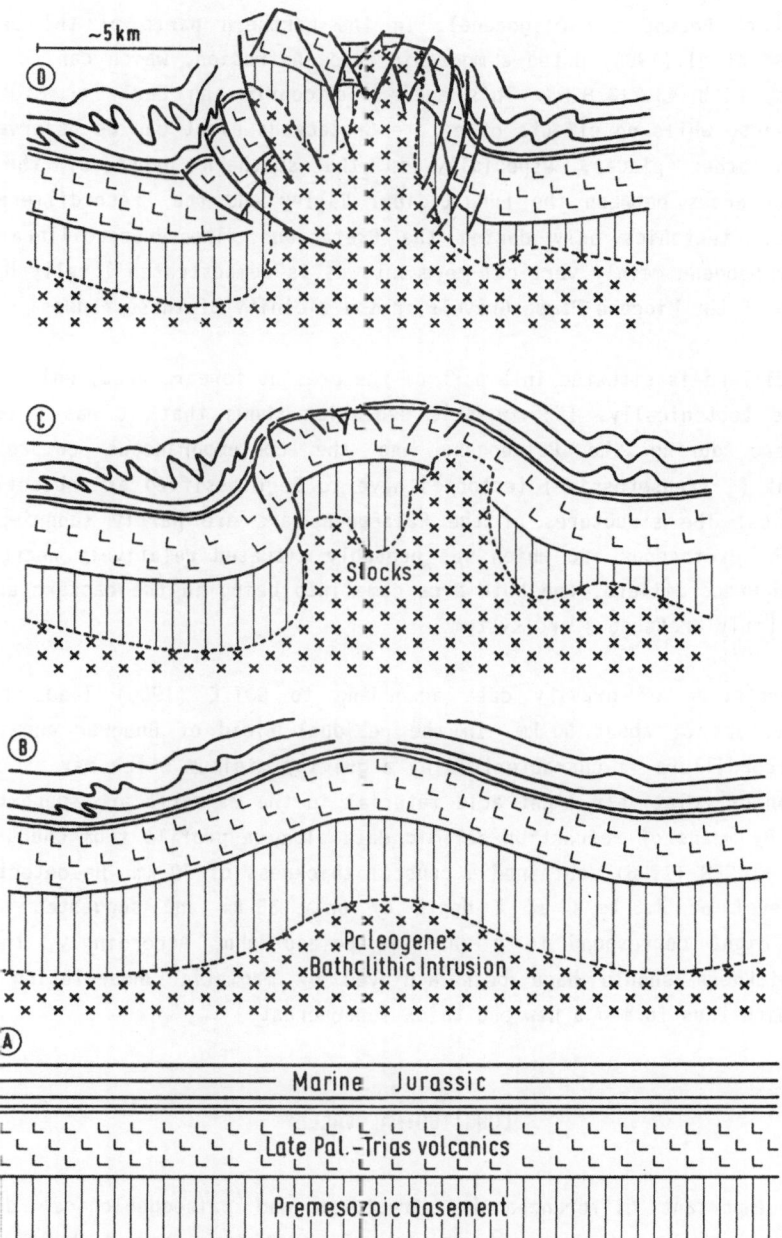

fig. 5: Hypothetical model showing the development of arc-related compressional structures in the upper crust of the Precordillera: <u>Stage A</u>: Plate convergence produces tangential stress in the crust. Due to the lack of horizons suitable for detachement, no deformation occurs. <u>Stage B</u>: Acid melts destabilize the level of intrusion and enable folding to occur in the rigid upper crustal level. <u>Stage C</u>: Increasing shortening steepens the flanks of the basement anticline giving way to further folding in the sedimentary cover on the limbs of the anticline. Diapiritic rise of granitic magma is thus possible. <u>Stage D</u>: Further shortening results in upthrusts on the flanks which, in turn, enhance the rising of the core. Further shortening occurs in the synclines on the flanks leading to the formation of special folds and cleavage.

(probably Late Eocene and Oligocene). In the northern parts of the segment (Q. Choja), DAMM et al.(1986) dated a monzodioritic instrusion, which can be considered syntectonic, with 43,7±3,8 Ma. This age would confirm an Early or Middle Eocene tectonic event, while no effects of a Late Cretaceous event can be observed in that region. In other places, especially in the southern parts of the segment, transitional areas between the Longitudinal Valley and the Precordillera suffered compressional tectonics only during the Cretaceous. The Precordillera was also affected by Neogene mainly vertical movements as is demonstrated locally by faulting and tilting of the Miocene Pampa Gravels in the vicinity of these ranges.

The Precordillera is situated in a part of the present forearc area, which is now not very active tectonically. Its magmatic evolution shows that it was a continental volcanic arc during the Palaeogene and the contemporaneous compressive and transcurrent (= transpressive) tectonics have to be classified as subduction linked arc tectonics. The structures of the Palaeogene arc are partly superimposed over those of the Cretaceous arc which was probably situated relatively nearby, to the west of the Precordillera, and both arcs came into being in the backarc area of the Jurassic - Early Cretaceous arc system.

The interpretation of gravity data according to GÖTZE (1986) leads to crustal thickness values of about 60 km. In the residual field of Bouguer anomalies, the Chilean Precordillera is characterized by a gravity minimum which may be due to the accumulation of relatively light acid material in the magmatic arc supported by arc tectonics. By means of refraction seismic data along a profile from Chuquicamata to the south WIGGER (1986), obtained a crustal thickness of 70 km. He detected a high velocity level of 7.3 km/s at a depth of only 35 km and suggested these high velocities might correspond to a Jurassic palaeo-Moho. Accordingly, the present crustal thickness should have been achieved by magmatic underplating of basic material which thus formed a new and thick lower crust.

LONGITUDINAL VALLEY

There are important differences in structures and palaeogeological development between the northern and the southern part of the Andean segment dealt with here. These differences are particularly evident in the Longitudinal Valley. In the northern part, the Pampa del Tamarugal represents a young tectonic depression morphologically separating the Coastal Range from the Chilean Precordillera, the debris of which accumulate in that basin. Tensional tectonics do not seem to play a role in the formation of that depression; tilting of the now rigid block of the Precordillera towards the west is more likely to be a kinematic motive. The Miocene and younger sediments are not very thick; elevations of Mesozoic and Palaeozoic rocks emerging from that peneplain show that this morphostructural unit was subject to strong tectonics and subsequent erosion prior to the Miocene. On comparsion with the southern part, the unit can probably be said to be of Cretaceous age.

To the east of Antofagasta, the mountains of the Coastal Range merge orographically into the Precordillera, i. e. a morphologically distinct Longitudinally Valley does not exist there. The situation changes south of $24^{0}30'$ S, where the Chilean Precordillera is separated from the Coastal Range by a 50 km wide hilly peneplain occupied by the Palaeogene volcanic Chile-Alemania Formation (CHONG, 1973). These basic and acid lavas, ignimbrites, tuffs and volcanic stocks unconformably overlie the folded and faulted transitional area between the bordering morphostrustural units, but the volcanic formation itself suffered almost no deformation except in the ambit of the Precordillera. The structures of the substratum, which can be observed in the mountaineous region east of Antofagasta, appear to be similar to those of the Precordillera, since anticlines with cores of Palaeozoic rocks exist. Therefore the age of the tectonics is older than the latest Cretaceous - possibly Mid-Cretaceous. Similarly to the Precordillera, compression within a destabilized crust may have been the reason for folding which, therefore, might have occurred in the ambit of a Mid Cretaceous magmatic arc whose location, however, is not well known.

According to the interpretation of seismics (WIGGER, 1986) and gravity data (GÖTZE, 1986), the crust of the Longitudinal Valley has a thickness of about 50 km. Neither the Bouguer gravity field nor the residual gravity point to an abnormal thickness of the sedimentary cover. The Longitudinal Valley is even controlled by positive residual gravity anomalies which cover both the Central Valley and the Coastal Range with values up to +80 mGals. Zones of low electric resistivity could not be detected in the magnetotelluric measurements (SCHWARZ et al., 1986). The spaceous and not intense young tectonics of that present forearc region coincide with these geophysical data.

COASTAL RANGE

During the Jurassic and Early Cretaceous the magmatic arc was situated in the Coastal Range (BUCHELT & TELLEZ, this issue). It consists of andesitic lavas, locally more than 10 km thick and of large plutons of mainly dioritic composition. The great thickness of the volcanics as well as their composition showing tholeiitic affinities in the early stages ("early basics", PICHOWIAK et al., 1988) indicate a geotectonic setting different to that of the later arc-systems. The volcanics overlie some Early Jurassic marine sediments, Triassic and Upper Palaeozoic deposits, Palaeozoic granitoids as well as rocks of the metamorphic basement (probably Cambrian, DIAZ et al., 1985, DAMM et al., 1986). Marine intercalations in the Jurassic volcanics indicate a depositional environment more or less at sea level. The extrusion of these volcanics was thus accompanied by a considerable crustal subsidence and the intrusion of huge dioritic batholites as early as Jurassic times. A second plutonic pulse took place in the Early Cretaceous, when smaller plutons of tonalitic to granodioritic composition intruded along N-S-trending faults.

Steep to nearly vertical faults are the most characteristic tectonic feature of the Coastal Range. Some of them are young and seem to be still active. The most important system of faults constitutes the N-S trending Atacama Fault Zone (AFZ) which can be traced over 1000 km from Iquique (19⁰S) to La Serena (30⁰S). The vertical displacements along these faults can be very important, e. g. a 12.5 km throw occurred along the El Way Fault (RÖSSLING et al. 1986) after the Barrêmian. The blocks between the main longitudinal faults are strongly inclined in some places, thus contrasting with the inclination of the neighbouring blocks (SCHEUBER et al. 1986).

The Post-Neocomian vertical displacements along the faults of the Coastal Range resulted in the phenomenon that rocks that formed in a deep crustal level are exposed over a large area. The rocks show features of metasomatism and partial melting (Bolfin Complex, RÖSSLING, 1987) as well as ductile shear deformation. SCHEUBER (1987) showed this shear deformation belonged to a Jurassic to Early Cretaceous period of wrenching along the AFZ. The deformation was closely related to the intrusion of plutonic bodies of that time. Petrological data indicate medium to high grade conditions for Jurassic shear zones and low grade conditions for Early Cretaceous mylonites with metamorphic pressures intermediate between low pressure and medium pressure series (35-70⁰C/km). The decreasing metamorphic grade indicates crustal uplift in the Coastal Range at least since the Early Cretaceous.

The sense of shear in the ductile shear zones is uniformly sinistral and corresponds to reconstructions of plate configurations and the directions of movement for the SE-Pacific (LARSON & PITMAN III, 1972, ENGEBRETSON et al., 1985, fig. 6). Because of the close relation between the arc magmatism of the Coastal Range and the wrenching along the AFZ, this fault zone can be viewed as an arc-related structure or as a "trench-linked strike-slip fault" sensu WOODCOCK (1986).

It can be presumed that the Jurassic to Early Cretaceous faults were reactivated later in the forearc stress regime when the subsequent magmatic arcs had developed farther to the east. The Neogene to recent activity of the faults of the Coastal Range does not reveal any strike-slip movements and reflects the vertical tectonics of the structural high at the outer rim of the forearc region. The faults, which cause the huge scarp along the coast (about 2000 m in the southern part of the segment) mark the transition to the inner trench slope which as a consequence of

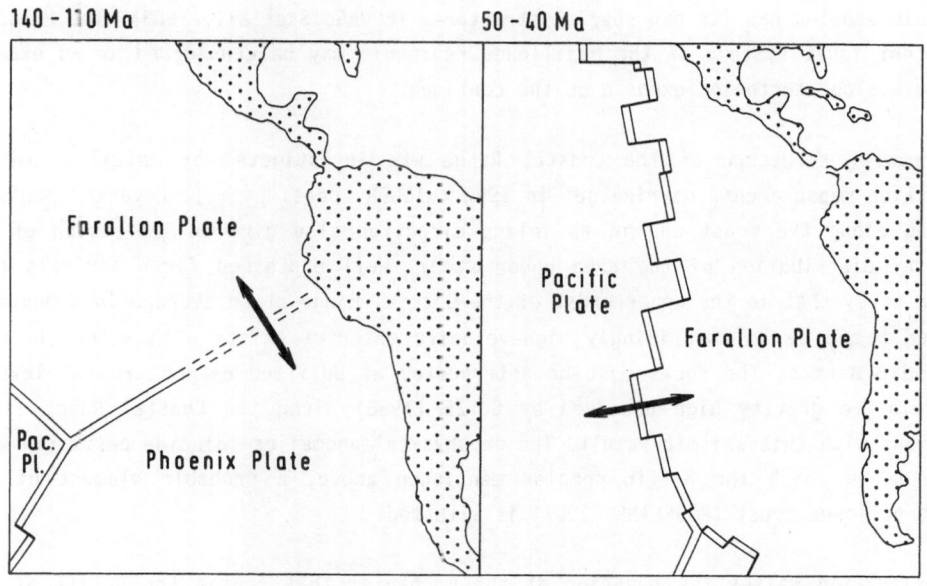

fig. 6: Reconstructions of plate configurations and directions of spreading. Left: Early Cretaceous, right: Palaeogene (Modified after LARSON & PITMAN III, 1972 (left) and WHITMAN et al., 1983 (right)).

tectonic erosion has its own special structures (BOURGOIS et al., 1988). The Pliocene to recent fault systems of the Mejillones Peninsula may be considered as an example of these slope tectonics exposed on the continent.

The crustal structure of the Coastal Range was investigated by detailed seismic refraction measurements carried out in 1987 (WIGGER, pers. comm.). Reversed profiles running along the coast and in an inland direction also give an impression of the velocity distribution of the area under study. Well expressed first arrivals show very clearly that in the upper crust of the Coastal Range at an average in a depth of between 5 and 15 km surprisingly high-velocity material exists with values between 6.5 and 6.8 km/s. The rocks must be interpreted as uplifted deeper crustal levels. The relative gravity high observed by GÖTZE (1986) along the Coastal Range is in agreement with this seismic result. The geophysical anomalies coincide perfectly with the area in which the Bolfin complex mentioned above, a probable element of the middle or lower crust (RÖSSLING, 1987) is situated.

This situation raises the question of the mechanism that caused the uplift of the Chilean Coastal Range with respect to the central units of the Andes. If the Bolfin complex represents a crustal element which formed a part of the lower (or middle) crust in Jurassic times, and which is now situated in the upper part of the continental crust of normal thickness, underplating must have occurred. Fission-track dating of apatites in a Jurassic amphibolite S of Antofagasta provided an age estimate of 118 ± 13 Ma (ANDRIESSEN, pers. comm.). This age designates the time when the temperature of the rock fell below 100°C. Thus, a great part of the uplift and consequently of the underplating probably took place in Early Cretaceous times and so this phenomenon may also be related to the arc tectonics.

CONCLUSIONS

In a converging plate system, the greatest amount of crustal shortening is accommodated in the subduction zone, i.e. at the interface between the upper plate and the downgoing plate, and in the subduction complex forming the inner slope between the trench and the structural high (DICKINSON & SEELY, 1979). At the active continental margin of the Central Andes, these structures are not exposed above sea level, perhaps with the exception of some blocks near the coast (e.g. Mejillones Peninsula). Due to the displacement of the arc system towards the east as a consequence of tectonic erosion of the continental border (RUTLAND, 1971; HILDE, 1983) the subduction complexes of the early stages of the Andean Cycle cannot have been preserved.

The visible structures of the Andes owe their existence to stresses which were, and still are, transmitted through the subduction zone into the continental crust of the upper plate. The tectonic mobilization of the crust is supposed to be achieved by its decoupling from the underlying mantle wedge by magmatic processes. The different settings of the arc configurations that developed during the Andean Cycle suggest that their formation was influenced not only by the inherited crustal conditions but also by the variable conditions of plate motion, e.g. convergence rate, obliqueness of plate motion relative to the trench axis, subduction dip and others. The structural development shows that intense tectonic activity not only affected the subduction complex, but also the area of the magmatic arc, where compressive or transpressive structures generated. During some stages in the backarc area crustal shortening was also accomodated in fold and thrust belts. The backarc outside the fold and thrust belt and especially the forearc between the structural high and the magmatic arc were relatively stable areas or subject only to slow vertical movements. Amounts and velocities of tectonic transport in the mobilized part of the continental crust of the upper plate are important, although certainly much less than those in the subduction zone and complex.

In the backarc area, according to the present situation, a huge crustal thrust system may be developed caused by underthrusting of the foreland under the mobilized crust of the central parts of the orogen. This A-subduction (BALLY, 1975) confers a certain bilateral symmetry to the orogenic system. According to its nature as a flat dipping shear zone, only dip slip movement is possible, hence strong crustal shortening can take place here.

The magmatic arc also shows effects of crustal shortening. Its crust is destablilized by intrusions, by which the upper still rigid part of the crust is enabled to react by folding and steeply dipping conjugate thrusts while the lower part may be deformed by more or less viscous flow. Important low angle thrusts are not necessarily developed and a thickening of the crust is achieved by an internal sort of pure shear deformation and not by crustal underthrusting. The weak crust and the tectonic structures of the magmatic arc also allow the accommodation of stresses parallel to the trench axis resulting from oblique subduction (WOODCOCK, 1986). As fossil Jurassic and Palaeogene structures show (SCHEUBER, 1987), the magmatic arc can be affected by longitudinal almost vertically dipping strike slip faults and secondary structures pertaining to this type of deformation.

In the segment under consideration, four different stages of formation of continental arc systems, one developing after the other, can be recognized for the time from the Jurassic to the present (fig. 7). As far as the magmatic arc is concerned, each stage involved the formation of a volcanic chain, intrusion of plutonic bodies and deformation; it was followed by a period of tectonic and magmatic quiescence with

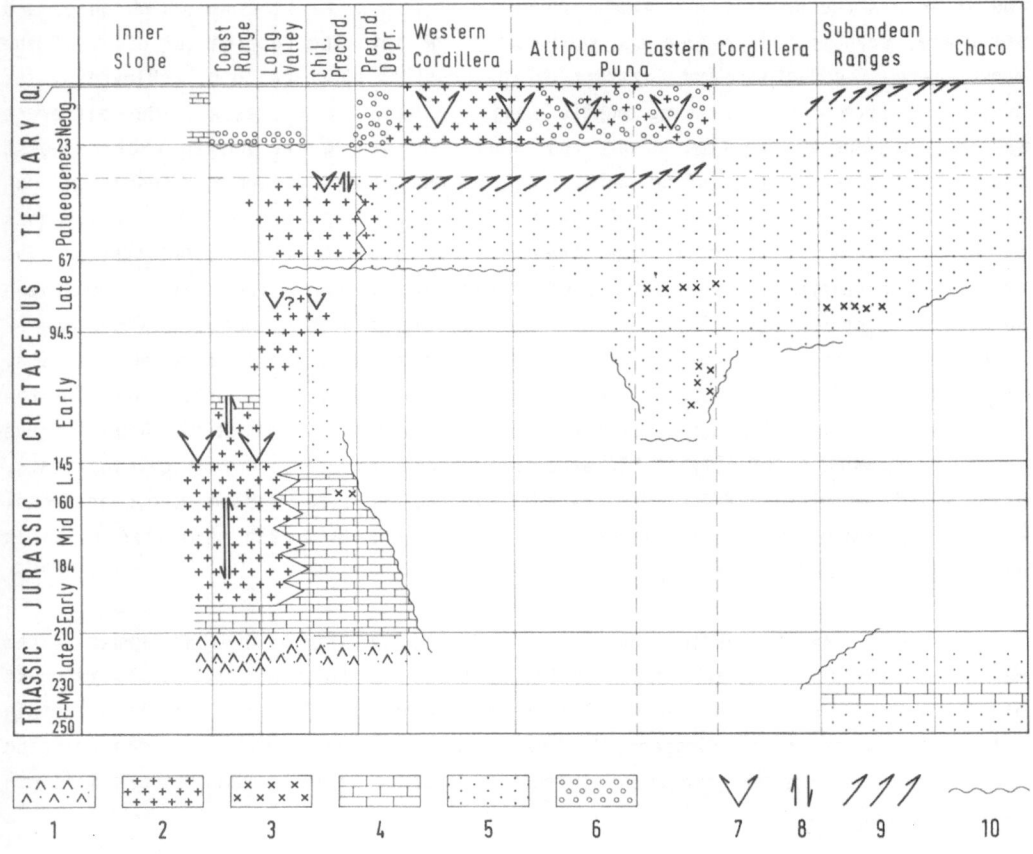

fig. 7: Schematic presentation of the Central Andes structural evolution since the Jurassic (Andean Cycle). 1. Triassic volcanics and continental deposits, 2. arc magmatism, 3. backarc magmatism, 4. marine deposits (backarc), 5. continental backarc sedimentation, 6. intramontane deposits, 7. compressional arc tectonics, 8. wrenching, 9. fold and thrust belt tectonics, 10. unconfirmity.

uplift and erosion before, once again, a new magmato-tectonic stage started with a volcanic chain in a new site, to the east of the preceding chain. If the volcanic activity is regarded in isolation, these stages are (1) Jurassic, (2) Mid-Cretaceous, (3) Latest Cretaceous - Eocene, and (4) Neogene-Quarternary (see COIRA et al., 1982). Not only the sites but also the configurations of these arc systems as well as the accompanying deformations differed from each other (COIRA et al., 1982 also indicated differences of the magmatic products). Nevertheless, there seems to be a regular trend of evolution in the arc systems and the continental crust where they developed from a marine environment to a continental environment and, finally, to high plateau conditions.

The Jurassic magmatic arc, the first of the Andean Cycle, was installed within a subsiding ensialic marine basin in the area of the present Coastal Range. It may have divided the basin into a marine forearc region and a likewise marine backarc basin. These marine conditions may have been a consequence of some crustal spreading and flattening under the load of the volcanics. Nothing is known about the Jurassic forearc, but if the subduction system that gave rise to the Andean Cycle came into being after a long period (Permian and Triassic ?) without plate convergence, it is possible that oceanic crust was adhered by the newly formed forearc. A backarc A-subduction in the ensialic backarc basin cannot be infered from the structures of the exposed marine Jurassic sediments. On the contrary, some volcanic extrusions attest local tensional regimes. However, the basin was limited to the east by a zone of crustal uplift, the connection of which with possible Jurassic compressional tectonics cannot a priori be excluded. Intense tectonic activity took place in the magmatic arc, where, as a consequence of oblique subduction (LARSON & PITMAN III, 1972), transpression caused important left lateral strike slip motion (SCHEUBER, 1987) and compressional warping and block tilting.

As radiometric age determinations in the considered segment are scarce, only little is known about a Cretaceous magmatic arc that is supposed to have developed to the east of the Jurassic arc. Some calc-alkaline volcanic formations and plutonic rocks of the Sierra de Moreno have been dated as "Mid" Cretaceous and may therefore be attributed to an arc of that time. There is no evidence of a Cretaceous A-subduction in the backarc, but pre-Maastrichtian uplift, leading to the erosion of Jurassic sediments in the area of the present Preandean Depression and the Western Cordillera may be seen in this context.

Starting in the latest Cretaceous and during the Paleocene and Eocene, a new important magmatic arc developed in the area of the present Chilean Precordillera (Cordillera Domeyko, Sierra de Moreno), somewhat to the east of the supposed Cretaceous arc. Due to the migrating arc tectonics, in the western part latest Cretaceous and Palaeogene lavas overlie folded Jurassic backarc sediments and rocks belonging to the supposed Cretaceous arc. In the eastern part of the new arc, the volcanics overlie or interfinger with Palaeogene continental backarc sediments. South of 24°S, the volcanics of this age extend into the Longitudinal Valley and even into the Coastal Range, which formed the forearc of this arc system. Consequently these volcanics were not affected by the well developed arc tectonics of the Precordillera. Oblique subduction in a northeastern direction (cf. WHITMAN et al., 1983, fig. 6) during the Palaeogene is revealed by folds with vertical axial plunge and Z-arrays due to N-S directed right lateral slip displacements in that magmatic arc. Maastrichtian marine sediments in the backarc area show that there, probably throughout the Palaeogene, lowland conditions prevailed. The mostly continental sediments of the backarc were intensely folded during the Late Eocene and Oligocene (Incaic Phase) in the region comprising the present Preandean Depression to the Eastern Cordillera. These tectonics certainly led to the formation of a fold and thrust belt, similar to that described by MEGARD (1984) from the northwestern Altiplano, although there is no proof of a significant A-subduction system (fig. 3).

The present arc regime which has taken over since the Early Miocene upon folded Palaeogene backarc rocks developed several new features. In the segment studied, volcanism extends from the virtual magmatic arc of the Western Cordillera for more than 200 km into the backarc. There, an important and, with respect to crustal shortening, effective A-Subduction is active and, finally, the Andes are much more upwarped than during the previous arc stages, which is certainly a consequence of an enormous crustal thickening. Also the area of arc tectonics, characterized by conjugate inverse fault systems (SCHWAB, 1985), is probably broader than that of the preceding arcs, as it comprises the young structures of the Preandean Depression, the Western Cordillera and the Altiplano-Puna region (fig. 7). This crustal shortening within the upper plate is probably much more effective than before. All these effects may be due to a higher plate convergence velocity than before with no or only slight obliquity (PARDO CASAS & MOLNAR, 1987). Furthermore, for the present configuration some final remarks about the subduction complex at the inner slope of the trench can be made. According to the postulated tectonic erosion of the continental border, no accretionary wedge exists, however, blocks separated by faults are tilted and subside until their remainders are underthrust together with the subducting slab (ARABASZ, 1971; EMAIS et al., 1988; BOURGOIS et al., 1988). Thus tectonic erosion in the outer forearc and uplift in the inner forearc seem to be interdependent.

The foregoing considerations show that throughout the Andean Cycle the continental crust on which the Andean arc systems were installed underwent intense deformations. One of the main effects was the shortening of the continental crust and, as pointed out, it occurred in three tectonically active parts of the upper plate: (1) the subduction complex, (2) the magmatic arc, and (3) the backarc. The tectonic erosion of the continental margin by subduction since the Jurassic may be aproximately 200 km

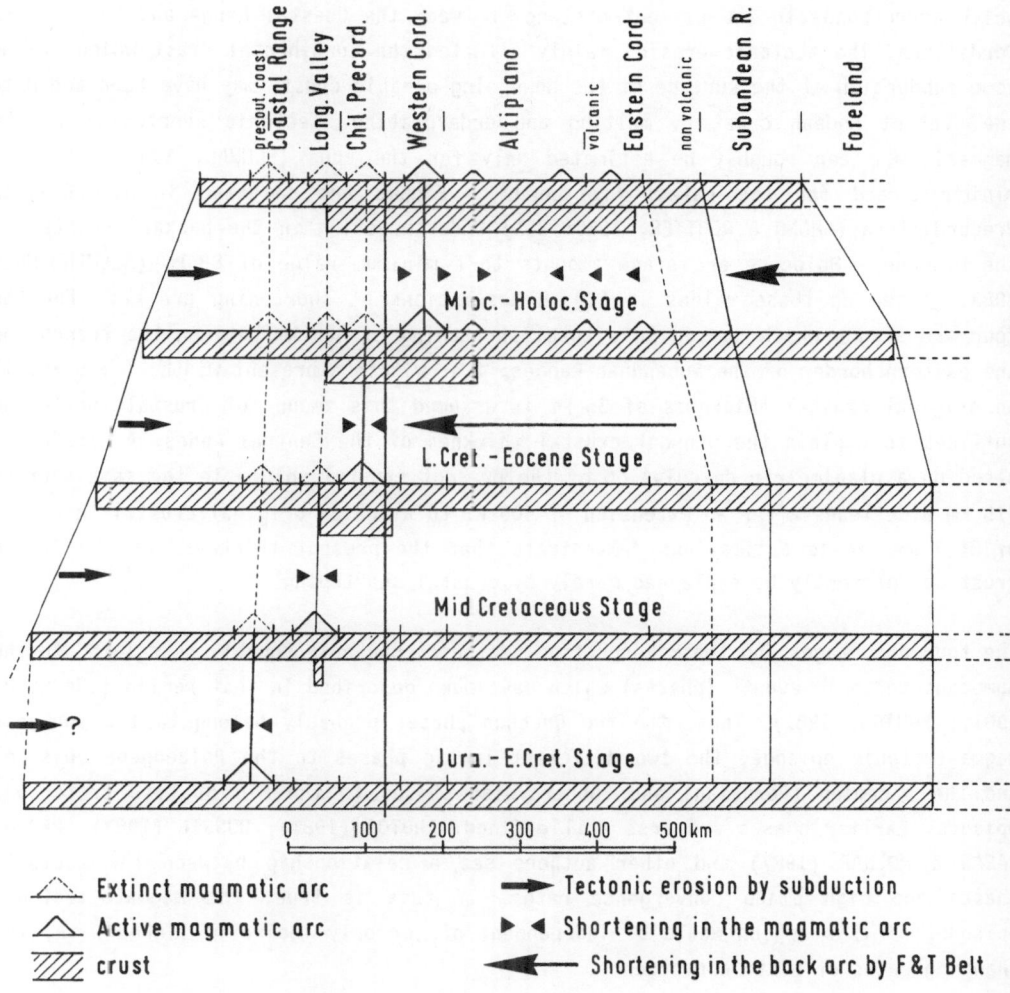

fig. 8: Diagram illustrating crustal shortening during the four arc stages since the Lias in a section through the Central Andes from the trench to the eastern border of the Subandean Ranges. Shortening is achieved by tectonic erosion in the subduction zone, compression in the subsequent magmatic arcs and, during the last two stages, by the installation of fold and thrust belts in the backarc areas. Tectonic erosion is considered to amount to 200 km corresponding to the distance between the Jurassic and the present volcanic arcs. shortening during the first three stages is estimated with 10 km in each case (about 17%) and with 40 km between the eastern border of the Chilean Precordillera and the center of the Eastern Cordillera (about 11%). Crustal shortening in te backarc fold and thrust belt of the presently active arc system is calculated with 80 km (ALLMENDINGER, 1983), while for the Late Cretaceous - Eocene stage an amount of 50 km is postulated. The resulting crustal thickness is conformable with that obtained from geophysical research.

which corresponds to the present distance between the Coastal Range and the Western Cordillera. The tectonic erosion mainly affected the continental crust which, after some subduction at the surface of the downgoing oceanic crust, may have been added to the present Andean crust by melting and underplating. Tectonic shortening in the magmatic arc can roughly be estimated only for the Puna (SCHWAB, 1985: 14% as a minimum), and for the Latest Cretaceous - Eocene arc situated in the Chilean Precordillera (CHONG & REUTTER, 1985: 25%) Underthrusting in the backarc related to the Miocene - Holocene arc stage amounts to a minimum value of 80 km (ALLMENDINGER 1983). Based on these values in fig. 8 assumptions of shortening are made for the four arc stages which sum up to a total shortening of 400 km between the trench and the eastern border of the Subandean Ranges, i.e. within a present width of 775 km. If an original crustal thickness of 35 km is assumed this amount of crustal shortening suffices to explain the present crustal thicknes of the Central Andes. A crosscheck based on a planimetric calculation of the present crustal volume in the same section 775 km wide leads after an extension of 400 km to a medium original crustal thickness of 36,7 km. These estimations demonstrate that the present thickness of the Andean crust can perfectly be explained merely by crustal shortening.

The four arc stages and their respective tectonics can only roughly be related to the numerous tectonic events (phases) which have been described in that region (CHARRIER, 1981,; FRUTOS, 1981). Thus, the two Quechua phases probably belong to the youngest magma-tectonic episode, the two different Incaic phases to the Palaeogene episode, and the Laramien (Peruvian) and Subhercynian phases to the Mid or Late Cretaceous episode. Earlier phases are less well defined. FRUTOS (1981), BUSSEL (1983), PARDO-CASAS & MOLNAR (1987) and other authors see a relationship between the tectonic phases and high plate convergence rates. If this is true, the magmato-tectonic episodes of this region would be independent of, or only indiretly linked with, the changing rates of plate convergence.

REFERENCES

ACEÑOLAZA, F.G. & TOSELLI, A.J. (1976) Consideraciones estratigraficas y tectonicas sobre el Paleozoico inferior del Noroeste Argentino.- Mem. II. Congr. Latinoamer. Geol. (1973), 755-763, Caracas.

ACEÑOAZA, F.G. & TOSELLI, A.J. (1981): Geología del noroeste Arentino.- Univ. Nac. de Tucuman, Publ. N°1287, pp 212, Tucuman.

ALLMENDINGER, R.W. (1986): Tectonic development, southeastern border of the Puna Plateau, northwest Argentine Andes.- Geol. Soc. Amer. Bull., 97: 1070-1082, Boulder.

ALLMENDINGER, R.W.; RAMOS, V.A.; JORDAN, T.E.; PALMA, M. & ISACKS, B.L. (1983): Paleogeography and Andean structural Geometry, Northwest Argentina.- Tectonics, 2, 1-16, Washington D.C..

ARABASZ, W.J. (1971): Geological and geophysical studies of the Atacama fault zone in northern Chile.- unpubl. Diss. Calif. Inst. Techn., pp 264, Pasadena/Calif..

ARANEDA, M.; CHONG, G.; GÖTZE, H.-J.; LAHMEYER, B.; SCHMIDT, S. & STRUNK, S. (1985): Gravimetric modelling of the Northern Chilean lithosphere (20°-26° Lat. South).- 4. Congr. Geol. Chileno Actas, 1: 2/18 - 2/34, Antofagasta.

BALLY, (1975): A geodynamic scenario for hydrocarbon occurences.- 9th world petroleum congress, Tokyo 1975, 33-44, London.

BOURGOIS, J.; PAUTOT, G.; BANDY, W.; BOINET, T.; CHOTIN, P.; HUCHON, P.; MERCIER DE LEPINAY, B.; MONGE, F.; PELLETIER, B.; SOSSON, M. & V.HUENE, R. (1988): Seabeam and seismic reflection imaging of the tectonic regime of the Andean continental margin off Peru ($4^{0}S$ to $10^{0}S$).- Earth Planet. Sci. Lett., 87: 111-126, Amsterdam.

BREITKREUZ, C. (1986): Das Paläozoikum in den Kordilleren Nordchiles.- Geotekt. Forsch., 70: pp 88, Stuttgart.

BUNESS, F.; WETZIG, E. & WIGGER, P. (1986): Seismologische Studien in den zentralen Anden.- Berl. geow. Abh. A 66, 5-33, Berlin.

BUSSEL, M.A. (1983): Timing of tectonic and magmatic events in the Central Andes of Peru.- J. geol. Soc. London, 140: 279-286, London.

CHARRIER, R. (1981): Mesozoic an Cenozoic stratigraphy of the Central Argentinian-Chilean Andes (32^{0}-$35^{0}S$) and chronology of their tectonic evolution.- Zbl. Geol. Paläont. Teil I, 1981: 344-355, Stuttgart.

CHONG, G. (1973): Reconocimiento geológico del área Catalina-Sierra de Varas y estratigrafía del Jurásico del Profeta.- unpubl. Thesis, pp 294, Santiago.

CHONG, G. & REUTTER, K.-J. (1985): Fenomenos de tectonica compresiva en las Sierras de Varas y de Argomedo, Precordillera Chilena, en el ambito del paralelo 25^{0} sur.- IV. Congr. Geol. Chileno Actas 2: 2-219 - 2-238, Antofagasta.

COIRA, B.; DAVIDSON, J.; MPODOZIS, C. & RAMOS, V. (1982): Tectonic and magmatic evolution of the Andes of northern Argentina and Chile.- Earth-Sc. Rev., 18: 303-332, Amsterdam.

DALMAYRAC, B.; LAUBACHER, G. & MAROCCO, R. (1977): Caractères generaux de l'evolution géologique des Andes peruviennes.- Thesis U.S.T.L., pp 36, Montpellier.

DALMAYRAC, B.; LAUBACHER, G.; MAROCCO, R.; MARTINEZ, C. & TOMASI, P. (1980): La chaine hercynienne d' Amerique du sud, structure et évolution d' un orogene intracratonique.- Geol. Rdsch., 69: 1-21, Stuttgart.

DAMM, K.-W.; PICHOWIAK, S. & TODT, W. (1986): Geochemie, Petrologie und Geochronologie der Plutonite und des metamorphen Grundgebirges in Nordchile.- Berliner geowiss. Abh., A 66: 73-146, Berlin.

DIAZ, M.; CORDANI, U.G.; KAWASHITA, K.; BAEZA, L.; VENEGAS, R.; HERVE, F. & MUNIZAGA, F. (1985): Preliminary radiometric ages from the Mejillones Peninsula, Northern Chile.- Comunicaciones, 35: 59-67, Santiago.

DICKINSON, W.R. & SEELY, D.R. (1979): Structure and stratigraphy of forearc regions.- Bull Am. Assoc. Pet. Geol., 63: 2-31, Tulsa.

DONATO, E.O. & VERGANI, G. (1976): Geología del Devonico y Neopaleozoico de la zona del Cerro Rincón, Provincia de Salta, Argentina.- IV Congr. Geol. Chileno, Actas, I, 1-262 - 1-283, Antofagasta.

EMAIS, K.-C.; SUESS, E. & WEFER, G. (1988): Tektonik und Paläozeanographie im Vorland der Anden.- Die Geowissenschaften, 6: 1-7, Weinheim.

ENGEBRETSON, D.C.; COX, A. & GORDON, R.G. (1985): Relative motions between oceanic and continental plateaus in the Pacific Basin.- Geol. Soc. Am., Special Paper, 206: 1-59, Boulder.

FROIDEVAUX, C. & ISACKS, B.L. (1984): The mechanical state of the lithosphere in the Altiplano-Puna segment of the Andes.- Earth Planet. Sci. Lett., 71: 305-314, Amsterdam.

FRUTOS, J. (1981): Andean tectonics as a consequence of sea-floor spreading.- Tectonophysics, 72: T21-T32, Amsterdam.

GALLISKI, M.A. & VIRAMONTE,J.G. (in press): Cretaceous paleorift in northwestern Argentina - petrological approach.- Journal of South-American Earth Sciences, Amsterdam.

GILL, J. (1981): Orogenic andesites and plate tectonics.- pp 390, Springer, Heidelberg, Berlin, New York.

GÖTZE, H.-J. (1986): Schweremessungen und deren Interpretation im mittleren und östlichen Teil der Anden-Geotraverse. Unpublished final report, DFG Project Go 380/1, Bonn.

GÖTZE, H.-J.; LAHMEYER, B.; SCHMIDT, S.; & STRUNK, S. (1987): Gravity field and megafault-system of the Central Andes (20°-26° L.S.) Abstracts, Terra Cognita, 7: 57, Cambridge.

HERVE, M.; MARINOVIC, N.; MPODOZIS, C. & PEREZ DE ARCE, C. (1985): Geocronologia K-Ar de la Cordillera de la Costa entre los 24° y 25° latitud sur. Antecedentes prelimnares.- IV Congreso Geologico Chileno, Resumenes, 4.19: 158, Antofagasta.
HILDE, T.W.C. (1983): Sediment subduction versus accretion around the Pacific.- Tectonophysics, 99: 381-397, Amsterdam.
HILLEBRANDT, A.v.; GRÖSCHKE, M.; PRINZ, P. & WILKE, H.-G. (1986): Marines Mesozoikum in Nordchile zwischen 21° und 26°S.- Berliner geowiss. Abh., A 66: 169-190, Berlin.
JORDAN, T. & ALONSO, R. (1987): Cenozoic stratigraphy and basin tectonics of the Andes mountains, 20°-28° south latitude.- Am. Ass. Petrol. Geol. Bull., 71: 49-64, Tulsa.
KUSSMAUL, S.; HÖRMANN, P.K.; PLOSKONKA, E. & SUBIETA, T. (1977): Volcanism and structure of southwestern Bolivia.- J. Volcanology and Geotherm. Res., 2: 73-111, Amsterdam.
LAHSEN, A. (1982): Upper Cenozoic volcanism and tectonism in the Andes of Northern Chile.- Earth Sci. Rev., 18: 285-302, Amsterdam.
LARSON, R.L. & PITMAN III, W.C. (1972): World-wide correlation of Mesozoic magnetic anomalies, and its implications.- Geol. Soc. Am. Bull.,83: 3645-3662, Boulder.
LEHMANN, B. (1978): A Precambrian core sample from the Altiplano/Bolivia.- Geol. Rdsch., 67: 270-278, Stuttgart.
LYON-CAEN, H.; MOLNAR, P. & SUAREZ, G. (1987): Gravity anomalies and flexure of the Brazilian shield beneath the Bolivian Andes.- Geodynamique Andes Centrales, Seminaire 14-16 janvier 1987, résuménes; 83, Bondy.
MARINOVIC, S. & LAHSEN, A. (1984): Hoja Calama.- Serv. Nac. Geol. Min.: Carta Geol. de Chile N°·58, pp 140, Santiago.
MARTINEZ, C. (1978): Géologie des Andes Boliviennes.- Travaux et documents de l'O.R.S.T.O.M., 119: pp 352, Paris.
MARTINEZ, C.; TOMASI, P.; SUBIETA, T. & BOTELLO, R. (1973): Mapa tectónico de Bolivia.- Univ. Mayor S. Andrés, Serv. Geol. Bol., La Paz.
MEGARD, F. (1984): The Andean orogenic period and its major structures in central and northern Peru.- J. Geol. Soc. London, 141: 893-900, London.
MENDEZ, V.; NAVARINI, A.; PLAZA, D. & VIERA, V. (1973): Faja Eruptiva de la Puna Oriental.- Actas V° Congr. Geol. Arg., IV: 89-100, Córdoba.
MINGRAMM, A.; RUSSO, A.; POZZO, A. & CAZAU, L. (1979): Sierras subandinas.- Segundo Simp. Geol. reg. Argent.,Córdoba, Acad. nac. Cienc., 1: 95-137, Buenos Aires.
MUÑOZ, N (1986): Estratigrafia de las Hojas Baquedano y Pampa Union al oeste de la Quebrada del Rio Seco Región de Antofagasta.- Taller de titulo II Univ. de Chile (unpubl.), pp 96, Santiago.
PALMA, M.A.; PARICA, P.D. & RAMOS, V.A. (1986): El granito Archibarca: su edad y significado tectónico, Provincia de Catamara.- Asoc. Geol. Argentina, Rev., 41: 414-419, Buenos Aires.
PARDO-CASAS, F. & MOLNAR, P. (1987): Relative motion of the Nazca (Farallon) and South American plates since Late Cretaceous time.- Tectonics, 6: 233-248, Washington D.C..
PICHOWIAK, S.; BUCHELT, M. & DAMM, K.-W. (1988): Mesozoic magmatic activity and tectonic setting in the N-Chile Central Andes Region: granitoid magmagenesis and relations to volcanic activity during early stages of the Andean Cycle.- Geol. Soc. Am. Memoir, manuscript submitted.
PRICE, R.A. (1981): The Cordilleran foreland thrust and fold belt in the southern Canadian Rocky Mountains.- In. McClay, K.R. & Price, N.J. (eds.): "Thrust and Nappe Tectonics": 427-448, London.
RAMIREZ, C.F. & GARDEWEG, M. (1982): Hoja Toconao.- Serv. Nac. Geol. Min.: Carta Geol. de Chile N°·54, pp 121, Santiago.
RIVANO, S.; SEPULVEDA, P.; HERVE, M. & PUIG, A. (1985): Geocronologia K-Ar de las Rocas Intrusivas Entre los 31°-32° Latitud Sur, Chile, Rev. Geol. Chile, No. 24, 63-74, Santiago.
ROEDER, D. 1986: Andean-Age structure of Eastern Cordillera (Province of La Paz, Bolivia).- Latin Amer.Coll. FU Berlin; Denver, Colorado.
ROGERS, G. (1985): A geochemical traverse across the North Chilean Andes.- Ph.D. thesis, Dept. Earth Sc., Open University, Milton Keynes.
RÖSSLING, R. (1987): Petrologie in einem tiefen Stockwerk des jurassischen magmatischen Bogens in der nordchilenischen Küstenkordillere südlich von Antofagasta.- Unpubl. Diss. FU Berlin, pp 165, Berlin.

RÖSSLING, R.; SCHEUBER, E. & REUTTER, K.-J. (1986): Strukturen und petrologische Einheiten der nordchilenischen Küstenkordillere zwischen Antofagasta und Paposo.- 10. Geowissenschaftl. Lateinamerika-Kolloquium Berlin, 19.-21.11.1986, Kurzfassungen der Beiträge. Berliner geowiss. Abh., A Sonderband: 174-178, Berlin.

RUTLAND, R.W.R. (1971): Andean orogeny and sea floor spreading.- Nature, 233: 252-255, London.

SALFITY, J.A. (1985): Lineamentos Transversales al Rumbo Andina en el Noroeste Argentino.- IV Congr. Geol. Chileno, 2: 119-137, Antofagasta.

SALFITY, J. & MARQUILLAS, R. (1981): Las Unidades Estratigraficas del Norte de la Argentina.- Comité Sudamericano del Jurásico, 1: 303-317, Buenos Aires.

SCHEUBER, E.; RÖSSLING, R. & REUTTER, K.-J. (1986): Strukturen in der chilenischen Küstenkordillere zwischen Paposo und Antofagasta.- Berliner geowiss. Abh., A 66: 209-224, Berlin.

SCHEUBER, E. (1987): Geologie der nordchilenischen Küstenkordillere zwischen 24°30' und 25°S - unter besonderer Berücksichtigung duktiler Scherzonen im Bereich des Atacama-Störungssystems.- Unpubl. Diss. FU Berlin, pp 157, Berlin.

SCHWAB, K. (1985): Basin formation in a thickening crust - the intermontane basins in the Puna and the Eastern Cordillera of NW-Argentina (Central Andes).- IV Congr. Geol. Chileno, 2: 138-158, Antofagasta.

SCHWARZ, G.; HAAK, V.; MARTINEZ, E. & BANNISTER, J. (1984): The electrical coductivity of the Andean crust in northern Chile and southern Bolivia as inferred from magnetotelluric measurements.- J. Geophys. 55, pp 169-178, Heidelberg.

SCHWARZ, G.; MARTINEZ, E. & BANNISTER, J. (1986): Untersuchungen zur elektrischen Leitfähigkeit in den zentralen Anden.- Berliner geowiss. Abh., A 66: 49-72, Berlin.

SKARMETA, J. & MARINOVIC, N. (1981): Hoja Quillagua.- Serv. Nac. Geol. Min.: Carta Geol. de Chile N°.51, pp 63, Santiago.

STRUNK, S. (1985): Auswertung gravimetrischer Messungen und deren 3-D Interpretation im Bereich der andinen Subduktionszone Nordchiles. unpubl. thesis.- Institut f. Geophysik; TU Clausthal.

WHITMAN, J.M.; HARRISON, C.G. & BRASS, G.W. (1983): Tectonic evolution of the Pacific Ocean since 74 Ma.- Tectonophysics, 99: 241-249, Amsterdam.

WIGGER, P (1986): Krustenseismische Untersuchungen in Nord-Chile und Süd-Bolivien.- Berliner geowiss. Abh., A 66: 31-48, Berlin.

WOODCOCK, N.H. (1986): The role of strike-slip fault systems at plate boundaries. - Phil. Trans. R. Soc. Lond., A 317: 13-29, London.